PAUL B. ZBA … OELKER

INDUSTRIAL ELECTRONICS

A TEXT-LAB MANUAL

FOURTH EDITION

McGRAW-HILL PUBLISHING COMPANY

New York Atlanta Dallas St. Louis San Francisco
Auckland Bogotá Caracas Hamburg Lisbon
London Madrid Mexico Milan Montreal New Delhi
Paris San Juan São Paulo Singapore
Sydney Tokyo Toronto

OTHER BOOKS BY PAUL B. ZBAR

Basic Electricity: A Text-Lab Manual
Basic Television: Theory and Servicing, A Text-Lab Manual (with P. Orne)
Electricity-Electronics Fundamentals: A Text-Lab Manual (with J. Sloop)
Basic Electronics: A Text-Lab Manual (with A. Malvino and M. Miller)

Sponsoring Editor: Paul Sobel
Editing Supervisor: Allen Appel
Design and Art Supervisor: Joseph Piliero
Production Supervisor: Kathy Porzio

Text Designer: Keithley Associates
Cover Designer: BC Graphics

Library of Congress Cataloging-in-Publication Data

Zbar, Paul B.,
Industrial electronics : a text-lab manual / Paul B. Zbar, Richard L. Koelker.—4th ed.
p. cm.
ISBN 0-07-072822-4
1. Electronics—Laboratory manuals. I. Koelker, Richard L. II. Title.
TK7818.Z23 1990 89-28171
621.381′078—dc20 CIP

Industrial Electronics: A Text-Lab Manual, Fourth Edition

1 2 3 4 5 6 7 8 9 0 SEM SEM 8 9 6 5 4 3 2 1 0 9

ISBN 0-07-072822-4

CONTENTS

Note on Experiment Content

Each of the experiments described below is set up in the following manner:
OBJECTIVES The objectives are enumerated and clearly stated.
INTRODUCTORY INFORMATION The theory and basic principles involved in the experiment are stated concisely.
SUMMARY Salient points in the introductory information are summarized.
SELF-TEST Based on the material in the introductory information, a self-test helps students evaluate their understanding of the principles involved in the experimental procedure. The self-test should be taken both before the experiment is undertaken and after the experiment is completed, for comparison. Answers to the self-test questions are given at the end of each experiment.
MATERIALS REQUIRED All the materials required for the experiment, including test equipment and components, are specified.
PROCEDURE Instructions on how to perform the experiment are given in step-by-step details.
QUESTIONS A series of pertinent questions help students draw conclusions from their experimental data.

Experiments

SERIES PREFACE

Electronics is at the core of a wide variety of specialized technologies that have been developing over several decades. Challenged by a rapidly expanding technology and the need for increasing numbers of technicians, the Consumer Electronics Product Group Service Committee of the Electronic Industries Association (EIA) and various publishers, has been active in creating and developing educational materials to meet these challenges.

In recent years, a great many consumer electronic products have been introduced and the traditional radio and television receivers have become more complex. As a result, the pressing need for training programs to permit students of various backgrounds and abilities to enter this growing industry has induced EIA to sponsor the preparation of an expanding range of materials. Three branches of study have been developed in two specific formats. The tables list the books in each category; the paragraphs following them explain these materials and suggest how best to use them to achieve the desired results.

THE BASIC ELECTRICITY-ELECTRONICS SERIES

Title	Author	Publisher
Electricity-Electronics Fundamentals	Zbar/Sloop	McGraw-Hill Publishing Company
Basic Electricity	Zbar/Rockmaker	McGraw-Hill Publishing Company
Basic Electronics	Zbar/Malvino/Miller	McGraw-Hill Publishing Company

The laboratory text-manuals in the Basic Electricity-Electronics Series provide in-depth, detailed, completely up-to-date technical material by combining a comprehensive discussion of the objectives, theory, and underlying principles with a closely coordinated program of experiments. *Electricity-Electronics Fundamentals* provides material for an introductory course especially suitable for preparing service technicians; it can also be used for other broad-based courses. *Basic Electricity* and *Basic Electronics* are planned for 270-hour courses, one to follow the other, providing a more thorough background for all levels of technician training. A related instructor's guide is available for each course.

THE TELEVISION-AUDIO SERVICING SERIES

Title	Author	Publisher
Television Servicing with Basic Electronics	Sloop	Howard W. Sams & Co., Inc.
Advanced Color Television Servicing	Sloop	Howard W. Sams & Co., Inc.
Audio Servicing—Theory and Practice	Wells	McGraw-Hill Publishing Company
Audio Servicing—Text-Lab Manual	Wells	McGraw-Hill Publishing Company
Basic Television: Theory and Servicing	Zbar and Orne	McGraw-Hill Publishing Company
Cable Television Technology	Deschler	McGraw-Hill Publishing Company

The Television-Audio Servicing Series includes materials in two categories: those designed to prepare apprentice technicians to perform in-home servicing and other apprenticeship functions, and those designed to prepare technicians to perform more sophisticated and complicated servicing such as bench-type servicing in the shop.

Television Servicing with Basic Electronics (text, student workbook, instructor's guide) covers the basics, the math, and the test equipment required in Television Servicing.

Advanced bench-type diagnosis servicing techniques are covered in *Advanced Color Television Servicing* (text, student workbook, instructor's guide). Written primarily for color television servicing courses in schools and in industry, this set follows the logical diagnostic troubleshooting approach consistent with the manufacturers' approach to bench servicing.

Audio Servicing (theory and practice, text-lab manual, and instructor's guide) covers each component of a modern home stereo with an easy-to-follow block diagram and a diagnosis approach consistent with the latest industry techniques.

Basic Television: Theory and Servicing provides a series of experiments, with preparatory theory, designed to provide the in-depth, detailed training necessary to produce skilled television service technicians for both home and bench servicing of all types of televisions. A related instructor's guide is also available.

Basic laboratory courses in industrial control and computer circuits and laboratory standard measuring equipment are provided by the Industrial Electronics Series and their related instructor's guides. *Industrial Electronics* is concerned with the fundamental building blocks in industrial electronics technology, giving the student an understanding of the basic circuits and their applications. *Electronic Instruments and Measurements* fills the need for basic training in the complex field of industrial instrumentation. Prerequisites for both courses are *Basic Electricity* and *Basic Electronics*.

The foreword to the first edition of the EIA-cosponsored basic series states: "The aim of this basic instructional series is to supply schools with a well-integrated, standardized training program, fashioned to produce a technician tailored to industry's needs." This is still the objective of the varied training program that has been developed through joint industry-educator-publisher cooperation.

Peter McCloskey, President
Electronic Industries Association

THE INDUSTRIAL ELECTRONICS SERIES

Title	Author	Publisher
Industrial Electronics	Zbar	McGraw-Hill Publishing Company
Electronic Instruments and Measurements	Zbar	McGraw-Hill Publishing Company

PREFACE

The field of industrial electronics has been strongly affected by the remarkable degree to which computers and digital technology have been introduced in the control of machines and processes. The low cost of digital devices and their ever-increasing usage suggest a need for the electronic technician to be able to deal with them as "components" in much the same way as individual transistors and other electronic devices were treated a short time ago. The approach taken, therefore, in the fourth edition of *Industrial Electronics* has been to reduce the emphasis on discrete components and stress modular building blocks wherever possible.

The format of previous editions has been retained. The objectives of each experiment have been clarified, and new material has been added to the text, the self-test, and the set of questions at the end of each experiment.

The early text and experiments laid the groundwork for the material in the last chapter, an introduction to the programmable controller. While the technician may continue to see machine controls with discrete components, mechanical relays, and timers in older equipment, new designs will usually involve applications of the modern and exceptionally flexible programmable controller.

To obtain the greatest value from the use of this edition, students should have a good background in basic electrical and electronic theory, as well as a mathematical background which includes algebra and trigonometry. Calculus is not required. Where necessary, new mathematical tools are introduced.

Over the years many students have studied the Zbar text-lab manuals as a solid foundation for a successful career in some phase of the vast field of electricity/electronics. We hope that this edition of *Industrial Electronics* will continue that tradition, and we are happy to have had the opportunity to make a small contribution.

Paul B. Zbar
Richard L. Koelker

SAFETY

Electronic techniques work with electric and electronic devices, motors, and other rotating machinery. They are often required to use hand and power tools in constructing prototypes of new devices or in setting up experiments. They use test instruments to measure the electrical characteristics of components, devices, and electronic systems. In short, electronics technicians are involved in any of a dozen different tasks.

These tasks are interesting and challenging, but they may also involve certain hazards if the technician has careless work habits. It is therefore essential that student technicians learn the principles of safety at the very start of their career and that they practice these principles.

Safe work requires a careful and deliberate approach to each task. Before undertaking a job, the technicians must understand what to do and how to do it. They must plan the job, setting out on the workbench in a neat and orderly fashion, tools, equipment, and instruments. Extraneous items should be removed, and cables should be securely fastened.

When working on or near rotating machinery, loose clothing should be anchored, ties firmly tucked away.

Line (power) voltages should be isolated from ground by means of an isolation transformer. Powerline voltages can kill, so these should *not* come in contact with the hands or body. Line cords should be checked before use. If the insulation on line cords is brittle or cracked, these cords must *not* be used. TO THE STUDENT: Avoid direct contact with any voltage source. Measure voltages with one hand in your pocket. Wear rubbersoled shoes or stand on a rubber mat when working at your experiment bench. Be certain that your hands are dry and that you are not standing on a wet floor when making tests and measurements in a live circuit. Shut off power before connecting test instruments in a live circuit.

Be certain that line cords of power tools and nonisolated equipment use safety plugs (polarized 3-post plugs). Do not defeat the safety feature of these plugs by using ungrounded adapters. Do not defeat any safety device, such as fuse or circuit breaker, by shorting across it or by using a higher amperage fuse than that specified by the manufacturer. Safety devices are intended to protect you and your equipment.

Handle tools properly and with care. Don't indulge in horseplay or play practical jokes in the laboratory. When using power tools, secure your work in a vise or jig. Wear gloves and goggles when required. Exercise good judgment and common sense and your life in the laboratory will be safe, interesting, and rewarding.

FIRST AID

If an accident should occur, shut off the power immediately. Report the accident at once to your instructor. It may be necessary for you to give emergency care before a physician can come, so you should know the principles of first aid. You can learn the basics by taking a Red Cross first-aid course.

Some first-aid suggestions are set forth here as a simple guide.

Keep the injured person lying down until medical help arrives. Keep the person warm to prevent shock. Do not attempt to give water or other liquids to an unconscious person. Be sure nothing is done to cause further injury. Keep the injured one comfortable until medical help arrives.

ARTIFICIAL RESPIRATION

Severe electric shock may cause stoppage of breathing. Be prepared to start artificial respiration at once if breathing has stopped. The two recommended techniques are:

1. Mouth-to-mouth breathing, considered the most effective
2. Schaeffer method

These techniques are described in first-aid books. You should master one or the other so that if the need arises you will be able to save a life by applying artificial respiration.

These safety instructions should not frighten you but should make you aware that there are hazards in the work of an electronics technician—as there are hazards in every job. Therefore you must exercise common sense and good judgment, and maintain safe work habits in this, as in every other job.

LETTER SYMBOLS

As noted in the Author's Preface, primary emphasis in this manual has been placed on semiconductor (solid-state) devices and circuits. However, vacuum tubes and their associated circuits are also treated, making it desirable to use letter symbols that have the same meaning throughout the text for both solid-state and vacuum-tube circuits. Accordingly, the IEEE (Institute of Electrical and Electronics Engineers) Letter Symbols for Semiconductor Devices (IEEE Standard #255) were used, with modifications for vacuum tubes.

The following summary of symbols for electrical quantities is intended to clarify their use throughout the text.

QUANTITY SYMBOLS

1. Instantaneous values of current, voltage, and power, that vary with time, are represented by the lowercase letter of the proper symbol.
   Examples: i, v, p
2. Maximum (peak), average (direct current), and root-mean-square values of current, voltage, and power are represented by the uppercase letter of the appropriate symbol.
   Examples: I, V, P

SUBSCRIPTS FOR QUANTITY SYMBOLS

1. Direct-current values and instantaneous total values are indicated by uppercase subscripts.
   Examples: i_C, I_C, v_{EB}, V_{EB}, p_C, P_C
2. Alternating-component values are indicated by lowercase subscripts.
   Examples: i_c, I_c, v_{eb}, V_{eb}, p_c, P_c
3. Symbols to be used as subscripts:
   E, e emitter terminal
   B, b base terminal
   C, c collector terminal
   A, a anode terminal
   K, k cathode terminal
   G, g grid terminal
   P, p plate terminal
   M, m maximum value
   Min, min minimum value

   Examples:
   I_E emitter direct-current (no alternating current component)
   I_e rms value of alternating component of emitter current
   i_e instantaneous value of alternating component of emitter current
4. Supply voltages may be indicated by repeating the terminal subscript.
   Examples: V_{EE}, I_{CC}, V_{BB}, V_{PP}, V_{GG}
   The one exception to this system is the occasional use of $V+$ for the plate supply voltage of a tube. Note that $V+$ replaces the more usual $B+$.
5. The first subscript designates the terminal at which current or voltage is measured with respect to the reference terminal, which is designated by the second subscript.

EXPERIMENT

1

CHARACTERISTICS OF A SILICON RECTIFIER

OBJECTIVES

1. To determine the voltage drop across a silicon rectifier
2. To study the operation of a silicon rectifier with a resistive load R and with a series RL load

BASIC INFORMATION

In communications electronics, we were concerned with the characteristics and circuit arrangements of solid-state diodes, of transistors, and to a lesser extent, of high-vacuum tubes. These devices operated with currents on the order of microamperes to milliamperes and required relatively low voltages. For industrial applications, devices are required which can supply currents in the range of amperes to thousands of amperes and higher voltages. The use of silicon diodes operating as efficient *rectifiers* can reliably provide this greater power-handling capability. In this experiment we will study the characteristics of silicon diodes and their use in elementary rectifier circuits.

A silicon diode is useful because it permits current to flow more readily in one direction than in the other. When it is forward-biased (the anode is at a positive potential with respect to the cathode), it acts like a closed switch, or nearly so. When it is reverse-biased (the anode is at a negative potential with respect to the cathode), it acts like an open switch. Figure 1-1 demonstrates this concept. A perfect diode would have zero resistance when forward-biased and infinite resistance when reverse-biased. When a practical silicon diode is conducting, its internal resistance is very low and the voltage drop across it is very low, usually less than 1 volt (v). It is therefore almost a lossless switching device and can convert large amounts of ac power to dc with high efficiency.

Diffused Junction Silicon Diodes

Silicon diodes have several advantages over other types, such as selenium, germanium, and copper oxide diodes used in the past. Silicon diodes can operate at higher temperatures than other types and can be constructed to withstand high reverse-bias voltages with low leakage currents. Compact diodes are available which can handle thousands of amperes.

Figure 1-1*b* shows the volt-ampere characteristic of a silicon diode. The graph shows the diode has very low forward resistance R_F, which equals the voltage at any point divided by the corresponding current: $R_F = V_F/I_F$. At 0.6 V, for example, $R_F = 0.6/0.2 =$ 3 ohms (Ω). In reverse bias the diode has very high resistance: R_R, the reverse resistance at 300 V = 300/0.4 microamperes (μA), = 750 megohms (Ω).

Although silicon diodes are more tolerant of heat than germanium diodes, for example, they are still heat-sensitive. The forward-bias voltage drop decreases as temperature increases, and the reverse-bias leakage current increases also as the temperature of the diode increases. If the maximum operating temperature is exceeded [175°C (347°F)], the diode will eventually fail. To avoid excessive diode temperature, the larger types are mounted on a "heat sink," the chassis or a special finned aluminum extrusion which can transmit the heat generated within the diode to the surroundings.

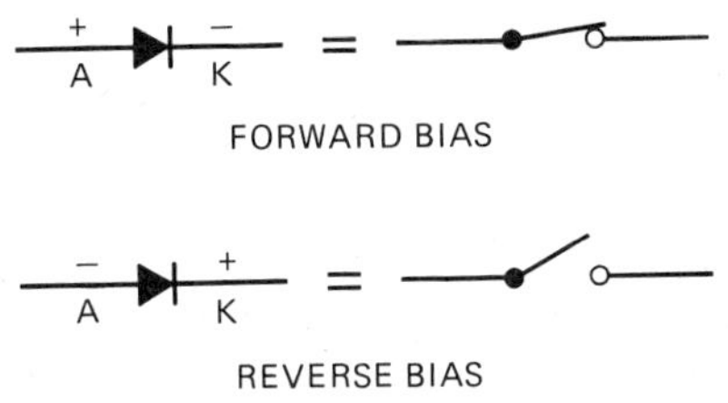

Fig. 1-1a. Forward bias and reverse bias.

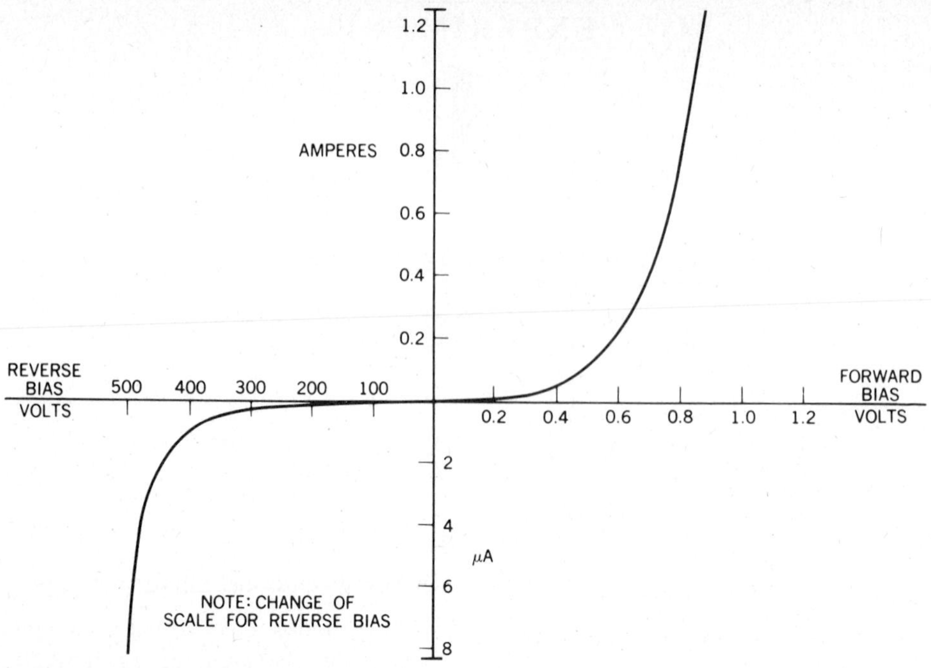

Fig. 1-1b. Volt-ampere characteristic of a silicon rectifier.

Silicon Diode Ratings

Diode characteristics usually supplied by the manufacturer include (at specified temperatures):

1. The maximum reverse bias voltage (V_{RM}) which can be applied to the diode without excessive reverse current.
2. Average half-wave rectified current with resistive load, at a specified temperature.
3. Peak recurrent forward current.
4. Maximum forward-biased voltage drop at a given current.
5. Maximum reverse-biased leakage current at a given voltage.
6. Operating and storage temperatures.
7. Derating factors as a function of temperature.

The physical size and internal construction of these diodes determine their characteristics. For very high voltages, diodes are connected in series, and for very large currents, they are connected in parallel. Some diodes are mounted by their leads, as shown in Fig. 1-2, and some have threaded studs for mounting on a heat sink. The full current and power ratings of stud-mounted diodes cannot be achieved unless the diode is mounted on a heat sink with a specified thermal resistance.

Half-wave Rectifier with Resistive Load

In some industrial applications of diodes as rectifiers, the pulsating rectifier output is used directly without filtering. Battery chargers, electroplating, and some chemical finishing processes do not require pure direct current. Similarly, the windings of dc motors are often provided with the unfiltered dc output current from a rectifier circuit.

A basic half-wave rectifier circuit with resistive load is shown in Fig. 1-3. Because the ac input voltage makes point *A* positive with respect to point *C* for one half-cycle and makes *A* negative with respect to *C* for the next half-cycle, the diode is alternately forward-biased and then reverse-biased. The resulting waveforms across the transformer, diode, and load resistor are shown in Fig. 1-4 and may be observed with an oscilloscope.

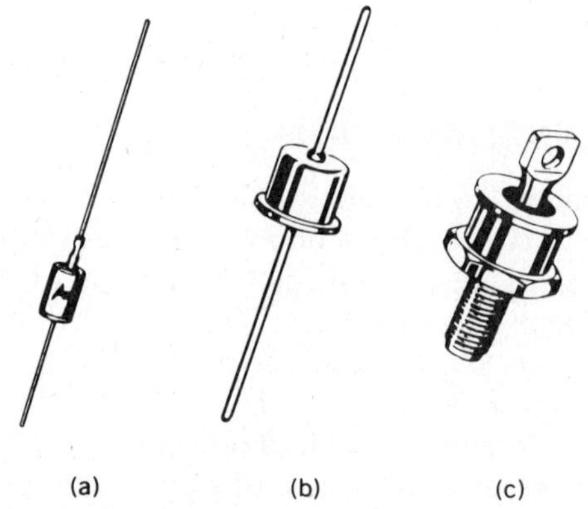

Fig. 1-2. Silicon rectifier types.

Half-wave Rectifier with *RL* Load

Where the load contains inductance in addition to resistance, with motor windings, for example, the oper-

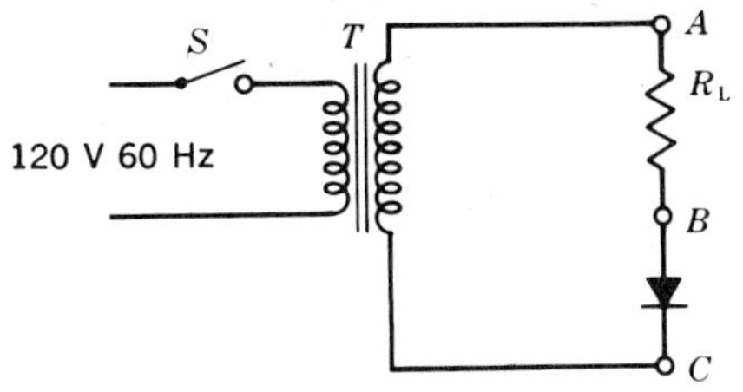

Fig. 1-3. Half-wave silicon rectifier with resistive load.

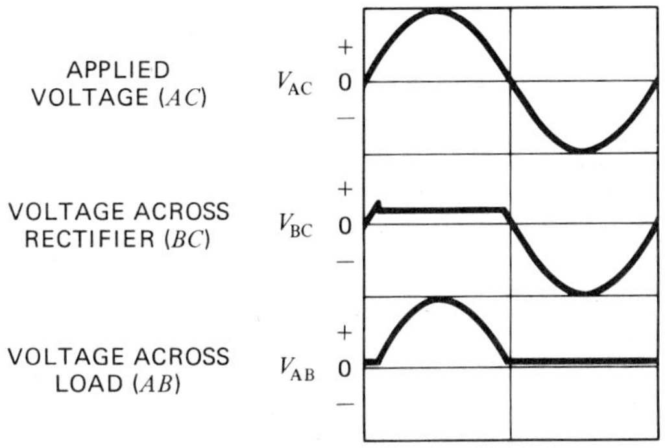

Fig. 1-4. Voltage waveforms in a silicon rectifier circuit with resistive load.

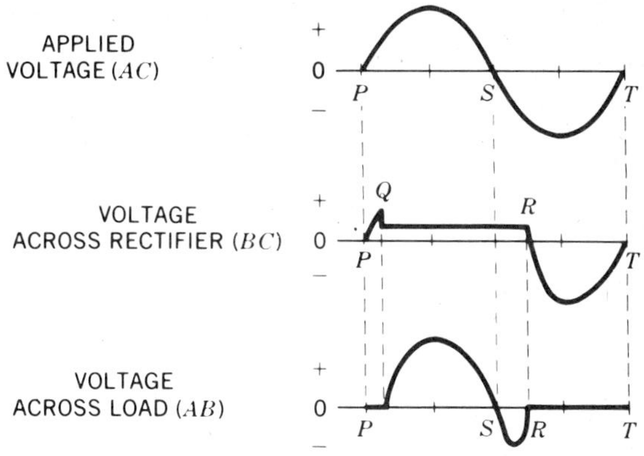

Fig. 1-6. Voltage waveforms in silicon half-wave rectifier with *RL* load.

ation of the circuit is slightly different. Figure 1-5 shows that because an inductance stores energy while current flows through it and releases it when the current is interrupted, the diode appears to conduct when the applied voltage is going negative. However, it is the collapsing magnetic field that creates an additional voltage which, when added to the transformer voltage, causes the diode to be forward-biased and conduct during the period *SR* in Fig. 1-6. The length of this interval increases, within limits, as the inductance value *L* is increased and *R* is decreased (longer time constant = *L*/*R*)

The Oscilloscope in Industrial Electronics

In signal tracing through an amplifier, the waveforms observed are typically ac sine waves, and dc components which may be present are not considered. In industrial applications, however, waveforms are encountered which often have both ac and dc levels, and a "dc-coupled" oscilloscope is required. A dc scope will show both the ac level of interest and any dc value it may be superimposed upon. An ac scope will only show the ac value and is, therefore, not suitable in some cases. Modern oscilloscopes are typically dc-coupled and have provisions to observe only the ac waveform, if that is necessary. In this experiment it is necessary to use an oscilloscope with dc coupling so that the waveforms are shown properly with respect to ground and to each other.

"Floating" the Oscilloscope

One terminal of the input to the oscilloscope is the metallic case of the oscilloscope in most cases. In addition, test equipment often has a three-wire power cable with the third, grounded wire, connected to earth at one end and the oscilloscope case at the other. In other words, when using the oscilloscope, you must be careful not to place the ground side of the oscilloscope input on a "hot" part of the circuit. When measurements must be made in which both sides of the voltage are above ground, it is necessary to "float" the oscilloscope to remove the ground reference placed on the case by the three-wire grounded power cable (Fig. 1-7). To accomplish this, place a 1:1 isolation transformer between the oscilloscope and the power receptacle as shown in Fig. 1-8.

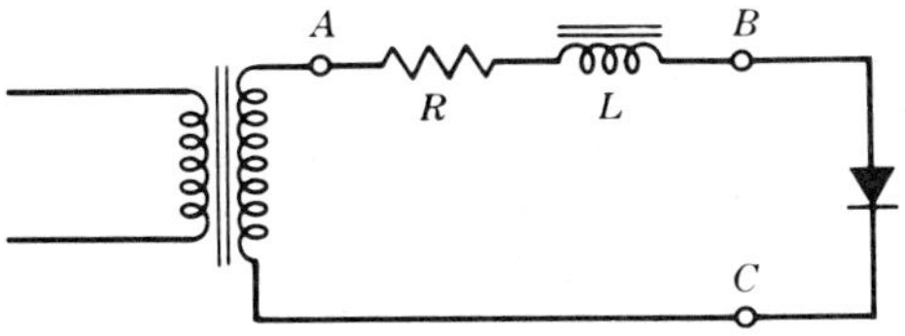

Fig. 1-5. Silicon rectifier with *RL* load.

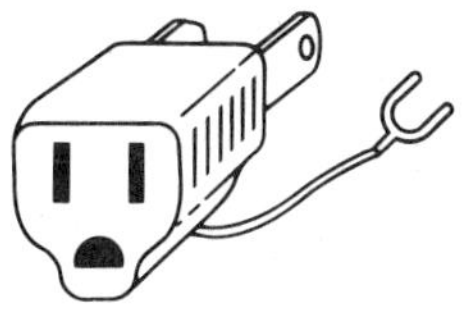

Fig. 1-7. Three-prong to two-prong line cord adapter.

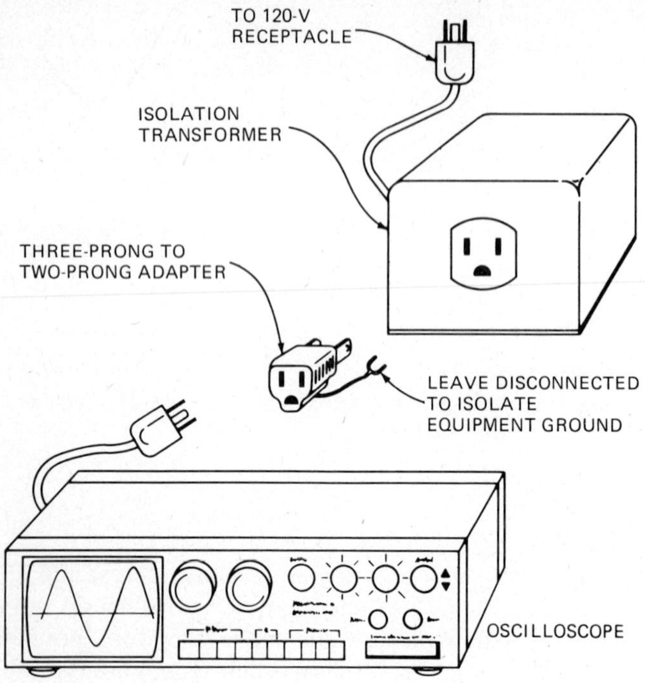

Fig. 1-8. "Floating" an oscilloscope. Plug isolation transformer. Now plug oscilloscope line cord into adapter. Leave the ground wire on adapter disconnected.

SUMMARY

1. Industrial electronics deals with the control, measurement, and monitoring of industrial processes, machinery, and equipment.
2. A silicon diode, used as a rectifier to convert alternating current to direct current, acts as a switch which is closed when forward-biased and open when reverse-biased.
3. Unfiltered direct current having a significant amount of "ripple" is commonly found in many industrial applications of electrical energy.
4. The most basic rectifier circuit uses one diode in a half-wave circuit.
5. To study the operation of circuits with an oscilloscope, it is desirable that the scope have dc-coupled vertical amplifiers.
6. A piece of equipment is "floating" when none of its terminals is connected to earth ground.

SELF-TEST

Check your understanding by answering these questions.

1. A diode conducts when its anode is at a ___________ (positive/negative) potential with respect to its cathode.
2. A reverse-biased diode is like a/an ___________ circuit.
3. A ___________ is used to remove heat from diodes to limit their temperature rise.
4. The voltage drop across a conducting silicon diode is usually about ___________ V.
5. A silicon diode acts as a ___________ (high/low) efficiency rectifier.
6. The forward-biased resistance of a diode = ___________ .
7. The device used to "float" an oscilloscope or other piece of test equipment is called a/an ___________ .

MATERIALS REQUIRED

- Power supply: Variable dc source (0 to 50 V)
- Equipment: Oscilloscope with dc vertical amplifiers; electronic voltmeter (EVM) or volt-ohm-milliammeter (VOM) (20,000 Ω/V)
- Resistors: 10-watt (W): 2500-, 10,000-Ω
- Transformer: Power *T*, Triad R22A, or equivalent
- Rectifier: 1N5625 or equivalent
- Miscellaneous: Inductance, 8 henrys (H) at 85 mA; single-pole single throw (SPST) switch

PROCEDURE

Characteristics of a Silicon Rectifier

1. Connect the circuit shown in Fig. 1-9. V_{AA} is a variable source of dc voltage. Switch S_1 is open. Set the output of the dc supply to 0 V.

2. Close S_1. Power ON. Adjust V_{AA} for 0.1 V. Measure the voltage V_{AB} across the load resistor R_L and record in Table 1-1. Measure also and record the voltage V_{BC} across the rectifier.

TABLE 1-1. Voltage Measurements in Silicon-Rectifier Circuit

Applied Voltage, dc (+)	V_{AB}	V_{BC}	R_{INT}
0.1			
0.2			
0.3			
0.4			
0.5			
0.6			
0.7			
0.8			
0.9			
1.0			
1.5			
2.0			
3.00			
5.00			
10.00			
20.00			
30.00			
40.00			
50.00			

3. Repeat step 2 for each of the applied voltages indicated in Table 1-1.

4. Calculate and record the internal resistance of the rectifier for each value of applied voltage, by substituting the measured values of V_{BC} and V_{AB} in the formula

$$R_{INT} = \frac{V_{BC}}{I_{RECT}} = \frac{V_{BC}}{\frac{V_{AB}}{10{,}000}} = V_{BC}V_{AB} \times 10{,}000\ \Omega$$

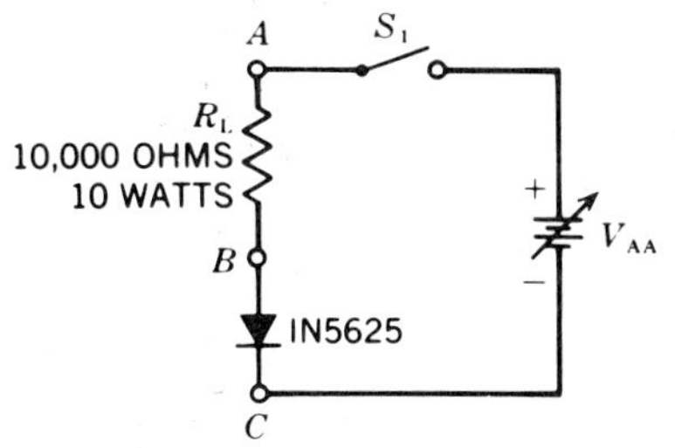

Fig. 1-9. Circuit for determining the characteristics of a silicon rectifier.

5. Power OFF.

Silicon Rectifier with Resistive Load

6. Connect the circuit shown in Fig. 1-10. Power OFF.

7. Calibrate the vertical dc amplifiers of your oscilloscope for a deflection sensitivity of 200 V/cm.

8. Close S_1. Power ON. Set the oscilloscope on line or external sync/triggering. If you are using external sync, connect point *A* (Fig. 1-10) through a 0.002-microfarad (μF) capacitor to the external sync jack of the scope.

9. Connect the scope vertical input leads across points *AC* in Fig. 1-10, positive lead to *A*, negative to *C*. Adjust sweep, sync/triggering, and positioning controls until the applied voltage sine wave is properly centered about the vertical and horizontal axes on the reticule of the cathode-ray tube (CRT) and the applied voltage waveform appears as in Fig. 1-11. The horizontal axis through the center of the waveform is the 0-V reference baseline. Measure and record in Fig. 1-11 the peaks of the positive and negative alternations. NOTE: *Do not vary sweep, sync/triggering, or positioning controls for the remainder of this section.*

10. Now connect the oscilloscope across the rectifier, points *BC*, positive lead at *B*. Observe, mea-

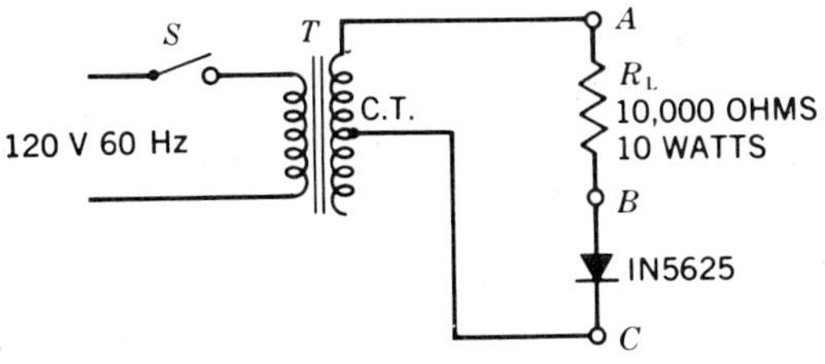

Fig. 1-10. Experimental silicon half-wave rectifier with resistive load.

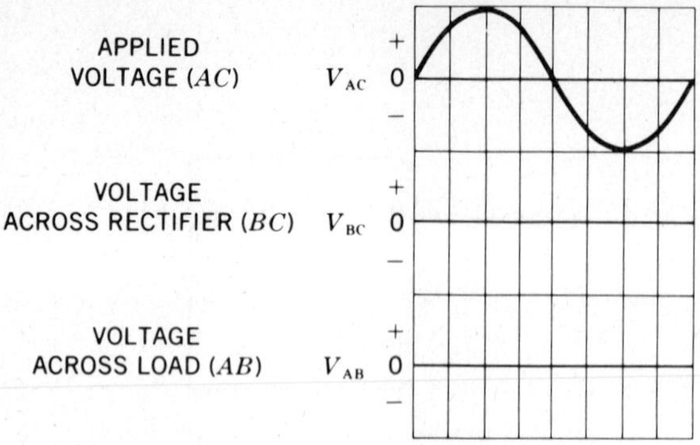

Fig. 1-11. Voltage waveforms in experimental silicon half-wave rectifier with resistive load.

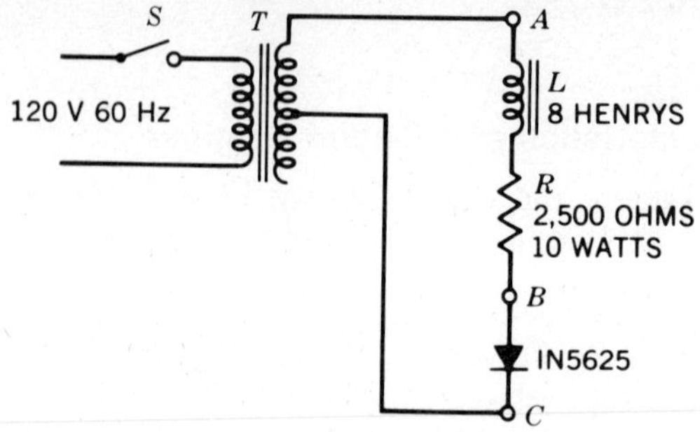

Fig. 1-12. Experimental silicon rectifier with *RL* load.

sure, and record in Fig. 1-11 the waveform across the rectifier, V_{BC} showing the O-V reference line, the amplitude of the voltage drop across the rectifier, and the peak of the negative alternation. The waveform should be drawn in proper time phase with the reference.

11. Connect the vertical scope leads across the load resistor, points *AB,* positive lead on *A*. Observe, measure, and record the waveform in proper time phase with the reference. Show the peak voltage across the load resistor. Power OFF.

Silicon Rectifier with *RL* Load

12. Connect the circuit shown in Fig. 1-12.
13. Repeat steps 8 through 11. Use Fig. 1-13 for your measurements.

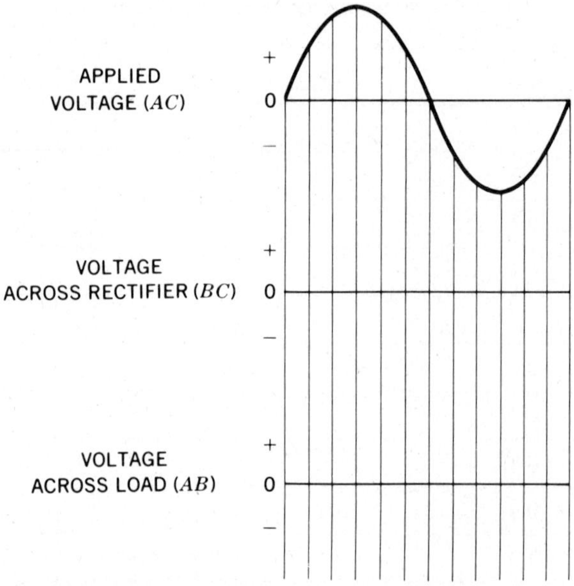

Fig. 1-13. Voltage waveforms in experimental silicon rectifier with *RL* load.

QUESTIONS

1. How did you calculate the rectifier current shown in Fig. 1-9?
2. What do the data in Table 1-1 show about (a) the voltage drop across a silicon rectifier when it is ON and (b) the internal resistance of the rectifier when it is ON?
3. Why does the use of Eq. 1-1 in the experimental procedure yield the value of internal resistance of the rectifier?
4. What do the waveforms in Fig. 1-11 reveal about the time the rectifier is ON; OFF? Explain.
5. How do the waveforms in Fig. 1-13 differ from those in Fig. 1-11? Why do they differ?
6. What indication would you have in Fig. 1-13 that the diode was shorted or open?
7. How would the waveforms in Fig. 1-13 be affected if the diode were reversed in the circuit?

Answers to Self-Test

1. positive
2. open
3. heat sink
4. less than 1 V (0.7)
5. high
6. V_F/I_F
7. isolation transformer

EXPERIMENT

2

THE SILICON-CONTROLLED RECTIFIER (SCR)

OBJECTIVES

1. To become familiar with the SCR (construction, characteristics, and terminology)
2. To study the dc gate-current control characteristics of an SCR

BASIC INFORMATION

The silicon-controlled rectifier (SCR) is a four-layer NPNP solid-state device which has three electrodes: an anode, a cathode, and a gate which serves as a control element. The circuit schematic symbol for an SCR and its four-layer representation are shown in Fig. 2-1. The SCR differs from the two-element diode in that it will not pass an appreciable current when forward-biased until the anode to cathode voltage equals or exceeds a value called the *forward breakover voltage* V_{BRF}. When V_{BRF} is reached, the SCR is switched ON and becomes highly conductive. The value of V_{BRF} can be controlled by the amount of gate current. The gate requires a much smaller current than the anode or load current it controls, and interfacing with other parts of a control system is therefore easily accomplished.

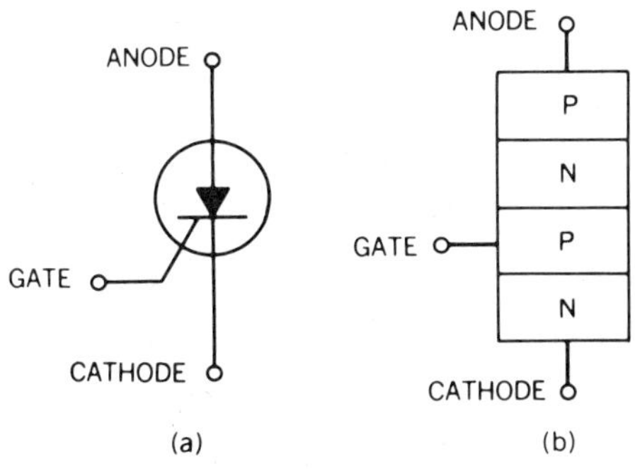

Fig. 2-1. (*a*) SCR circuit symbol; (*b*) four-layer representation.

SCRs with a wide range of voltage and current capacities are available. Small SCRs in TO-5 or TO-92 packages operate at an anode current of less than 1 A and may be used to control indicator lights or small relays. High-voltage, high-current SCRs capable of controlling thousands of amperes with hundreds of volts from anode to cathode may resemble stud-mounted silicon diodes with an added terminal (Fig. 2-2).

Voltage-Current Characteristic

Figure 2-3 shows the voltage-current characteristic of an SCR whose gate is not connected (open). When the anode-cathode circuit is reverse-biased, there is a slight reverse leakage current called the *reverse blocking current*. This current remains small until the *peak reverse voltage* V_{ROM} is exceeded. At that point the reverse avalanche region begins, and the current through the SCR increases sharply. As with two-terminal diodes, SCRs are not designed to conduct when reverse-biased and the anode-to-cathode voltage normally supplied should be less than V_{ROM}.

When the SCR is forward-biased, there is a small forward leakage current, the *forward blocking current,* which remains small until the forward breakover voltage is reached. At the forward avalanche point the current jumps rapidly to a much larger value, the anode-to-cathode resistance of the SCR becomes very small, and the SCR acts as a closed switch. At this point the voltage across the SCR drops to a very small value (about 1 V), and nearly all the source voltage appears across the load which is connected in series with it (Fig. 2-4 on page 9). It is the external load, then, which must limit the current through the SCR and hold it within its ratings. The maximum ratings of V_{ROM} and V_{BRF} are usually the same. The amount of gate current determines V_{BRF} but does not affect V_{ROM}.

The two operating states of an SCR correspond, then, to the two states of an on-off switch. When the anode-to-cathode voltage is less than the breakover voltage,

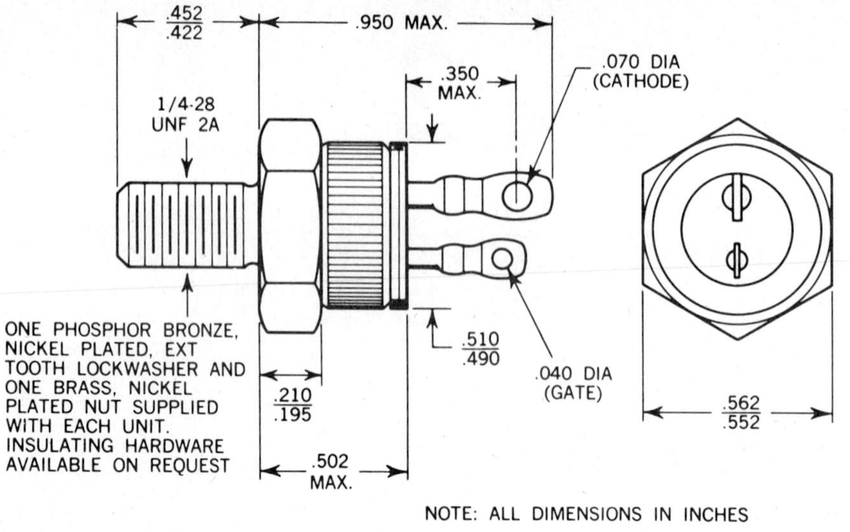

Fig. 2-2. A stud-mounted SCR, GE-type C20; the stud is the anode.

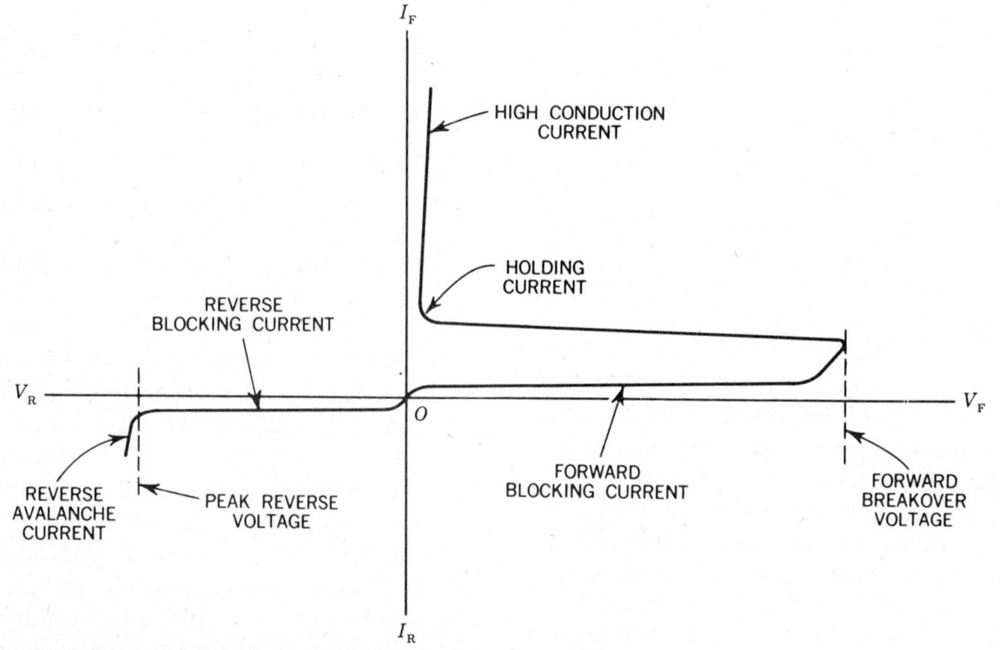

Fig. 2-3. Voltage-current characteristic of an SCR whose gate is open.

the switch is OFF. When the voltage increases to a value equal to or greater than the breakover voltage, the SCR is turned ON. It remains ON (in its conducting state) as long as the current stays above a certain value called the "holding current." When the voltage from anode-to-cathode drops to a value too low to sustain the holding current, the SCR turns OFF.

Gate Control of Forward Breakover Voltage

When the gate-cathode junction is forward-biased, the SCR is turned ON at a lower anode-to-cathode voltage than with the gate open. That is, *forward bias of the gate-cathode junction reduces the value of the forward breakover voltage*. The effect of increasing the value of the gate current is shown in Fig. 2-5. The maximum value of forward breakover voltage V_{BRF0} occurs when the gate is open and $I_{GO} = 0$. With gate current I_{G1} the forward breakover voltage V_{BRF1} is less than V_{BRF0}. Similarly, V_{BRF2}, determined by I_{G2}, is less than V_{BRF1}. *With sufficient gate current, the SCR acts essentially like an ordinary silicon diode.*

After the SCR has been turned ON by gate current, the gate loses control; reducing the value of gate current has no effect on anode current. (This behavior is

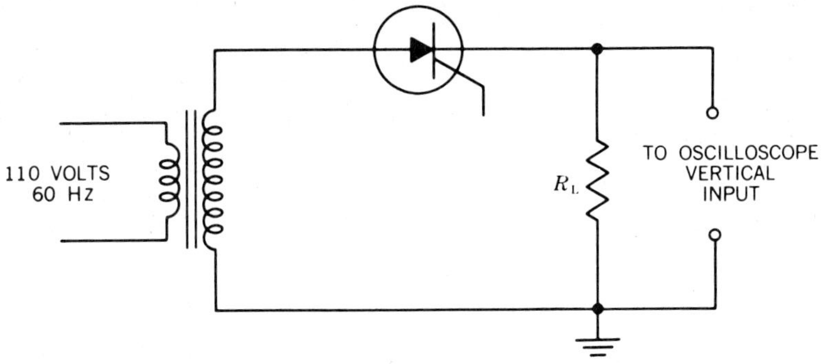

Fig. 2-4. AC source feeding a silicon-controlled rectifier and load.

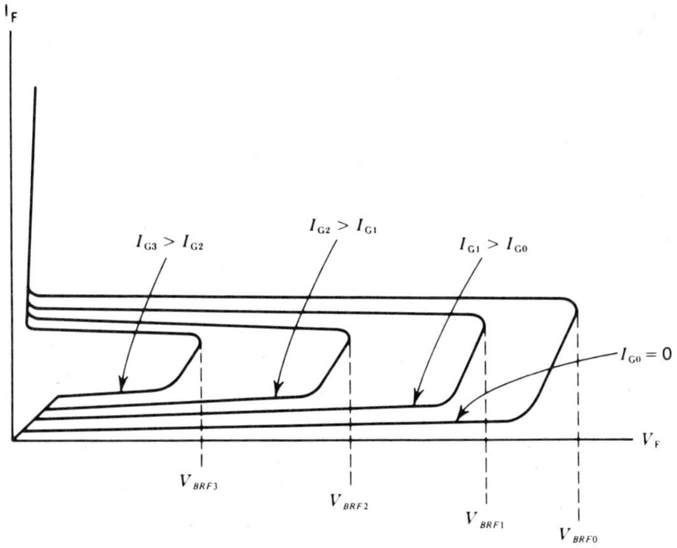

Fig. 2-5. Effect of gate current on forward voltage breakover level.

similar to the operation of a vacuum-tube Thyratron.) The SCR remains ON until the anode voltage is removed or falls below the level required to maintain the *holding current,* which may be a few milliamperes or less. When the anode source is alternating current, the SCR turns OFF during the negative alternation, when the anode-cathode is reverse-biased.

Since the anode-to-cathode voltage at which conduction occurs is determined by the value of forward gate bias, SCRs are operated well below the maximum forward breakover point and the average current through the SCR can be controlled by the amount of gate current or the time at which it is applied.

SCR Ratings

To assure trouble-free operation, the manufacturer's maximum ratings should not be exceeded or permanent damage to the SCR may result. Different manufacturers may use somewhat different symbols or ratings, but the following are typical:

- PFV—repetitive peak forward voltage, gate open
- I_F—root mean square (rms) forward current, ON state
- I_{Fav}—average (dc) forward current, ON state
- V_{ROM}—peak reverse voltage, gate open
- P_{GM}—peak gate power dissipation
- P_{Gav}—average gate power dissipation
- V_{GRM}—peak reverse gate voltage
- T_{stg}—storage temperature
- T_j—operating junction temperature
- *di*/*dt*—maximum rate of rise of anode-cathode current
- *dv*/*dt*—maximum rate of rise of anode-cathode voltage

If the anode-to-cathode voltage increases too rapidly, the SCR will be turned ON without the appropriate value of gate current. To prevent turn ON by exceeding the

dv/dt rating, special networks are connected in parallel with the SCR.

DC Gate-Current Control

Figure 2-6 shows a circuit with separate dc sources for the anode-to-cathode and gate circuits. To study the control characteristics of an SCR with this circuit, both the anode V_{AA} and gate V_{GG} supplies are made variable so that gate current can be made independent of the anode voltage. The load resistance R_L must be large enough to limit anode current to a value within the SCR's ratings. A_1 is the ammeter used to measure anode current, and A_2 measures gate current. Voltmeter V_1 measures anode-to-cathode voltage, and voltmeter V_2 measures the gate-to-cathode voltage. V_{GG} is a variable-voltage source used to set the level of gate current. Resistors R_1 and R_2 establish a specified resistance between the gate and cathode so that the SCR is operated under standard conditions.

Figure 2-7 shows an SCR circuit employing the same dc source for both the anode and gate circuits. Instead of a separate source V_{GG} as in Fig. 2-6, a potentiometer R_3 across V_{AA} in Fig. 2-7 is used to set the value of gate current.

Until the SCR is turned ON, the anode-cathode resistance is very high (a small leakage current may flow). After the SCR is turned ON, the anode-cathode resistance drops sharply to a low value and the gate loses control. Reducing the gate current to zero should not switch the SCR OFF. The SCR is switched OFF when anode-cathode forward bias is removed, when the anode-cathode circuit is broken, or when the anode current is reduced below the holding current value.

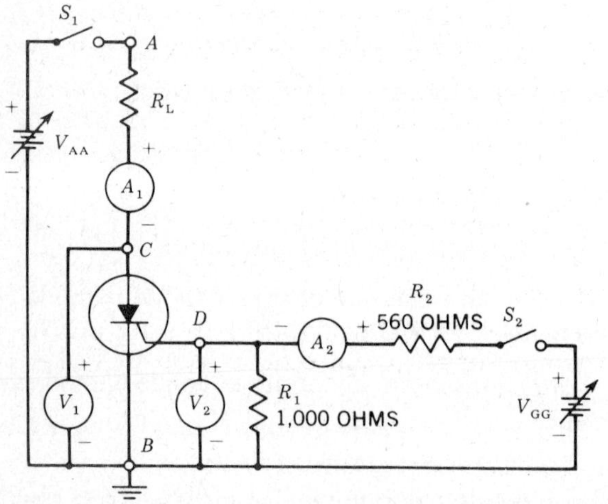

Fig. 2-6. SCR circuit employing separate dc anode and gate source.

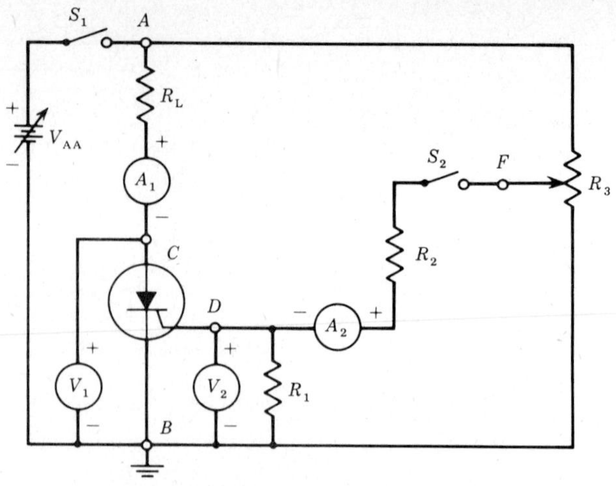

Fig. 2-7. SCR circuit employing same dc anode and gate source.

SUMMARY

1. A silicon-controlled rectifier (SCR) is a three-element, four-layer (NPNP) solid-state device which permits current to flow through it in one direction only, when its anode-cathode circuit is forward-biased by a voltage equal to or greater than the forward breakover voltage V_{BRF}.
2. When an SCR conducts, its anode-cathode resistance is very small. It acts as a switch which turns ON when it is sufficiently forward-biased and is OFF when it is reverse-biased, when forward bias is removed, or when the anode-cathode circuit is broken.
3. Once an SCR is turned ON, it continues to conduct as long as it is forward-biased and the current through it is equal to or greater than the holding current. The holding current is the minimum anode current which will keep the SCR turned ON.
4. The control element of an SCR is the gate. The value of gate current determines the value of the anode-to-cathode voltage required to cause conduction (V_{BRF}).
5. Increasing the value of gate current reduces the value of the forward breakover voltage.
6. When gate current is sufficiently increased, SCR action approaches that of an ordinary silicon diode.
7. Once an SCR is turned ON, the gate loses control and the value of the anode current is determined by the size of the external load and the voltage provided by the source.

SELF-TEST

Check your understanding by answering these questions.

1. The control element of an SCR is the ____________. The forward-biased breakover voltage ____________ as the ____________ current increases.

2. Physically, the SCR is a __________ -layer device having __________ PN junctions.

3. Once an SCR is turned ON, the __________ loses control and it continues to conduct until it is turned OFF.

4. Figure 2-3 shows that the voltage drop across the SCR is very __________ in the high conduction current region.

5. Once an SCR is turned ON, its resistance is very __________ and the current through it is limited by the size of the __________ __________.

6. When the peak reverse-bias voltage V_{ROM} is exceeded, reverse __________ current flows which may damage the SCR.

7. An SCR is much like a switch: it is either __________ or __________.

MATERIALS REQUIRED

- Power supply: Variable regulated 400-V dc source; variable regulated low-voltage dc source
- Equipment: EVM; 0- to 10-mA dc milliammeter; 0- to 100-mA dc milliammeter (or comparable current scales on a VOM or a multirange milliammeter)
- Resistors: ½-W, 560-, 1000-Ω; two 20-W, 1250-Ω
- Semiconductors: GE-type C20D or C22D or C220D or the equivalent; 2N1596 or the equivalent
- Miscellaneous: Two SPST switches

PROCEDURE

Nonconducting Anode-Cathode Resistance of SCR

1. Connect the circuit shown in Fig. 2-6. S_1 and S_2 are OFF. Turn V_{AA} and V_{GG} outputs down to 0 V. Load resistance R_L equals 2500 Ω and consists of two series-connected 1250-Ω 20-W resistors. The SCR used is a C20D or C22D or C220D, or the equivalent. V_{AA} is a variable 0- to 400-V supply. V_{GG} is a variable 0- to 40-V supply.

2. Close S_1, applying power to the anode-cathode circuit. S_2 remains OFF. Adjust V_{AA} until V_1 measures 50 V. This is V_{CB}. Measure and record anode current I_F in Table 2-1.

TABLE 2-1. Nonconducting Anode-Cathode Resistance of an SCR

	V_{CB}, V				
	50	*100*	*150*	*200*	*250*
I_F, A					
R_{CB}, Ω					

Compute the value of anode-cathode resistance R_{CB} by substituting your measured values of V_{CB} and I_F in the formula

$$R_{CB} = \frac{V_{CB}}{I_F}$$

Record R_{CB} in Table 2-1.

3. Repeat step 2 for every value of anode voltage in Table 2-1.

Characteristics of Medium-Current SCR

NOTE: The C20D or C22D is rated at 7.4-A root mean square (rms), maximum anode current at 80°C [176°F], and this value should not be exceeded. The gate current required to turn this SCR ON may vary widely with individual units. Thus one unit may turn ON at a gate current of 2 mA, another at a gate current of 12 mA, and so on. According to the specification sheet, this SCR (C20D/C22D) will definitely fire at 25-mA gate current.

In this part of the experiment, you will first determine the minimum gate current required to fire the SCR when its anode-cathode voltage is set at 100 V. You will then determine the holding current I_H required to keep the SCR ON, first with gate current ON and then with gate current OFF. You will then repeat this procedure for other specified values of applied voltage.

The experimental technique is carefully set forth in the following instructions. These should be followed precisely, particularly the sequence of opening and closing of circuit switches.

4. S_1 and S_2 are both OPEN. Set V_{AA} at 100 V, as measured across the power supply. Do not vary V_{AA} again until you are instructed to do so. S_1 is still OPEN. Set V_{GG} at O V. Close S_2 and adjust V_{GG} until milliammeter A_2 measures a gate current of 1 mA.

5. Now close S_1. Milliammeter A_1, will measure anode current, and voltmeter V_1 will measure anode-cathode voltage. If the SCR has fired, the voltage across the SCR will be very low (about 1.5 V)

and the anode current I_F will be determined by V_{AA} and the size of R_L. Thus

$$I_F = \frac{V_{AA}}{R_L} (approximately)$$

6. If the SCR has not fired, open S_1. Increase gate current to 2 mA and repeat step 5. Continue this process, increasing gate current by 1 mA each time. Note that *switch S_1 must be opened each time before gate current is increased.* After gate current is set, repeat step 5 for each trial.

7. After the SCR fires, open S_1 and *reduce* gate current 0.25 mA from its last level. Now close S_1. If the SCR no longer fires, open S_1. Bring gate current up 0.25 mA to the last value at which the SCR fired. Close S_1. The SCR should fire again. *You have now determined* the value of gate current required to turn the SCR ON when the applied voltage V_{AA} is 100 V. Record this value of gate current I_G in Table 2-2. Record also the value of anode current I_F, the voltage across the rectifier V_{CB}, and the voltage across the load V_{ac}.

8. Open S_2. Observe and record in Table 2-2 the effect on anode current as gate current falls to zero. Again close S_2.

TABLE 2-2. Characteristics of Medium-Current SCR (C20D)

V_{AA}, V	I_G, mA	V_{CB}, V	V_{AC}, V	I_F, mA	I_H, mA
100					
150					
200					
250					

Effect on anode current, after SCR is turned ON, of opening gate switch S_2: ____________

9. Determine the holding current I_H as measured on A_I by gradually reducing V_{AA} until the SCR turns OFF. I_H is the value of I_A just before the SCR turns OFF. Record I_H in Table 2-2. NOTE: As long as anode current remains equal to or greater than I_H, the voltage V_{CB} across the SCR remains very low. Reduction in anode current below I_H will cause V_{cb} to jump suddenly to the applied voltage V_{AA} and will also cause anode current to drop sharply.

10. Open switch S_1. Increase V_{AA} to 150 V. Back off (reduce) gate current about 1 mA from its last setting. Close S_1 and note if the SCR fires. If it does, repeat steps 7 and 8. If the SCR does not fire, open S_1 and *increase* gate current 0.25 mA. Close S_1 again and note if the SCR fires. Repeat this process in 0.25-mA increases of gate current until it does fire. You have now determined the value of gate current I_G required to turn the SCR ON with 150 V V_{AA}. Record I_G. Also determine and record I_H.

11. Repeat steps 9 and 10 turn for V_{AA} = 200 V and 250 V.

Characteristics of Low-Current SCR

12. Power OFF (Open S_1 and S_2). Replace the C22D SCR with 2N1596. This is a low-current SCR, which is rated at 1.6-A rms anode current. Replace R_L with a 1250-Ω 20-W resistor, so that now R_L = 1250 Ω

13. Set the output of the anode supply V_{AA} at 20 V. Now, determine and record the characteristics of the 2N1596 following a procedure similar to that in steps 4 through 11. However, instead of setting gate current at 1 mA as in step 4, set it at 0.5 mA. Increase it to 1 mA, as in step 6, if the SCR does not turn ON. Continue the process of increasing I_G by 0.5mA until the SCR fires. Then determine I_G, the gate current required to fire the SCR, in a manner similar to that in step 7. Record in Table 2-3.

TABLE 2-3. Characteristics of Low-Current SCR (2N1596)

V_{AA}, V	I_G, mA	V_{CB}, V	V_{AC}, V	I_F, mA	I_H, mA
20					
40					
60					
80					

Effect on anode current, after SCR is turned ON, of opening gate switch S_2: ____________

14. Repeat step 13 for every value of V_{AA} in Table 2-3.

QUESTIONS

1. What are the differences between an ordinary silicon rectifier and an SCR?
2. Refer to the data in Table 2-2. For what value of gate current did the medium-current SCR turn ON when (*a*) V_{AA} = 250 V? (*b*) V_{AA} = 100 V?
3. Refer to the data in Table 2-2. After the SCR is turned ON, what is the effect, if any, on anode current of opening gate switch S_2 and reducing gate current to zero?
4. Refer to the data in Table 2-2. Did the value of holding current I_H appear to be affected by the level of gate current? Explain the difference, if any.
5. Compute the anode-cathode resistance of the medium-current SCR when it is just turned ON for V_{AA} = 250 V. Show your computations.
6. How does the value of resistance determined in answer to question 5 compare with the nonconducting resistance of the SCR in Table 2-1?
7. From your data in Table 2-1, does an SCR have a constant resistance in the reverse direction?
8. From Tables 2-2 and 2-3, what are the differences, if any, you observed between the low- and medium-current SCRs?
9. Refer to Fig. 2-7. Compute the maximum current that may be drawn by R_L when V_{AA} = 200 V.
10. After the rectifier turned ON, what was the effect on anode current of reducing anode voltage? Why?
11. How does forward breakover voltage vary with gate current? Refer specifically to your data to confirm your answer.

Answers to Self-Test

1. gate; decreases; gate
2. four, three
3. gate
4. small/low
5. small/low; load resistance
6. avalanche
7. ON/OFF, OFF/ON

EXPERIMENT

3

RECTIFICATION CHARACTERISTICS OF AN SCR

OBJECTIVES

1. To study the operation of an SCR connected as a gate-controlled ac rectifier
2. To observe the effects of varying gate current on the firing point of an SCR connected as an ac rectifier.

BASIC INFORMATION

SCR Used as a Rectifier

SCRs are useful in ac circuits as rectifiers whose average output current can be controlled by the amountof gate current used or the point in time at which gate current is applied. Examples of such applications include the use of SCRs to control the torque or speed of a dc motor or the average voltage or current supplied to a dc load from an ac source, as in electroplating, anodizing, or other electrochemical metal finishing process.

In Fig. 3-1 the SCR is connected as a half-wave rectifier supplying current to R_L. The SCR switches ON when the forward breakover voltage is reached, at point V_{BR1} on the positive alternation of the applied sine wave, which depends on the setting of the gate current control (see Fig. 3-2). During the interval V_{BR1} through V_X, the SCR conducts. When the anode-cathode voltage drops below V_X, the holding potential, the SCR turns OFF and remains OFF during the negative half-cycle, like a two-terminal diode would, provided that the maximum reverse bias rating of the SCR is not exceeded. Current flows through the load during the interval V_{BR1} through V_X. The supply voltage V_{AC}, the anode-to-cathode voltage V_{BC}, the load voltage V_{AB}, and the current through the resistive load are shown graphically in Fig. 3-2.

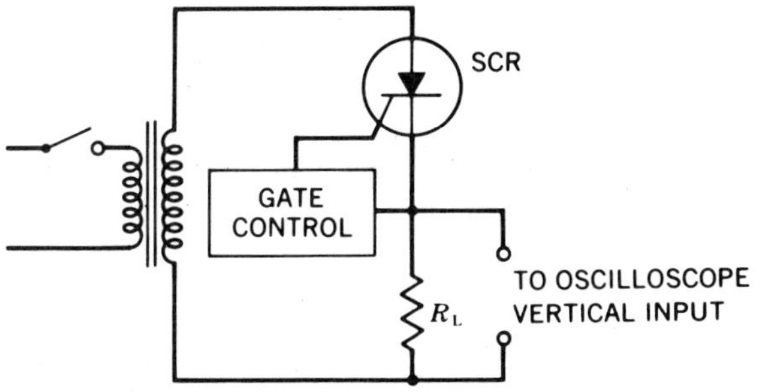

Fig. 3-1. AC source feeding a silicon-controlled rectifier. Gate control determines the firing point.

Effect of Varying Gate Current

In Fig. 3-3, the voltage at which the SCR turns ON is controlled by the amount of dc gate current supplied by V_{GG}. With greater gate current, the SCR begins to conduct at a lower anode-to-cathode voltage (V_{BR2}), and the period during which the load draws current, V_{BR} to V_X, is increased. Therefore the average current through the load is determined by the setting of the gate current control.

The amount of gate current required to fire the SCR is relatively small compared with the anode current. In Fig. 3-3, a dc milliammeter A_2 is used to measure the gate current, and a meter with a higher range A_1 measures the anode current. The average load current through R_L may be many amperes, depending on the amplitude of the applied ac voltage and the size of R_L.

Although dc gate current control of anode current is shown in Fig. 3-3, it is more common to employ ac control of the firing point of the SCR, as shown in Fig. 3-4. In this circuit both the anode-to-cathode voltage and the voltage applied to the gate circuit are supplied by the transformer T. In Fig. 3-4, diode D_1 operates as a half-wave rectifier supplying a positive voltage to the gate-cathode circuit. The amount of gate current is adjusted by the setting of R_3. Since the gate-to-cathode voltage is in phase with the anode-to-cathode voltage, the gate is forward-biased at the same time the anode is positive with respect to the cathode. If the amount

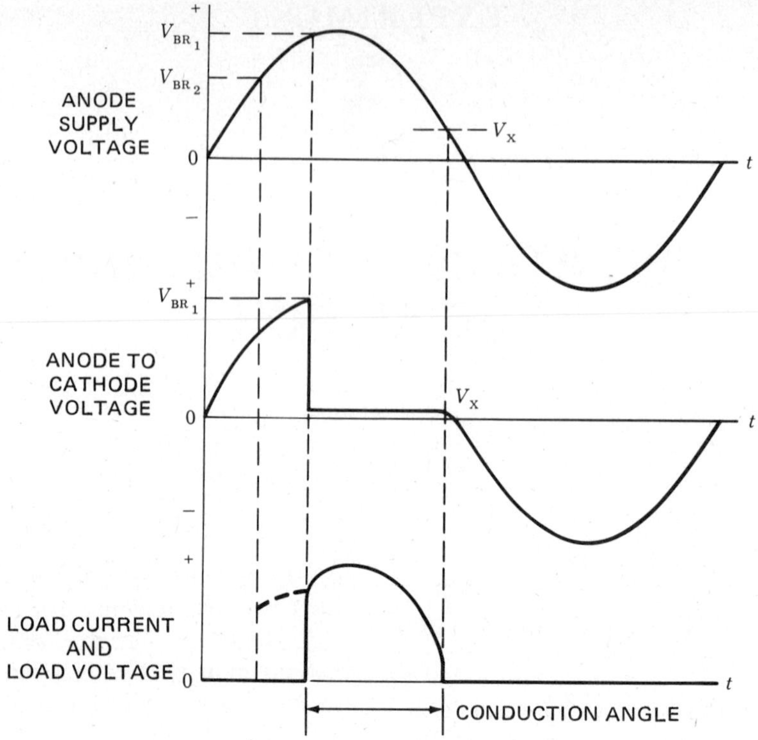

Fig. 3-2. Waveforms in SCR circuit.

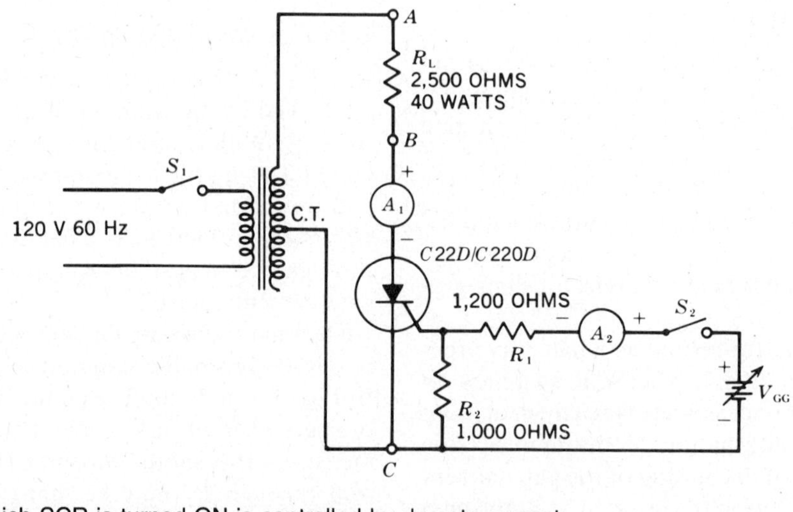

Fig. 3-3. Voltage at which SCR is turned ON is controlled by dc gate current.

of gate current is sufficient, anode current will flow. The length of time that anode current flows depends on the point on the positive half-cycle at which the SCR fires, and this point is determined by the amount of gate current. Increasing the amount of gate current causes the SCR to fire sooner and the average current through the load to increase.

SUMMARY

1. An SCR can be used to control the amount of power supplied to a load.
2. Because a small amount of gate power can control a large amount of power in the anode-cathode circuit, the SCR is particularly useful in industrial electronics.
3. An SCR operating as a half-wave rectifier from an ac source into a resistive load conducts only during the positive alternation (half-cycle) of the applied ac wave (Figs. 3-1 and 3-2). The point at which the SCR starts to conduct depends on the amount of gate current.
4. The waveform across the load shows at which points the SCR starts to conduct, and when con-

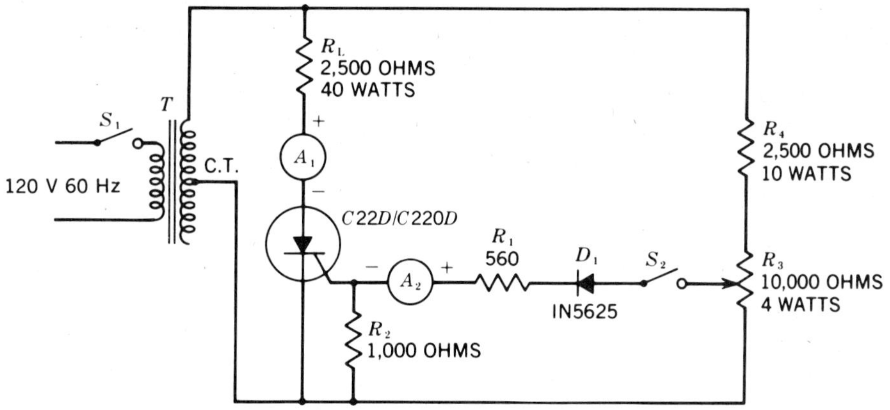

Fig. 3-4. Experimental SCR circuit controlled by ac gate current.

duction stops, with respect to the input waveform. Such waveforms, which are fractions of a sine wave, may generate large amounts of radio frequency interference (RFI) in the vicinity of the equipment in which they are used.

5. Gate current may be supplied by either a dc or an ac source.

SELF-TEST

Check your understanding by answering these questions.

1. In Fig. 3-1, increasing the amount of gate current would make the width of the load current pulse __________ .

2. In Fig. 3-2, the voltage V_{GG} causes the gate-to-cathode circuit to be __________ -biased.

3. In Fig. 3-2, increasing the setting of V_{GG} should cause the indication of A_1 to __________ and the indication of A_2 to __________ .

4. V_{AC}, Fig. 3-3, observed with an oscilloscope, is 500 V peak to peak. With S_1 and S_2 closed and power applied, the anode-to-cathode voltage V_{BC} is also a 500-V peak-to-peak sine wave. Under these conditions, the SCR __________ (is/is not) conducting.

5. If the waveform across the load in Fig. 3-3 or 3-4 is almost a complete half-cycle, the gate current is __________ .

6. The maximum conduction time through the SCR in Fig. 3-3 or 3-4 is __________ .

MATERIALS REQUIRED

- Variable, regulated dc voltage source (0 to 40 V dc)
- DC-coupled oscilloscope
- Digital multimeter or equivalent
- DC milliammeters: 0- to 1- to 0- to 10-mA and 0- to 25-mA ranges
- Resistors: 560, 1000, 1200 Ω (½ W); 2500 Ω (10 W); two 1250 Ω (20 W)
- Capacitor: 0.002 μF, 400 V
- SCR: C220D/C22D or equivalent; 1N5625 diode or equivalent
- Transformer: 120-V/60-Hz primary; 250-0-250–V/100-mA secondary
- Two SPST switches
- Potentiometer: 10k/4 W

PROCEDURE

CAUTION: *This experiment involves measurement of hazardous voltages. Perform the setup and measurements as described and observe appropriate safety precautions.*

DC Gate Current Control and AC Anode Source

1. Connect the circuit shown in Fig. 3-3. S_1 and S_2 are both open. Note that only half of the secondary winding of T is used. Adjust V_{GG}, the gate-current control, to zero output.

2. Externally synchronize the oscilloscope by connecting a 0.002-μF capacitor from point A, the top of the secondary winding of T, to the external synchronizing input to the oscilloscope. NOTE: Because the waveforms to be observed are at the line frequency, 60 Hz, *line sync* may be used instead of

external sync. Turn S_1 ON. S_2 is still OFF.

Observe the waveform from A to C. Adjust the oscilloscope controls until this *reference* waveform appears as in Table 3-1. Measure and record its peak positive amplitude.

Connect the ground lead of the oscilloscope to the cathode of the SCR and the input lead of the oscilloscope to the anode. Observe the waveform across the SCR. (The SCR should not be conducting.)

3. Close S_2. Slowly adjust V_{GG} over its range and observe the resulting effect on the waveform across the SCR. In Table 3-1, draw, in proper time phase with respect to the reference waveform, the anode-to-cathode waveform for the maximum conduction period (waveform 1) and for the minimum conduction period (waveform 2). For both conditions, measure the peak positive amplitude of the waveform, the load current, the gate current, and the conduction angle (the duration of the ON interval in any one cycle, in degrees). Record the data in Table 3-1.

AC Gate Control and AC Anode Source

4. Power source disconnected. Modify the preceding experimental circuit to conform with Fig. 3-4. Note that the dc gate source has been replaced by an ac source.

5. Power source connected, oscilloscope synchronized as before, S_1 and S_2 closed. With the oscilloscope connected across the SCR, anode to cathode, observe the effect on the conduction angle as R_3, the gate current control, is varied from minimum to maximum. Record the minimum and maximum conduction angles, and the corresponding load current I_L and gate current I_G in Table 3-2.

TABLE 3-1

Waveform Number	*Waveform*	*Volts, Peak Positive*	I_L, mA	I_G, mA	*Conduction Angle*
Reference	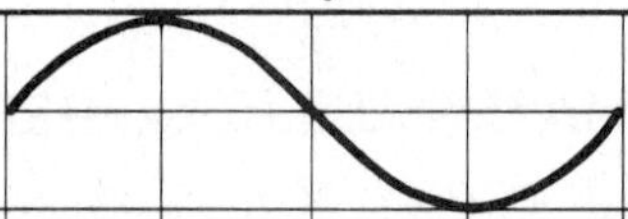		X	X	X
1					
2					

TABLE 3-2

	Conduction Angle	I_L, mA	I_G, mA
Minimum			
Maximum			

QUESTIONS

1. Under what conditions will an SCR conduct current?
2. What would happen if the voltage supplied to the gate in Fig. 3-4 were 180° out of phase with the anode-to-cathode voltage?
3. From your data, comparing the circuits of Fig. 3-3 and 3-4, which has the greatest *range* of control: (*a*) dc gate current or (*b*) ac gate current?
4. Assume R_L is a light bulb or electrical heater: What are the limitations of the circuits in this experiment?
5. What is meant by "synchronizing" the oscilloscope? Would *internal* synchronization work as well as *line sync?*

Answers to Self-Test

1. wider/greater
2. forward
3. increase; increase
4. is not
5. high/large/maximum
6. less than 1/120 second/ [8.3 milliseconds (ms)]

EXPERIMENT

4

PHASE-SHIFT BRIDGE CIRCUIT

OBJECTIVES

1. To study the properties of *RL* and *RC* phase-shift bridge circuits
2. To observe the waveforms and extent of phase shift provided by *RL* and *RC* phase-shift bridge circuits

BASIC INFORMATION

Experiment 3 demonstrated that the firing point of an SCR can be controlled by the amount of current supplied to the gate. But this method only provides about 90° of control: the SCR comes ON at the peak of the anode-to-cathode voltage, and increasing the gate current causes the SCR to fire progressively sooner. The range of control is approximately 0 to 90°. In many applications it is necessary to delay the firing point beyond the positive peak of the anode-cathode voltage for smooth and complete control. One method of accomplishing this is to use a circuit which shifts the phase of the gate current relative to the anode-to-cathode voltage.

Simple *LC* or *RC* phase-shift networks (Fig. 4-1) provide a maximum phase shift of 90° and have the undesirable property that the amplitude of the output voltage decreases as the amount of phase shift increases.

The phase-shift bridge circuit, Fig. 4-2, provides an output whose amplitude is constant and whose phase shift is adjustable over approximately 180° relative to the input.

LR Phase-Shift Bridge

The elements of an *LR* phase-shift bridge circuit are shown in Fig. 4-2. Center-tapped transformer *T* supplies two equal voltages which are out of phase with each other by 180° relative to the center tap *A*. The output voltage is taken from *B* to *A*. The operation of this circuit is described by the phasor diagram, Fig. 4-2*b*. It will be shown that, as *R* is varied over a wide range, the output voltage lags the input voltage by an angle between approximately 180 and 0° and the output voltage has a constant amplitude.

In the phasor diagram (a special form of vector diagram), Fig. 4-2*b*, the direction of the arrows indicates the relative phase of the voltages and the length represents their magnitude (voltage). Phasors **PA** and **AQ** lie on the same line because the two voltages are in phase, adding up to the total secondary voltage V_{PQ}. In an inductive circuit the current lags the voltage. Therefore, the current phasor **I** is shown at a lagging (clockwise) angle α with respect to the applied voltage **BQ.** The voltage drop across *R* is in phase with the current and is shown by phasor **PB.** Since the voltage across the inductor leads the current through it, the phasor representing this voltage drop **BQ** is shown at a right angle to **PB. PB** plus **BQ** always equals the applied voltage **PQ.** The angle α is determined by the value of *R* with respect to the inductive reactance X_L of the inductor.

In Fig. 4-2*b* the bold diagram *PAB* represents the case where *R* and X_L are nearly equal and the output

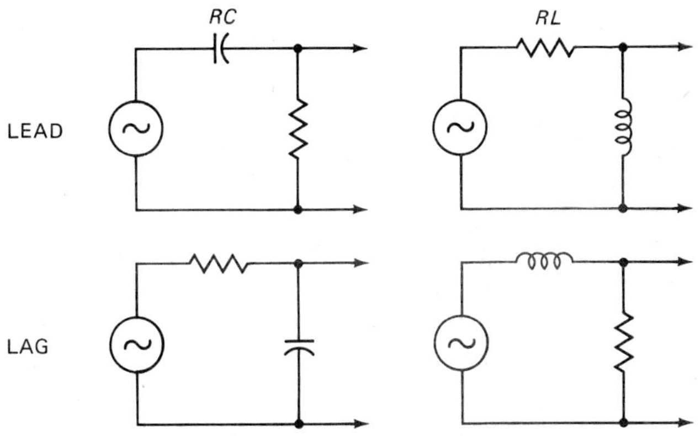

Fig. 4-1. Simple phase-shift networks.

age would lead the anode voltage and the conduction angle of the SCR could not be controlled.

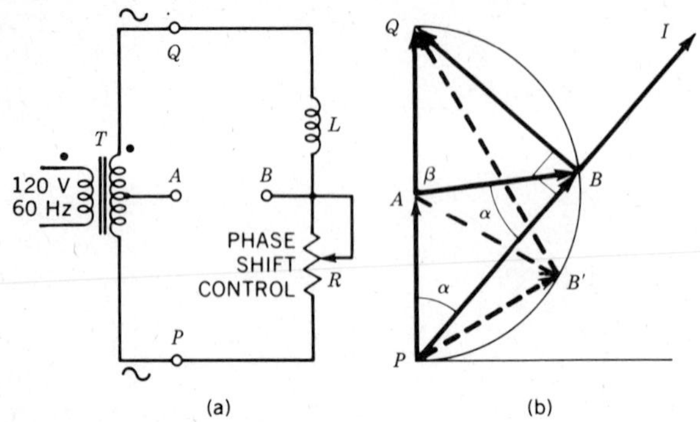

Fig. 4-2. *LR* phase-shift circuit and phasor diagram.

voltage **AB** lags the voltage **AQ** by nearly 90°. The dotted diagram PAB' occurs when R is smaller than X_L and the voltage **AB′** lags **AQ** by a larger angle. When R is zero, the output voltage V_{AB} lags V_{AQ} by 180°. When R is maximum, the two voltages are nearly in phase.

For any value of R, B' is a point on a semicircle whose center is A. Therefore, the length of phasor **AB** is constant as R is varied. Angle $\beta = 180° - \angle PAB$ and $\angle PAB = 180° - 2\angle\alpha$. Therefore, $\angle\beta = 2\angle\alpha$. Since **BQ** is proportional to X_L and **PB** is proportional to R, $\tan\alpha = BQ/PB = X_L/R$. When $R = 0$, $\tan\alpha = \infty$, $\angle\alpha = 90°$, and $\angle\beta = 180°$. If R were ∞, $\tan\alpha = 0$. Practical considerations limit the value of R to 10 to 20 times X_L. When $R = 20X_L$, $\alpha = 3°$ approximately, and $\beta =$ about 6°.

In summary, if R is adjustable from 0 to 20 times the reactance of L, the range of phase delay possible is from 6 to 180° and the amplitude is constant.

In Fig. 4-3 the anode-to-cathode voltage of the SCR is V_{QA} and the gate-to-cathode voltage is V_{BA}, a voltage whose phase lags V_{QA} by an angle determined by the value of R. As R is varied, the conduction angle of the SCR is adjustable over approximately 180°. It is essential that the gate voltage lag the anode voltage. If the positions of L and R were exchanged, the gate volt-

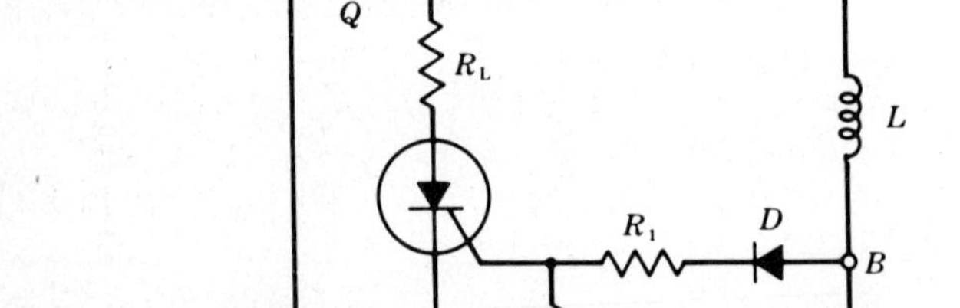

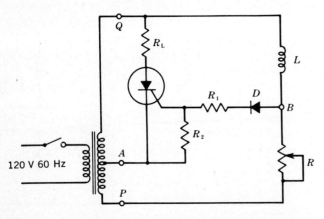

Fig. 4-3. SCR fired by controlling *LR* phase-shift circuit.

RC Phase-Shift Bridge

Inductors tend to be larger, more expensive, and less ideal than capacitors and unnecessary in many applications. If a capacitor is substituted for the inductor and the circuit revised as in Fig. 4-4, a circuit having properties similar to Fig. 4-2 results. In this case the current phasor **I** leads the voltage **PQ** by an angle determined by the value of R with respect to X_C. **BQ** is drawn parallel to **I** and represents the voltage drop across R. **PB**, the drop across the capacitor, is at 90° lagging with respect to the current **I.** By an analysis similar to that describing the *LR* circuit, it may be shown that the output voltage is constant as R is varied and that $\tan\beta/2 = R/X_C$. If R is varied from 0 to its maximum value ($20X_C$), the phase shift varies from 0 to 174°, approximately.

NOTE: *In Figs. 4-2 and 4-4 a center-tapped transformer provides the required two equal voltages of opposite phase. A similar result can be obtained by using two equal resistors to establish an electrical center tap. The circuit of Fig. 4-5 has properties similar to that of Fig. 4-4. The resistors providing the center tap should be relatively small.*

SUMMARY

1. The firing point of an SCR may be controlled by (*a*) a variable dc gate current or (*b*) an ac gate current whose amplitude or phase is variable.
2. A phase-shift bridge circuit provides an output voltage which is constant in amplitude and variable in phase over nearly a 180° range.
3. For proper control of the firing point of an SCR, the gate signal must *lag* the anode-to-cathode voltage.
4. Using a phase-shift bridge to provide variable-phase gate current, the firing angle of an SCR can be controlled over almost a 180° range. (The anode-to-cathode voltage must be derived from the same ac source or an equivalent.)
5. The phase-shift bridge may use either R and L or R and C components. Reversing the positions of the components changes the output phase from lagging to leading the input.
6. The properties of the phase-shift bridge circuit are due to the facts that (*a*) the current through an inductor lags the voltage across the inductor and (*b*) the current through a capacitor leads the voltage across the capacitor.

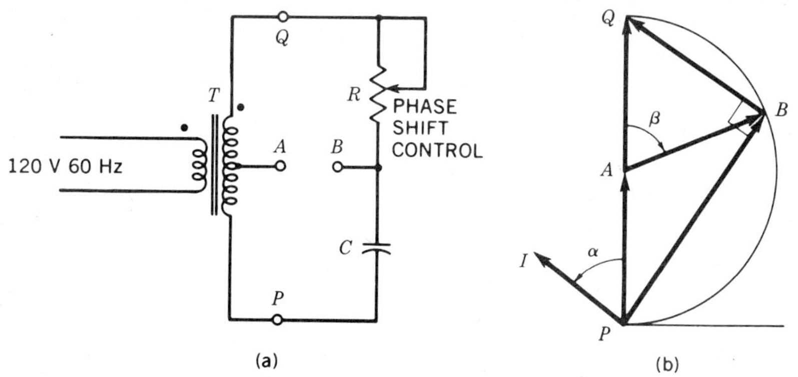

Fig. 4-4. *RC* phase-shift circuit and phasor diagram.

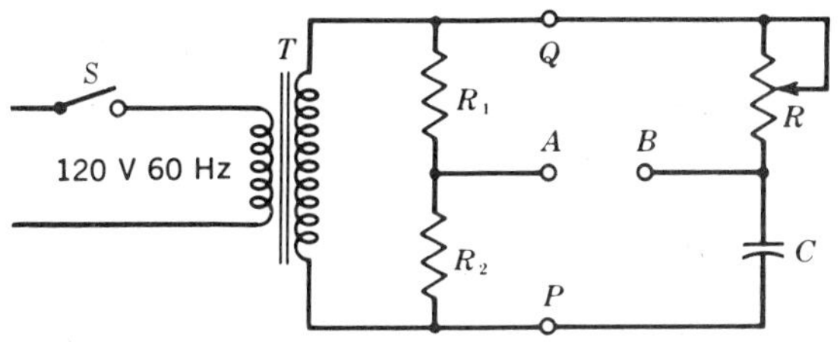

Fig. 4-5. Phase-shift circuit variation.

SELF-TEST

Check your understanding by answering these questions.

1. In the circuit shown in Fig. 4-2 the voltage from *B* to *P* should ___________ (lead/lag) the voltage from *Q* to *P*.
2. Relative to point *A*, the voltages from *Q* to *A* and *P* to *A* are ___________ and ___________ degrees out of phase.
3. The purpose of the diode *D* in Fig. 4-3 is to ___________ the gate ac voltage, thus permitting the gate to be ___________ -biased, only.
4. In Fig. 4-2, if $R = X_L$, the voltage V_{BA} would be shifted ___________ ° with respect to voltage V_{QA}.
5. In Fig. 4-4 maximum phase shift occurs when *R* is ___________ (minimum/maximum).
6. If the positions of *R* and *C* in Fig. 4-4 are reversed, the phase of the output voltage will change from lagging to ___________ .
7. With no load the output voltage from the phase-shift bridge remains ___________ as the phase shift varies.

MATERIALS REQUIRED

- Power supply: 120-V 60-Hz source
- Equipment: Oscilloscope
- Transformers: T_2, filament step-down transformer, 120-V primary, 25-V 1-A center-tapped secondary
- Miscellaneous: Inductor, 8 H at 85 mA; 500,000-Ω 2-W potentiometer; 50,000-Ω 4-W potentiometer; 0.1-μF capacitor at 400 V; SPST switch

PROCEDURE

RC Phase-Shift Circuit

1. Connect the circuit shown in Fig. 4-3. *T* is a step-down transformer whose primary receives 120 V ac. The secondary is a 25-V center-tapped winding. *R* is a 500,000-Ω 2-W potentiometer, and *C* is a 0.1-μF capacitor. **Do not apply power.**
2. Connect the ground lead of the oscilloscope to point *A* and the vertical input lead to point *Q*, the top of the secondary winding.
3. Power ON. Adjust the oscilloscope's sweep, sync, and vertical gain controls until one 60-Hz sine wave appears on the screen, with the polarity shown in Table 4-1, waveform number 1. The positive alternation leads the negative on the time axis. Use line sync or external sync with point *Q* connected to the external sync jack of the oscilloscope through a 0.001-μF capacitor. Adjust the vertical and horizontal gain controls so that the waveform is 4 cm high and 4 cm wide. Center it in the 4- by 4-cm square on the scope, using the guidelines on the etched faceplate. If you cannot secure waveform 1, reverse the ac power plug. You will get a stable waveform, as in Table 4-1, in one position or the other of the power plug. Once you have obtained the proper reference waveform 1, *do not* readjust the controls on the

TABLE 4-1. Waveforms—*RC* Phase-Shift Circuit

Waveform Number		*Waveform*	*Resistance*	*Waveform Number*		*Waveform*
1	V_{QA} + 0 −	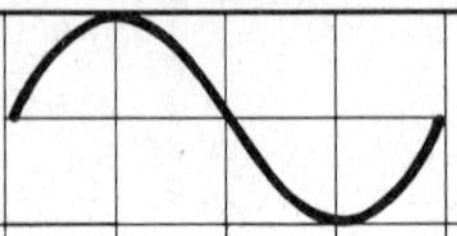		5	V_{QA} + 0 −	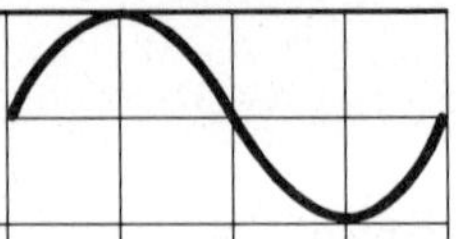
2	V_{BA} + 0 −		$R=$	6	V_{BA} + 0 −	
3	V_{BA} + 0 −		$R=$	7	V_{BA} + 0 −	
4	V_{BA} + 0 −		$R=$ (max)	8	V_{BA} + 0 −	

oscilloscope until you have completed checking waveforms 1 through 4.

4. Keep the ground lead of the oscilloscope at point *A* for the remainder of this exercise. Connect the vertical input lead of the oscilloscope to point *B*. Adjust the phase-shift control to minimum ($R = 0$). Observe and record this waveform 2 in Table 4-1, in proper time relationship with the reference voltage, waveform 1. Compare the amplitude of the waveform with the reference voltage.

5. Adjust *R* until there is a 90° phase shift between V_{BA} and V_{QA}. Draw this waveform 3 in Table 4-1. Do not disturb the setting of *R*. Compare the amplitude of the waveform with the reference voltage.

6. Power OFF. Measure the resistance of *R* and record in Table 4-1. This is the resistance value which causes a 90° phase displacement.

7. Power ON. Adjust *R* to maximum resistance. Observe and record waveform 4 in Table 4-1. Observe proper time relationship to reference waveform 1. Compare the amplitude of the waveform with the reference voltage.

8. Power OFF. Reverse the positions of *R* and *C* in Fig. 4-3. Otherwise, the circuit remains the same.

9. The ground lead of the oscilloscope remains connected at point *A*. Connect the vertical input lead to point *Q*. If necessary, readjust the oscilloscope controls until you get a waveform 5, identical with waveform 1.

10. Repeat step 4. Record this waveform 6 in Table 4-1, in proper time relationship with reference waveform 5.

11. Repeat step 5. Record this waveform 7 in Table 4-1. Power OFF.

12. Repeat step 6.

13. Repeat step 7. Record this waveform 8 in Table 4-1. Power OFF.

14. Connect the circuit of Fig. 4-1. *T* is the transformer previously used in this experiment. *L* is an 8-H 85-mA inductor. *R* is a 50,000-Ω 4-W potentiometer. **Do not apply power.**

15. Repeat step 2.

16. Repeat step 3. The waveform you observe is waveform 9 and is drawn for reference in Table 4-2.

17. Repeat step 4. The waveform you observe is waveform 10. Draw it in Table 4-2, in the proper time relationship to waveform 9.

18. Repeat step 5. The waveform you observe is waveform 11. Draw it in Table 4-2. Power OFF.

19. Repeat step 6, and record the value of *R* in Table 4-2.

20. Repeat step 7. The waveform you observe is waveform 12. Draw it in Table 4-2.

21. Power OFF. Reverse the positions of *R* and *L* in Fig. 4-2. Otherwise the circuit remains the same.

22. The ground lead of the oscilloscope remains connected at point *A*. Connect the vertical input lead to point *Q*. If necessary, readjust the oscilloscope controls until waveform 13 is identical with waveform 9.

23. Repeat step 4. Record this waveform 14 in Table 4-2, in proper time relationship with waveform 13.

24. Repeat step 5. Record waveform 15 in Table 4-2.

25. Repeat step 6, and record the value of *R* in Table 4-2.

26. Repeat step 7. Record waveform 16 in Table 4-2. Power OFF.

TABLE 4-2. Waveforms—*RL* Phase-Shift Circuit

Waveform Number		*Waveform*	*Resistance*	*Waveform Number*		*Waveform*
9	V_{QA} + 0 −			13	V_{QA} + 0 −	
10	V_{BA} + 0 −		$R = 0$	14	V_{BA} + 0 −	
11	V_{BA} + 0 −		$R =$	15	V_{BA} + 0 −	
12	V_{BA} + 0 −		$R =$ (max)	16	V_{BA} + 0 −	

QUESTIONS

1. In setting the firing point of an SCR, what is the advantage of using phase-shift control of gate current over dc or ac amplitude control?
2. Which of the waveform series in Table 4-1, 2 through 4 or 6 through 8, represents lagging voltages with respect to V_{QA}?
3. Calculate the reactance of a 0.1-μF capacitor at 60 Hz and compare this value with that obtained in Table 4-1, waveform 3. Explain.
4. Why is it necessary to properly synchronize the oscilloscope to an external reference in this experiment?
5. Why are the circuits in this experiment referred to as "bridge" circuits? Sketch the *RC* bridge and resistive center-tap circuit as a four-arm bridge.
6. Calculate the reactance of an 8-H inductor at 60 Hz and compare this with the resistance value obtained in step 18. Explain.

Answers to Self-Test

1. lag
2. equal/the same; 180°
3. rectify; forward
4. 90°
5. maximum
6. leading
7. constant/fixed/ the same

EXPERIMENT

5

PHASE-SHIFT CONTROL OF AN SCR

OBJECTIVES

1. To observe how a phase-shift network affects the firing point of an SCR
2. To study the requirements for phase-shift control

BASIC INFORMATION

In previous experiments we established that controlling the firing point of an SCR, whose anode-cathode is connected to a sinusoidal source of ac power, can be effected by varying the amplitude of gate current. Moreover, we found that gate control was possible by means of either dc or ac current. However, the range of firing control was limited to approximately 90° or one-half of the alternation, during which the anode is positive relative to the cathode (the only time in the ac cycle when the SCR can be turned ON). In this experiment we will observe that a properly polarized phase-shifted ac gate current can effect a much wider firing range than amplitude variation of the gate current.

Requirements for Phase-Shift Control of an SCR

In Experiment 3 we found that, for a specified ac anode-cathode voltage V_{AK}, there is a minimum level of gate-cathode voltage V_{GK} which will trigger the SCR. This can be represented by the graph in Fig. 5-1. Moreover, at the point of triggering, V_{GK} must be a positive voltage, like the reference waveform V_{AK}. The case where V_{GK} and V_{AK} are exactly in phase is shown in Fig. 5-2. When V_G reaches the level V_{GK} at point P_1, the SCR is triggered at the level V_1 on the anode-cathode voltage waveform. When V_G is 180° out of phase with V_A, the SCR cannot fire, as is evident from Fig. 5-3. For SCR triggering, then, the phase of V_G must lie between 0° (in phase) and 180° (out of phase) relative to V_A. Moreover, the phase of V_G in the phase-shift trigger (Fig. 5-4) must be lagging relative to V_A, in order to effect continuous control of anode current.

The frequency of the gate-cathode and anode-cathode waveforms must be the same. Figure 5-5 shows that a leading gate-cathode waveform V_{G1}, which has reached V_{GK} at the minimum level of anode-cathode voltage V_{min} which will support SCR triggering, will turn ON the SCR at the time t_1 in the cycle. Moreover, a waveform V_{GK} with a greater leading phase than V_{G1} will turn on the SCR at the same time t_1. Therefore, if the gate-cathode voltage has a leading phase relative to the anode-cathode voltage, the SCR will fire at the same time and there can be no continuous control of the firing point.

Figure 5-6 shows that a lagging gate-cathode voltage V_{G1}, which intersects V_{GK} at P_1, will trigger the SCR t_1 seconds or α_1° after the start of the positive alternation of the anode voltage. Similarly, V_{G2} will trigger the SCR t_2 seconds or α_2° after the start of the positive alternation of the anode voltage. It is evident that it should be possible to effect control of SCR triggering

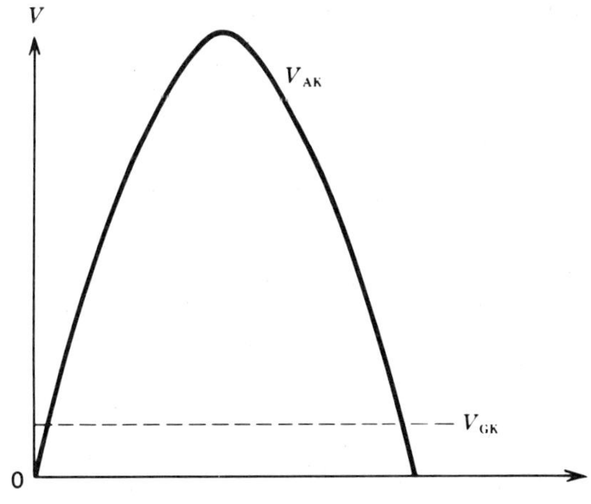

Fig. 5-1. *VGK* is the minimum gate-cathode voltage required to trigger an SCR whose anode-cathode voltage is *VAK*.

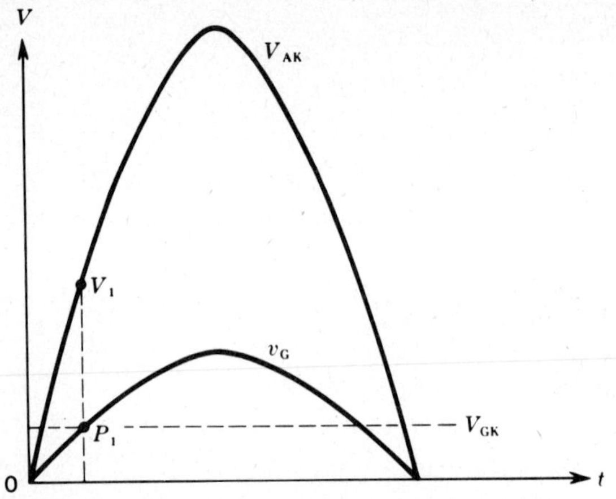

Fig. 5-2. When the voltage from gate to cathode reaches the level *VGK* at point *P*1, the SCR fires. At that point, its anode-cathode voltage is *V*1.

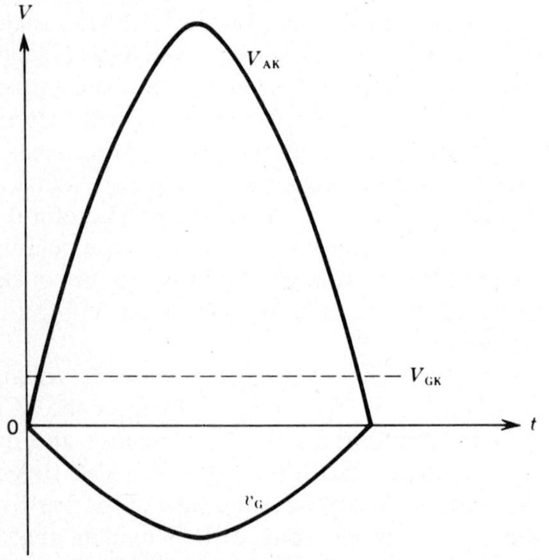

Fig. 5-3. When the gate-cathode voltage *VG* is 180° out of phase with the anode-cathode voltage *VAK,* the SCR will not turn ON.

over almost 180°, if we can vary the lagging phase of V_G over a range of 180°.

Phase-Shift Bridge Circuit Controlling an SCR

Fig. 5-7 is a phase-shift bridge circuit, identical with that which we studied in Experiment 4, whose output V_{HC} controls the firing of the SCR. Note that to achieve a lagging phase for control, the phase of voltage V_{AC} in the secondary of transformer T_1 must be the same as the phase of voltage V_{DF} across the secondary of transformer T_2. This is indicated by the dot (·) at the end of the secondary winding of T_1 and T_2.

We can study the operation of this circuit by observing the waveform from *H* to *C*, V_{HC}, and the waveform across the load resistor R_L, V_{AB}, and comparing these waveforms with the applied voltage V_{AC}, as R_3, the phase-shift control, is varied over its entire range. Not only can we observe when the SCR is triggered, but we can also study the effect on load current I_F for the various firing angles, as R_3 is varied over its range.

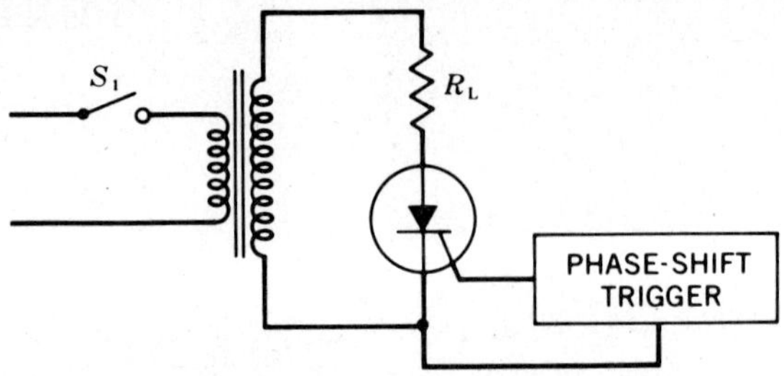

Fig. 5-4. The phase-shift trigger circuit must have lagging phase to effect continuous control of SCR firing.

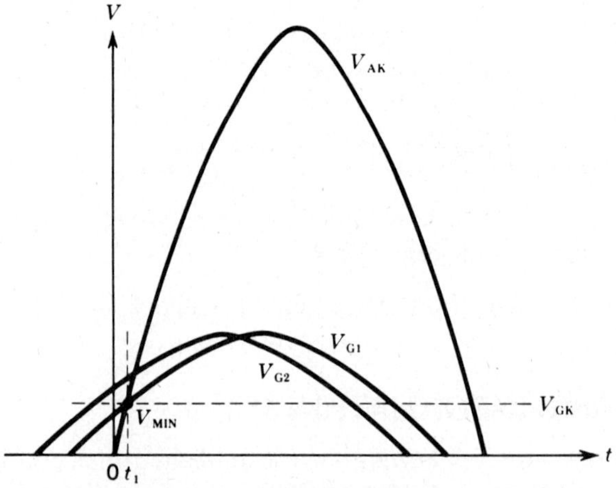

Fig. 5-5. A leading gate-cathode waveform *VG* will trigger the SCR at the same point *V*min.

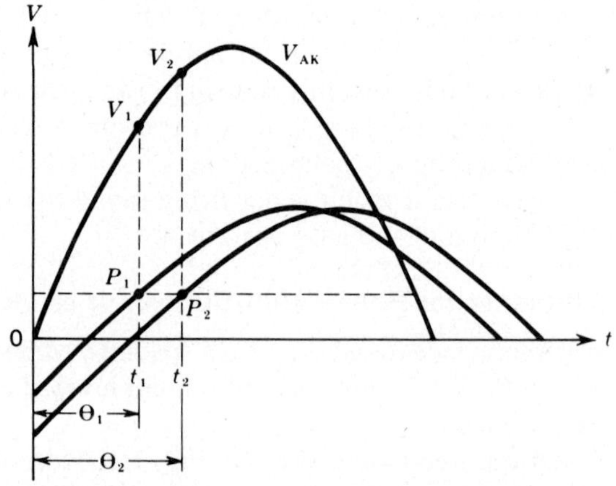

Fig. 5-6. A lagging gate-cathode voltage can be made to effect continuous triggering of the SCR.

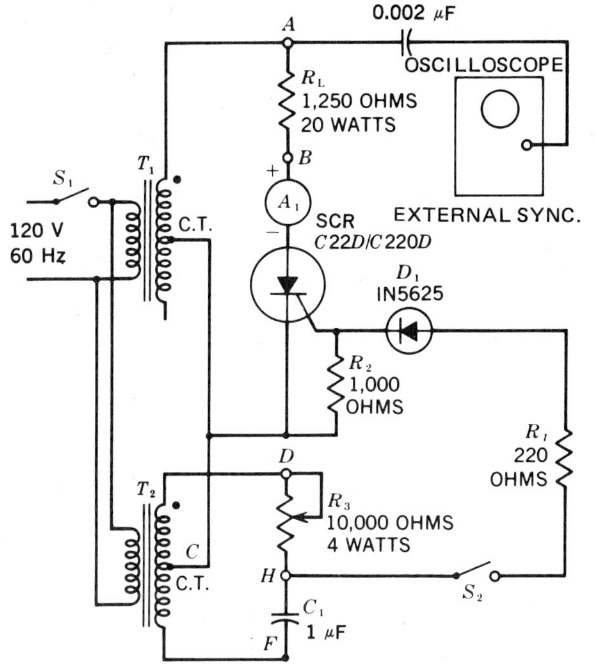

Fig. 5-7. An experimental phase-shift bridge circuit gives continuous control over SCR triggering.

SUMMARY

1. An SCR with a resistive load can be turned ON only during the positive alternation of the anode-cathode voltage waveform.
2. The phase of the ac waveform on the gate of an SCR must be *lagging* relative to the anode in order to achieve continuous control of anode current in the SCR.
3. A leading ac waveform in the gate of an SCR (leading relative to the anode) will not change the firing point.
4. A lagging ac waveform in the gate, whose phase relative to the anode can be varied over a range of 180° (approximately) will trigger the SCR over its range of variation, 180° (approximately).

SELF-TEST

Check your understanding by answering these questions.

1. An ac voltage on the gate, which is 180° out of phase with the anode voltage, __________ (will/will not) turn on the SCR.
2. In Fig. 5-2 if the maximum amplitude of V_g, the gate voltage, is lower than V_{GK}, the SCR __________ (will/will not) fire.
3. In order to achieve continuous (180° approximately) triggering of the SCR in Fig. 5-7, the phase of the waveforms at points *D* and *A* relative to *C* must be the __________ (same/opposite).
4. Phase relations in the circuit shown in Fig. 5-7 can be observed if the oscilloscope is on __________ or __________ sync.

MATERIALS REQUIRED

- Power supply: 120-V 60-Hz source
- Equipment: Oscilloscope: 0- to 100-mA milliammeter; EVM
- Transformers: T_1 power transformer; T_2 filament step-down, 120-V 60-Hz primary, 25-V 1-A center-tapped secondary
- Resistors: ½-W 220-, 1000-Ω; 20-W 1250-Ω
- Capacitors: 1-μF 400-V; 0.002-μF 400-V
- Semiconductors: C22D or C220D SCR or equivalent; IN5625 diode or equivalent
- Miscellaneous: Two SPST switches; resistor-decade box or 10,000-Ω 4-W potentiometer

CAUTION: *In the following procedure, high voltage is present. Observe safety precautions.*

PROCEDURE

1. Connect the circuit shown in Fig. 5-7. S_1 and S_2 are both open.
2. Close switch S_1 applying power to T_2 and T_1. S_2 remains open. Calibrate the vertical amplifiers of the oscilloscope for a deflection sensitivity of 100 V/cm and 10 V/cm.
3. Connect the ground of the oscilloscope to point *C* and the "hot" lead to point *A*. Set the oscilloscope on line sync. If line sync is not available, set scope on external sync, and through a 0.002-μF capacitor, connect point *A* to the external sync jack of the scope.
4. Adjust the oscilloscope sweep and sync controls until one 60-Hz sine waveform appears on the screen, with the polarity shown in Table 5-1. The positive alternative leads the negative on the time axis. Center the waveform, using the guidelines on the etched face plate.

   Once you have obtained the proper reference waveform 1, *do not* readjust the sync, sweep, and

TABLE 5-1. SCR Phase-Shift Control Data

Waveform Number		*Waveform*	*Volts, p-p*	*Load Current, I_F* mA	*Resistance, R_3* Ω
1	+ 0 −	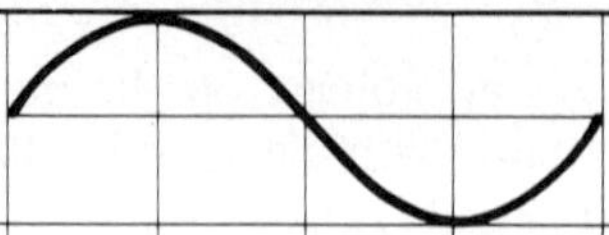		X	X
2	+ 0 −				
3	+ 0 −				
4	+ 0 −				
5	+ 0 −				
6	+ 0 −				
7	+ 0 −				

Effect of reversing secondary of T_2: ____________

centering controls on the oscilloscope until you have completed checking waveforms 1 through 7.

5. Measure the peak-to-peak (p-p) amplitude of waveform 1 and record in Table 5-1.

6. Connect the vertical input lead of the oscilloscope to point *D* with the ground lead remaining at point *C*. If the phase of the waveform observed on the scope is not the same as reference waveform 1, open switch S_1 and reverse the primary connections of transformer T_2.

   Close S_1. Observe waveform 2 between points *D* and *C* and draw it in Table 5-1, in proper time phase with waveform 1. Measure and record its p-p amplitude.

7. Close S_2. Connect the vertical input lead of the oscilloscope to point *H*, while the ground lead remains connected to point *C* (the center tap of the phase-shift transformer). Increase R_3 to maximum resistance and observe the sine wave. Draw this waveform 3 in Table 5-1, in proper time phase with the reference waveform. Measure and record in Table 5-1 the load current I_F and the p-p amplitude of waveform 3. Do *not* vary R_3, the phase-shift control, until instructed to do so. Open S_1. Measure and record in Table 5-1 the resistance of R_3.

8. Connect the vertical input lead of the oscilloscope to the SCR anode and ground lead to the cathode at point *C*. Close S_1. Observe and record waveform 4 across the SCR. Measure and record its p-p amplitude. Measure also and record the anode current I_F, if any.

9. With the vertical input leads of the oscilloscope still connected across the SCR, vary R_3 all the way from its setting in step 7. Observe the effect on the waveform and load current as R_3 is varied from its maximum to minimum resistance. With R_3 now set to its minimum resistance, observe and record waveform 5, across the SCR, in proper time phase with waveform 1, in Table 5-1. Measure and record its p-p amplitude. Also measure and record the load current I_F. Open S_1 and S_2. Measure and record the resistance of R_3.

10. Close S_1 and S_2. Connect the vertical input lead of the oscilloscope to point *H* and the ground lead to point *C*. Observe and record waveform 6 in Table 5-1, in proper time phase with the reference waveform. Measure and record in Table 5-1 its p-p amplitude.

11. Open S_1 and S_2. Reverse the connections of *D* and *F* on secondary of transformer T_2. Set R_3 at

10,000 Ω, maximum resistance. Close S_1 and S_2. Connect the oscilloscope vertical leads across the SCR and the ground lead to point C. Draw in Table 5-1 waveform 7 across the SCR. Measure and record its p-p amplitude. Also measure and record load current.

12. Vary R_3 over its entire range from maximum to minimum. Note and record in Table 5-1 the effect of reversing the secondary of T_2 on control of SCR firing.

Extra Credit (Optional)

13. Design a transformerless phase-shift circuit which may be used to control the firing of an SCR. Draw the circuit and indicate maximum range of control possible.

14. Connect the circuit and try it. Record your data in tabular form.

QUESTIONS

1. Does your experiment confirm that a lagging rather than a leading phase-shift voltage is required to effect continuous triggering control of an SCR? Refer specifically to the measurements which confirm this.
2. How is the firing angle of an SCR related to the load current?
3. From your observations in this experiment, what is the approximate range in control of a phase-shift bridge circuit over the triggering of an SCR? Refer specifically to your measurements, which confirm this answer.
4. What must be the phase relationship between the voltage V_{AC} and V_{DF} to effect continuous triggering control? Refer to the measurements in your experiment which confirm this.
5. What is the effect on the amplitude of the phase-shift voltage V_{HC} as R_3 is varied over its entire range?
6. What do you think would be the effect on triggering if R_1 were increased to 20 times its size? Why?
7. When is maximum firing angle achieved, with R_3 maximum or minimum? Why?

Answers to Self-Test

1. will not
2. will not
3. the same
4. external; line

EXPERIMENT

6

THE DIAC

OBJECTIVES

1. To study the electrical characteristics of a diac
2. To observe and measure current flow through a symmetrical diac

BASIC INFORMATION

A *diac* is a two-terminal, three-layer diode having negative resistance. It will conduct in either direction if the voltage applied to main terminal 1 (MT_1) is either positive or negative with respect to main terminal 2 (MT_2) and is sufficiently large.

The circuit schematic diagram symbol for the diac is shown in Fig. 6-1*a,* and the junction arrangement is shown in Fig. 6-1*b*. The diac resembles a bipolar (PNP) transistor in that it has two junctions, J_1 and J_2. However, unlike the bipolar transistor, the center N layer of the diac is not externally connected. A diac also differs from a bipolar transistor in that the doping at the two diac junctions is the same, whereas in a transistor the doping levels are different at the *BE* and *BC* junctions.

The equal doping level in the junctions of the diac helps explain the symmetrical current characteristics of some diacs (some are not symmetrical). Figure 6-2 is a graph of the current which flows through the diac as a function of the voltage across the diac. This graph shows that only a small current I_{BO} flows through the diac as MT_1 is made either more positive or more negative than MT_2. When the voltage across the diode reaches V_{BO} in either direction, avalanche breakdown occurs and current increases dramatically. At this point the diac is operating in a region of negative resistance because an increase in current through it is associated with a decrease in voltage across it. (Other devices, such as neon bulbs, also have regions of negative resistance in their volt-ampere characteristic curves.) In the negative resistance region, resistance in the external circuit limits the amount of current flow, as in Fig. 6-3.

The difference between the breakover voltage V_{BO} and the voltage drop in the negative resistance region is called the *breakback voltage change* $\Delta V;$ for a 1N5411 diac, this is at least 5 V. This change in voltage is an important consideration when the diac is used to trigger a triac or SCR.

Applications

Diacs are used in conjunction with SCR-like devices, called triacs, in a wide variety of light-, heat-, and motor-control circuits. Some triacs have built-in diacs, which reduce the number of components and connections in some cases. Diacs provide a high current pulse to the gate of an SCR or triac for reliable triggering.

SUMMARY

1. A diac is a two-terminal, three-layer solid-state device which has the property of negative resistance. It will conduct current in either direction.
2. A symmetrical diac will break down at the same voltage in either direction (GE ST2, for example). An asymmetrical diac (GE ST4) breaks down at different voltages when MT_1 is positive with respect to MT_2 than when it is negative.

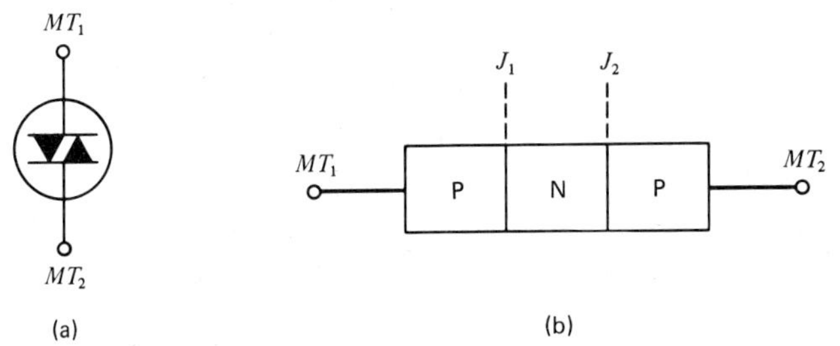

Fig. 6-1. *(a)* Diac symbol; *(b)* junction diagram.

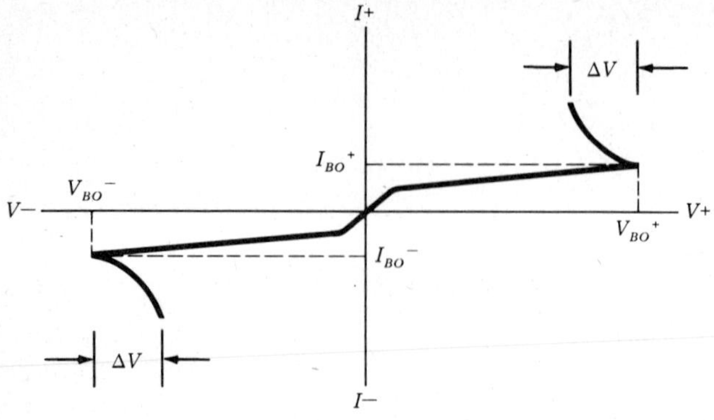

Fig. 6-2. Diac voltage-current characteristic.

3. A device has negative resistance when a decreasing voltage across the device causes an increase in current flow through the device.
4. The useful characteristic of a diac is the negative resistance region where avalanche breakdown occurs when V_{BO} is exceeded.
5. Resistance in the external circuit limits the amount of current flow to a safe value after avalanche breakdown is reached.

SELF-TEST

Check your understanding by answering these questions.

1. In the circuit shown in Fig. 6-3, current through R changes its __________ as the polarity of the ac input voltage changes.
2. Current flows through R __________ (continuously/in pulses).

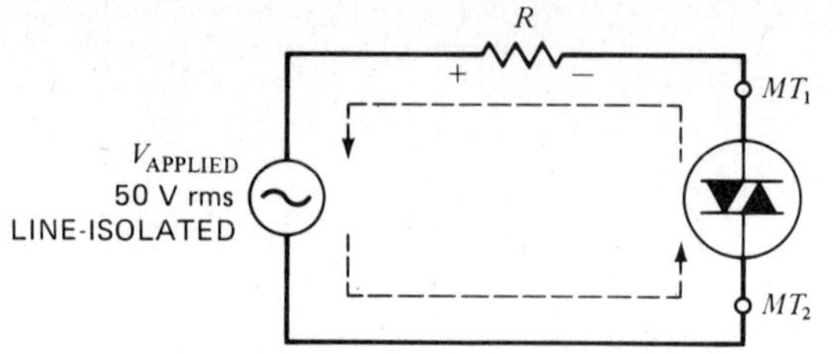

Fig. 6-3. When avalanche breakdown is reached, current in the diac is limited by the external resistance R.

3. In a symmetrical diac the doping of the two junctions is __________ .
4. If the current through a device increases as the voltage across it decreases, in that region the device has __________ __________ .
5. The terminals of a diac are the anode and cathode. __________ (true/false)

MATERIALS REQUIRED

- Power supply: Isolation transformer and variable autotransformer
- Equipment: Oscilloscope, EVM, or VOM (20,000 Ω/V)
- Resistors: ½-W 6.8-kilohm (kΩ)
- Capacitors: 0.002-μF 400-V (if external sync is used)
- Solid-state devices: Diac, with breakover voltage of about 30 V, GE ST2 or equivalent
- Miscellaneous: SPST switch

PROCEDURE

1. Connect the circuit shown in Fig. 6-4. Use an isolation transformer and a variable autotransformer for the 60-Hz source. Close switch S_1. Power ON. Set voltage to 50-V rms.

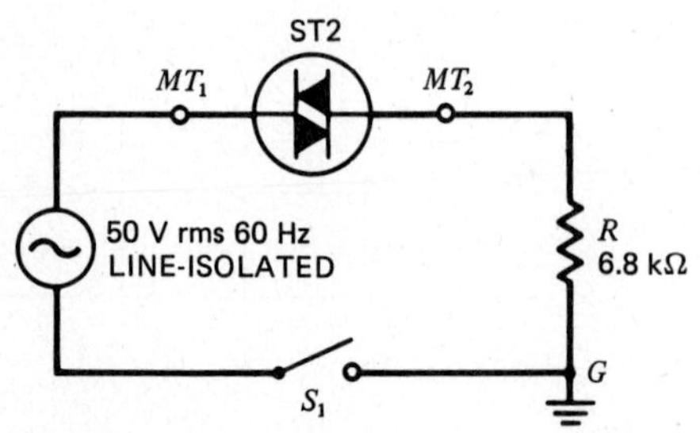

Fig. 6-4. Experimental diac circuit.

2. Externally synchronize the oscilloscope by connecting a 0.002-μF capacitor from MT_1 to the external sync post of the scope. NOTE: Line sync may be used instead of external sync.
3. Observe the waveform from MT_1 to G (ground). Adjust the oscilloscope controls until this reference waveform, properly centered, appears as in Table 6-1. Measure and record in Table 6-1 its p-p amplitude.
4. Observe the waveform 1, $V_{MT_1}MT_2$ across the diac. The hot lead of the scope is connected to point MT_1 and the ground lead to point MT_2. Record this waveform in Table 6-1 in proper phase with the reference waveform. Show the 0-V reference line. Also measure and record the peak voltage positive and negative. Identify on waveform 1 the periods

TABLE 6-1. Diac Measurements, 50-V rms

Waveform Number		*Waveform*	*Volts, p-p*	*Conduction Angle*
Reference	+ 0 V −	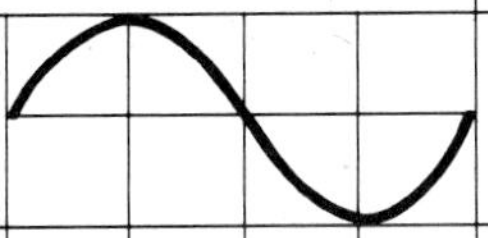		X
1	+ 0 V −			
2	+ 0 V −			X

during which the diac is conducting. Determine and record the conduction angle.

5. Observe waveform 2, $V_{MT_2}G$, across the resistor *R*. Measure and record this waveform in proper phase with the reference. Also record the peak voltage positive and peak voltage negative. Identify on waveform 2 the periods during which there is current in *R*. Determine and record the time when there is maximum current in *R*.

6. Increase the applied voltage to 75-V rms. Repeat steps 3, 4, and 5 and record your results in Table 6-2.

TABLE 6-2. Diac Measurements, 75-V rms

Waveform Number		*Waveform*	*Volts, p-p*	*Conduction Angle*
Reference	+ 0 V −	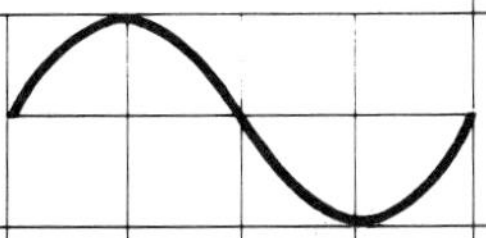		X
1	+ 0 V −			
2	+ 0 V −			X

QUESTIONS

1. Do the results of your experiment confirm the fact that the diac is a bidirectional current device? Refer specifically to the data in Tables 6-1 and 6-2 to substantiate your answer.
2. Which measurements in Tables 6-1 and 6-2 give the forward breakover voltage of the diac? Is this voltage the same when 50- and 75-V rms are applied?
3. What is the value of the peak current in the circuit when the input voltage is (*a*) 50 V, and (*b*) 75 V? Show your calculations.
4. From the data in Table 6-1, during the positive alternation, how long is the diode (*a*) ON and (*b*) OFF? Show your calculations.
5. Does the diac conduct equally during the positive and negative alternations? Substantiate your answer.
6. How do the conduction times of the diac compare during the 50-V and 75-V inputs? Explain any differences.

Answers to Self-Test

1. direction
2. in pulses
3. equal/the same
4. negative resistance
5. false: MT_1 and MT_2

EXPERIMENT 7

THE TRIAC

OBJECTIVES

1. To study the current characteristics of a triac
2. To observe and measure the performance of a triac triggered two different ways

BASIC INFORMATION

Triac Current Characteristics

A triac, like an SCR, is a gate-controlled device; but unlike the SCR, the triac has bidirectional current characteristics (it will conduct in either direction).

The triac has three terminals, designated MT_1, MT_2, and gate. Figure 7-1*a* is the schematic symbol for the triac and Fig. 7-1*b* is its junction diagram. Note that the triac symbol resembles that of a diac, but with a gate added. Either main terminal MT_1 or MT_2 can act as the current source.

The voltage-current characteristic of a triac (Fig. 7-2) is symmetrical. In the graph in quadrant 1, MT_2 is positive relative to MT_1 and acts like the anode of an SCR. In the third quadrant MT_2 is negative relative to MT_1 and acts like the cathode of an SCR. Like the SCR, the triac remains in an OFF state until the breakover voltage is reached, when it turns ON. In its ON state the voltage across the triac drops to a low value, and current is limited by the impedance of the external circuit. In the third quadrant, when the voltage polarity across MT_1 and MT_2 is reversed, current through the triac is also reversed.

As in the SCR, the breakover voltage can be controlled by changing the amplitude of the gate current. The breakover voltage is decreased by increasing gate current. Gate current to activate the triac can be either positive or negative.

Because of its construction, Fig. 7-1*b*, and voltage-current characteristics, the triac appears to be equivalent to two SCRs connected in parallel, but oriented in opposite directions, as in Fig. 7-3. This equivalent circuit also shows that the triac acts like an ac switch, first permitting current to flow in one direction, and then reversing the current as the polarity of the voltage across MT_1 and MT_2 changes.

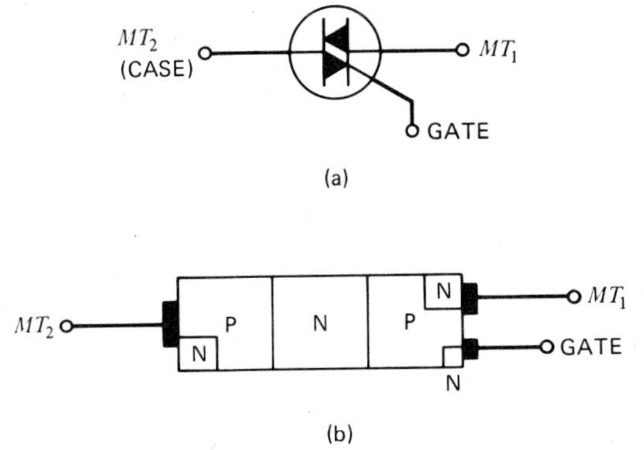

Fig. 7-1. *(a)* Triac schematic symbol; *(b)* junction diagram.

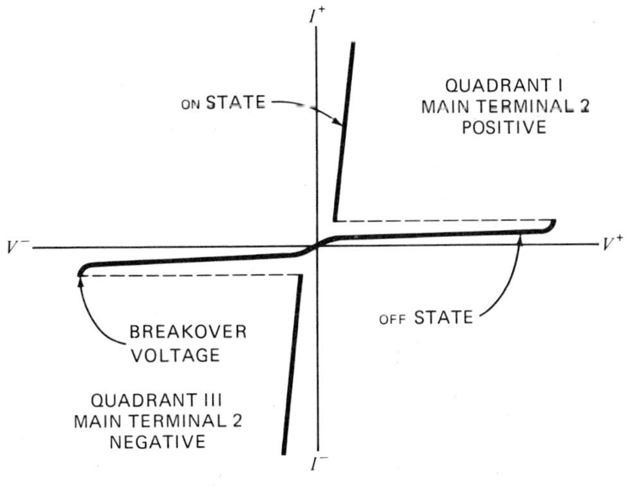

Fig. 7-2. Triac voltage-current characteristics are symmetrical.

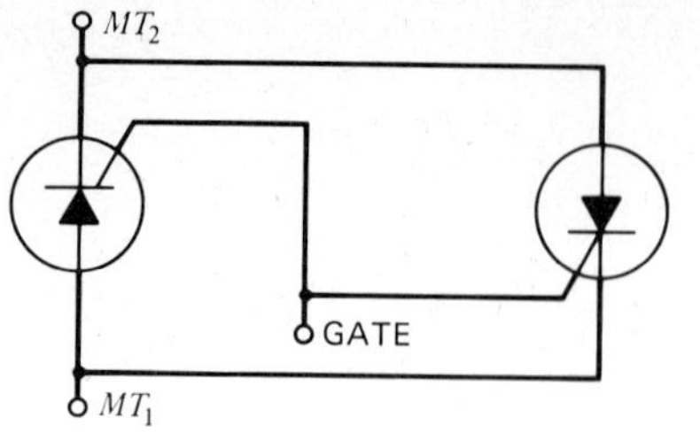

Fig. 7-3. Triac equivalent circuit; two SCRs in parallel, but oppositely oriented.

Gate-Triggering Requirements

Gate current will trigger a triac whether the gate is positive or negative relative to terminal MT_1 (all polarities in a triac are referenced to MT_1), regardless of the voltage polarity across terminal MT_1 and MT_2. So that in quadrant 1 (see Fig. 7-2), when MT_2 is positive relative to MT_1, the gate may be either positive or negative. Again in quadrant 3, when MT_2 is negative, the gate may be either positive or negative. This defines the four modes of operation shown in Table 7-1.

The most sensitive mode, that is, the one requiring the least gate current for turning the triac ON, is mode 1 (MT_2^+, gate$^+$, quadrant 1). The other modes require higher gate currents. Each operating mode needs a different gate current level to turn the triac ON. These requirements are given in the manufacturer's specifications. However, gate characteristics even for the same device will vary from unit to unit. The manufacturer therefore supplies the values (bands) of the limits of gate current and voltage required to trigger a device.

It should be noted that the level of gate current required to trigger any triac depends on the value of junction temperature. The higher the junction temperature, the lower the value of gate current required. This explains why higher gate voltages and currents are needed to turn ON a triac when it is cold than when it is warm.

Gate-Triggering Circuits

Gate circuits used to trigger a triac are very similar to circuits used for SCRs. For each device, the gate signal must be strong enough to turn it ON and keep it on during the full turn-on period. Trigger *pulses* are frequently used. These must have sufficient amplitude and duration [about 30 microseconds (μs)] for sustained firing. Higher-amplitude trigger pulses require less triac turn-on time. In the case of a triac the trigger pulse must have sufficient amplitude to turn on the device in the less sensitive modes of operation; otherwise the triac might fire in mode 1 but not in the other modes. The particular application usually determines the type of trigger circuit used.

TABLE 7-1. Triac Triggering Modes

MT_2	*Gate*	*Quadrant*	*Mode*
+	+	1	1
+	−	1	2
−	+	3	3
−	−	3	4

Two types of control circuits, deriving the trigger from the 60-Hz power line will be discussed here. The first (Fig. 7-4) is an example of amplitude control. A triac in series with a load R_L is connected across the 60-Hz power source. Current will flow through the load only when the triac is turned ON and will continue to flow as long as it remains ON. Current in the gate-control circuit is determined by the setting of rheostat R_v. The value of R_v is such that at maximum resistance there is insufficient gate current and the triac is OFF. As the resistance of R_v is reduced, thus increasing gate current, a point will be reached at which the triac fires. This will occur near the peak of the positive alternation on MT_2, about the 80 to 90° point, as shown in Fig. 7-4, and the triac will be ON for the time between 80° (90°) and 180°, approximately. Operation during this interval is in mode 1, because both the gate and MT_2 are positive. On the negative alternation, when both MT_2 and the gate are negative (mode 4), the triac may or may not turn ON, depending on the level of gate current. If gate current is insufficient to turn the device ON, the triac will not fire in mode 4; that is, it will not conduct during the negative alternation.

If gate current is increased further, the triac may be made to fire earlier, and still earlier with higher gate current. Eventually, a setting of R_v is reached at which the triac conducts on both the negative and positive alternations. That is, the triac can be made to conduct almost the full cycle by a proper setting of R_v. NOTE: With this trigger arrangement *delay* of the firing cannot be exercised during the second 90° (half) of each alternation. That is, control can be exercised for less than 90° of each alternation.

Figure 7-5 illustrates a second type of trigger control. Here R_v and C constitute a voltage divider and phase-shifting network; and the voltage across C, applied to the gate, has a lagging phase relative to the 60-Hz voltage. As in the case of Fig. 7-4, triggering of the triac is achieved by reducing the resistance of R_v, thereby increasing gate current and reducing the lagging phase angle. In Fig. 7-5, as in the preceding circuit, the triac is turned ON first on the positive alternation. As gate current is increased, firing of the triac occurs earlier in the cycle, until finally the triac fires on both the positive and negative alternations of the applied voltage. Control of firing is increased somewhat over amplitude control, but not substantially.

In Fig. 7-6 a double RC phase-shift circuit applies the control signal to the gate. This circuit also produces nonsymmetrical firing, although a slightly wider range of control is possible than with the single RC trigger.

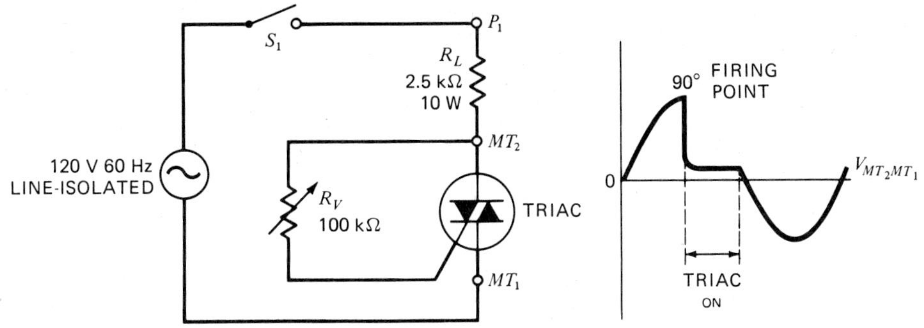

Fig. 7-4. AC amplitude control of a triac.

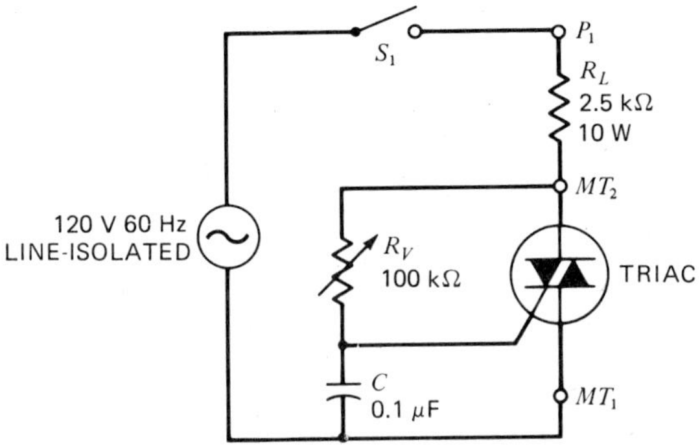

Fig. 7-5. Phase control used to determine conduction angle of a triac.

The widest range of control is made possible by introducing a diac in the gate circuit, as in Fig. 7-7. The resistance of R_v again controls the gate current by controlling the voltage across C_2. Here the voltage across C_2 first turns on the diac, which, in turn, applies enough voltage to the gate to turn on the triac. By properly matching the diac to the triac, symmetrical triggering of the triac on both the positive and negative alternations is accomplished. Moreover, the range of firing control is much wider, so that a turn-on delay of almost 180° is possible in each alternation.

Applications

Triacs are used in light-dimming control circuits, in motor controls, in heat-control circuits, and in other industrial applications.

SUMMARY

1. A triac is a special type of silicon-controlled rectifier whose main elements, MT_1 and MT_2, can act either as anodes or cathodes. When MT_1 acts as a cathode, MT_2 acts as the anode of the triac, and vice versa.
2. A triac can be turned ON when MT_2 is either positive or negative relative to MT_1.
3. A triac can be turned ON when the gate is either positive or negative relative to MT_1.
4. A triac can therefore be turned ON in any one of four modes. These are:

MT_2	Gate	Mode
+	+	1
+	−	2
−	+	3
−	−	4

The most sensitive is mode 1, because this mode requires the least amount of gate current to turn the triac ON.

5. In the other three modes of operation, higher gate currents are required to trigger the triac, each mode having its own unique requirements.
6. There are a variety of circuits used industrially to supply gate current for triggering a triac. These include resistive networks, *RC* networks, and diacs connected in conjunction with *RC* circuits. Diacs used to supply gate current are more effective than simple *RC* or resistive networks and help assure symmetrical triggering on both the positive and negative alternation of the input cycle.
7. The current in a triac is reversed as the triac turns ON when the polarity of the applied voltage is reversed across the main terminals.

SELF-TEST

Check your understanding by answering these questions.

1. The symbol for a triac is like that for a ____________ with a gate added.
2. A triac has ____________ current characteristics.
3. A triac has a clearly defined anode and cathode. ____________ (true/false)

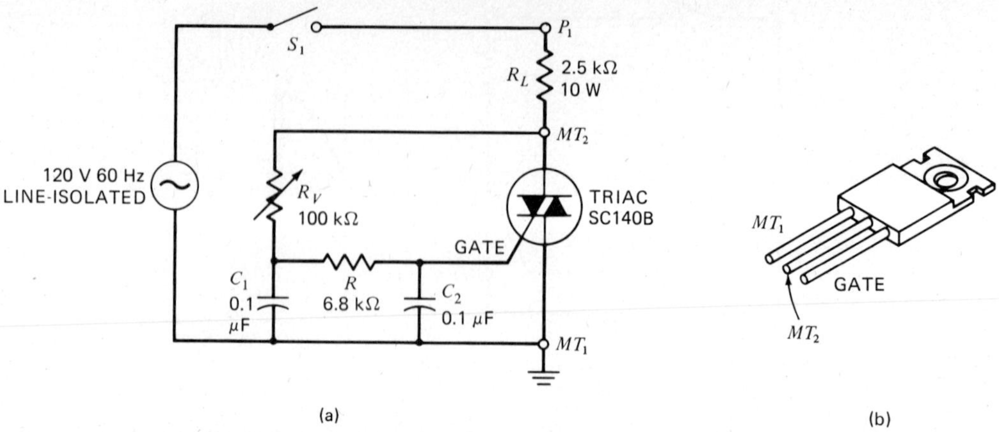

Fig. 7-6. *(a)* Double *RC* phase-shift circuit used to trigger a triac; *(b)* terminal identification for SC140B.

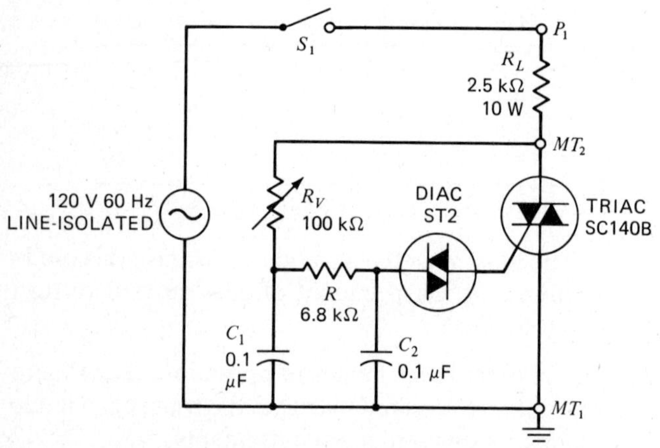

Fig. 7-7. Diac triggering of a triac produces wider and symmetrical control of firing.

4. The operating mode of a triac which requires the least gate current to turn it ON occurs when main terminal 2 is __________ with respect to terminal 1 and when the gate is __________ relative to terminal 1.

5. For stable triac triggering, a gate pulse with sufficient __________ and __________ is required.

6. Current through a triac always flows in the same direction when operated from an ac source. __________ (true/false)

MATERIALS REQUIRED

- Power supply: Isolation transformer
- Equipment: Oscilloscope
- Resistors: ½-W 6.8-kΩ; 10-W 2500-Ω
- Capacitors: Two 0.1-μF 400-V; 0.002-μF 200-V
- Solid-state devices: SC140B (triac) or equivalent; ST2 (diac) or equivalent
- Miscellaneous: 100-kΩ 2-W potentiometer; SPST switch

CAUTION: *Hazardous voltage present. Observe safety precautions.*

PROCEDURE

Double *RC* Time Constant in Gate Circuit

1. Connect the circuit shown in Fig. 7-6. Use the isolated output of an isolation transformer as the 120-V 60-Hz power source. Set R_v for *maximum* resistance. Close switch S_1. Power ON.

2. Externally synchronize your oscilloscope by connecting a 0.002-μF capacitor from P_1 to the external sync post of the oscilloscope. NOTE: Line sync, if available, is preferable to external sync.

3. Observe the waveform from P_1 to MT_1. Adjust the oscilloscope controls until this reference waveform, properly centered, appears as in Table 7-2. Use the dc amplifier (vertical) of your oscilloscope. Measure and record in Table 7-2 the peak positive and peak negative alternations of the wave.

4. Observe waveform 1, $V_{MT1,\ MT2}$ (ground lead to MT_1). Record this waveform in Table 7-2, in proper time phase with the reference. Also measure and record its peak positive and peak negative amplitude. Measure also and record the conduction angle in degrees, if the triac is ON, in each alternation. Confirm your conclusion by observing the waveform across R_L.

5. Gradually reduce the resistance of R_v just to the point where the triac fires. Observe and record waveform 2, $V_{MT1,\ MT2}$ in proper time phase with

TABLE 7-2. Triac Measurements, Double *RC* in Gate Circuit

Step	*Waveform Number*		*Waveform*	*Volts, Peak*		*Conduction Angle, Degrees*	
				+	−	+	−
3	Reference	+ 0 −	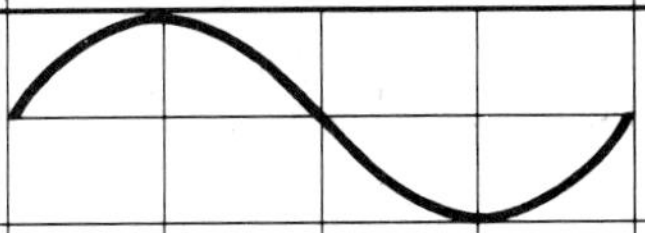				
4	1	+ 0 −					
5	2	+ 0 −					
6	3	+ 0 −					
7	4	+ 0 −					

the reference. Also measure and record its peak positive and negative amplitude. Indicate in Table 7-2 the conduction angle during the positive-negative alternation when the triac is ON.

6. Reduce the resistance of R_v gradually, just to the point where the triac fires on the negative (and positive) alternation. In Table 7-2 draw waveform 3 in proper time phase with the reference. Measure and record the peak positive and negative alternations. Measure and record the conduction angle during each alternation.

7. Reduce R_v to its minimum resistance. Measure and record waveform 4, $V_{MT1,\,MT2}$ in proper time phase with the reference. Record the peak positive and negative amplitude and the conduction angle in each alternation.

Diac in Gate Circuit

8. Connect the circuit shown in Fig. 7-7. Set R_v for *maximum* resistance. Close S_1. Power ON.

9. Repeat step 3. Use Table 7-3.

TABLE 7-3. Triac Measurements, Diac in Gate Circuit

Step	*Waveform Number*		*Waveform*	*Volts, Peak*		*Conduction Angle, Degrees*	
				+	−	+	−
9	Reference	+ 0 −	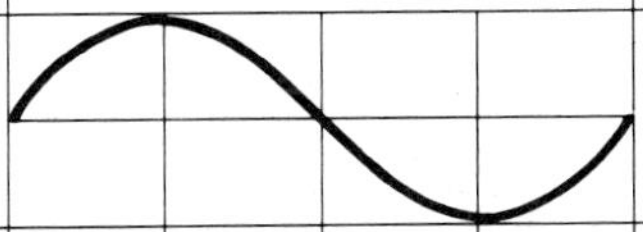				
10	1	+ 0 −					
11	2	+ 0 −					
12	4	+ 0 −					

10. Repeat step 4.

11. Repeat step 5.

12. Repeat step 7.

13. Power OFF.

QUESTIONS

1. Which experimental circuit provides symmetrical triggering of the triac [that is, triggering on both the plus (+) and minus (-) alternation of the applied voltage]? Refer specifically to your data to confirm your answer.

2. In the experimental circuits, does a value of $R_v = 100\ \text{k}\Omega$ permit triggering of the triac? Why?

3. (*a*) How can you determine what maximum value of R_v is required to trigger the triac in Fig. 7-4? (*b*) Will this value be the same for each of similarly numbered triacs? Why?

4. Which experimental circuit provides the greatest range of control of the triac? Refer specifically to your data to confirm your answer.

5. The conduction angles of a triac are 120° during the positive alternation and 100° during the negative. What is the total conduction time in seconds assuming a 120-V 60-Hz input?

Answers to Self-Test

1. diac
2. bidirectional
3. false
4. positive; positive
5. amplitude; duration
6. false

EXPERIMENT

8

CHARACTERISTICS OF PHOTOELECTRIC DEVICES

OBJECTIVES

1. To study the effect of an increase or decrease of light on the resistance of a photoconductive cell
2. To observe the effect of an increase or decrease of light on the output of a photovoltaic cell

BASIC INFORMATION

Photoelectric or light-sensitive devices are used in almost every branch of industry for control and safety, amusement and sound reproduction, and inspection and measurement.

In the first category we have, among other applications, door openers, burglar-alarm systems, smoke detectors, and automatic control of street lighting systems.

In the second class we find photoelectric devices used in theater sound systems, pinball games, and timers.

In the third class we find applications in pinhole detectors, daytime measurement of cloud heights, color measurement, counting devices, and other process-control systems.

There are many types of solid-state light-sensitive devices. In this experiment we will be concerned only with two, the photoconductive (photoresistive) cell and the photovoltaic cell. The photoconductive cell is a two-element device whose resistance decreases when light falls on it. Photovoltaic devices generate a voltage across the cell when they are exposed to light.

Nature of Light

It is not the purpose of this experiment to study the theories governing the nature and behavior of light. We merely mention that light behaves like an electromagnetic radiation, and associated with light are the characteristics of wavelength and frequency. The wavelength of light determines the color of that light. White light consists of many wavelengths, which may be separated by a prism. The human eye responds to the prismatic colors ranging from violet to red (approximately). The wavelengths of light which the eye can ''see'' are in the nanometer (nm) range of 400 nm (violet) to 700 nm (red). A nanometer is a 10^{-9} part of a meter.

Another unit frequently used in measuring the wavelength of light is the angstrom unit (Å). An angstrom unit is one-tenth of a nanometer. Visible light is therefore in the range of 4000 to 7000 Å. Although once a standard unit of light measurement, the angstrom is being discontinued in favor of the metric unit nanometer.

The average human eye exhibits selective properties within the range of the visible spectrum. Its response is similar to a series resonance curve, with maximum sensitivity at 5500 A, falling to minimum at 4000 and 7000 Å (see Fig. 8-1). This means that the response of the eye is not uniform within the range of the visible spectrum, and equal amounts of light of different wavelengths will not appear equal to the eye. Thus an object bathed in yellow-green light will appear brighter than an object illuminated by an equal amount of red light or violet light.

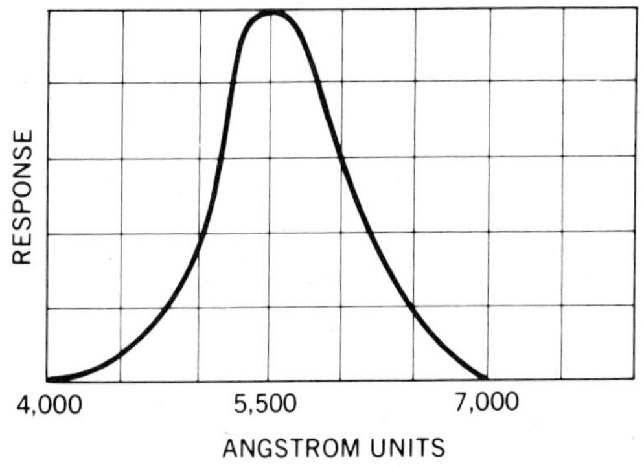

Fig. 8-1. Spectral response of the eye.

Like the human eye, photocells also exhibit unique spectral sensitivities. The sensitivity of a photocell to a particular color (frequency of light) depends on the nature of the material (semiconductor) from which the cell is constructed, and on the manner of its construction. Manufacturers specify the spectral sensitivities of their devices in the form of a frequency response curve.

Photometric Terms

In describing photo characteristics, photometric terms are often used. Let us define some of these terms.

- International candela: The standard light source used in defining light measurements. It is the unit of luminous intensity.
- Point Source of Light: A source whose dimensions are very small in comparison with the distance from which it is observed.
- Candlepower: The measure of the ability of a light source to produce a total amount of light. A large light surface which is not very bright may give off more light than a small bright source. Its candlepower rating would therefore be higher than that of the small bright source. This unit is being discontinued in favor of the SI (International System of Units) unit lumen.
- Footcandle: A measure of illumination on a surface. A footcandle is the amount of light a surface will receive from a source of 1 candlepower (cp) at a distance of 1 foot (ft) [0.305 m]. This unit is being replaced by the SI unit lumen per square meter or lux.
- Lumen: A measure of the total light flux (luminous flux) that either a light source gives off or a surface receives.

Photoconductive Devices

Photoconductive cells are made of a thin layer of semiconductor material such as cadmium sulfide, cadmium selenide, or lead sulfide. The semiconductor layer is enclosed in a sealed housing. A glass window in the housing permits light to fall on the active material of the cell. Figure 8-2 is a circuit symbol for a photoconductive cell.

Photoconductive cells exhibit the peculiar property that their resistance decreases in the presence of light and increases in the absence of light. The cell simply acts as a conductor whose resistance changes when illuminated.

A simple circuit arrangement for a photoconductive cell (*PC*) is shown in Fig. 8-3. Here the resistance of *PC*, in series with R_1, limits the amount of current I in the circuit. A is used to measure the current I. When no light falls on *PC*, its resistance is very high and the current I is very low. Hence V_1, the voltage drop across R_1, is relatively low. When *PC* is illuminated, its resistance becomes very low, I increases, and V_1 increases. We shall see that this simple circuit arrangement, slightly modified, will be found in control circuits employing photoresistive cells.

Photovoltaic Cells

These are also light-sensitive semiconductor devices, but they produce a voltage when illuminated which increases as the intensity of light falling on the semiconductor junction of this two-element cell increases. The usual basic material from which these cells are made today is silicon or selenium.

Photovoltaic cells convert light into electric energy, which may be used directly to supply small amounts of electric power for electrically powered devices. Because of the low levels of power which photovoltaic cells generate, they have been used in the past in low-power devices such as light meters and photographic exposure meters. However, with an improvement in the efficiency of these cells, more power has been produced, as in solar cells, which are photovoltaic devices. Solar cells appear destined to play a substantial role in the development of new sources of energy.

Photovoltaic cells consist of a single semiconductor crystal which has been doped with both N- and P-type materials. When light falls on the PN junction, which is the boundary of these dopants, a voltage appears across the junction. About 0.6 V is developed by the photovoltaic cell in bright sunlight. The amount of power the cell can deliver depends on the extent of its active surface. An average cell will produce about 30 milliwatts per square inch (30 mW/in^2) of surface, operating into a load of 4 Ω. To increase the power out-

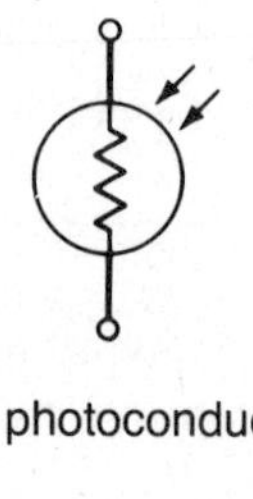

Fig. 8-2. Circuit symbol, photoconductive cell.

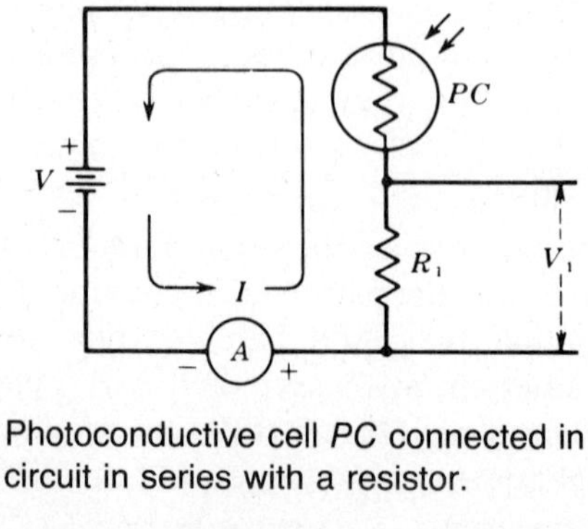

Fig. 8-3. Photoconductive cell *PC* connected in a simple circuit in series with a resistor.

put, large banks of cells are used in series and parallel combinations. An example is the use of solar cells to power the experimental circuits on lunar and space modules.

Photovoltaic cells may be used directly to energize sensitive relays, or where required their output may be amplified electronically.

The circuit symbol for a photovoltaic cell is shown in Fig. 8-4.

In addition to the spectral sensitivity of photoelectric cells, an important consideration for many applications is the speed of response of the photocell. Older cadmium sulfide cells exhibit a relatively slow change of resistance as the amount of light falling on them changes. Modern solid-state photodetectors respond very rapidly and are used in a wide variety of applications where speed, reliability, and ruggedness are important.

Photoconductive Amplifiers

The small currents of photoconductive cells must be amplified if these devices are to be used effectively in industrial control. For example, it is necessary to amplify the output of a photocell if we wish to actuate a relay or if we desire to employ it for automatic control of street lights. Transistors, SCRs, and other semiconductor devices are employed as amplifiers with solid-state photocells.

In the circuit shown in Fig. 8-5 the output of photoresistor *PC* is amplified by transistor *Q*. The circuit operates in the following manner: *PC* and R_1 constitute a resistive voltage divider which biases the transistor amplifier. When no light or little falls on *PC*, the resistance of *PC* is very high, and forward bias on *Q* is thus very low. Little current flows in the emitter-collector circuit of *Q*, which is in series with load resistor R_L. As the light intensity on *PC* increases, the resistance of *PC* drops, forward bias on *Q* increases, and current in *Q* (and R_L) increases. Finally, when *PC* is bathed in very bright light, the resistance of *PC* becomes very low, forward bias on *Q* becomes high, and current in *Q* becomes high. If R_L were the coil of a relay, the circuit would be so designed that the increase of current in *Q* would be sufficient to energize the relay. The circuit arrangement shown in Fig. 8-5, in which an increase of light on the photocell increases load current in R_L, is called the *forward* connection.

Figure 8-6 is the circuit of a photoresistor and amplifier in the *reverse* connection. In this, circuit amplifier current, and hence load current, is high when little or no light falls on *PC*. When light intensity on *PC* is high, there is little or no amplifier or load current. An explanation of circuit operation is left to you.

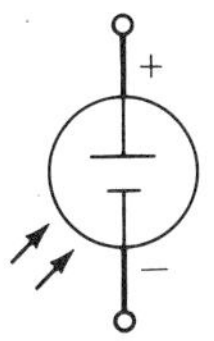

Fig. 8-4. Circuit symbol, photovoltaic cell.

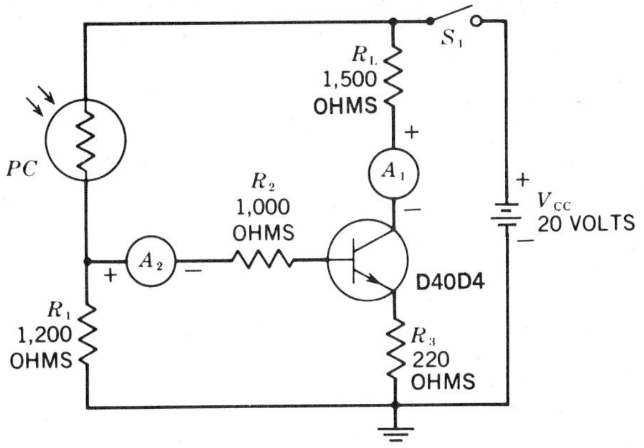

Fig. 8-5. Photoresistor and amplifier—forward connection.

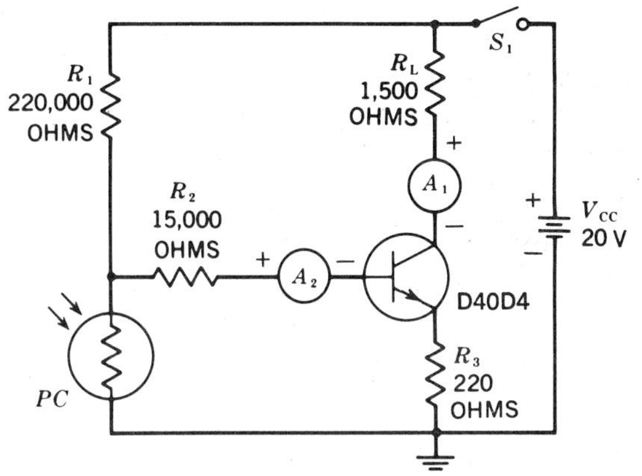

Fig. 8-6. Photoresistor and amplifier—reverse connection.

Experimental Techiniques

The purpose of this experiment is merely to familiarize the student with the operation and connections of photoelectric devices. Hence no attempt will be made to use precise light-measuring instruments. The following suggestions will be helpful in assuring successful results in this experiment. An ordinary 60-W lamp housed in a reflector may be used as a light source. A gooseneck desk lamp and a reflector will serve admirably.

A variable light source may be obtained by plugging the lamp into an ac outlet whose voltage may be adjusted. The use of a variable autotransformer simplifies the problem.

In order to minimize background illumination, a cardboard cylinder may be placed over the photoelectric device. A window, cut in the cardboard cylinder and properly oriented, will admit the desired light. The size of the window should correspond to the size of the light-sensitive element.

SUMMARY

1. Photoconductive (also called *photoresistive*) cells and photovoltaic cells are two-element semiconductor light-sensitive devices used in industrial control.
2. The resistance of photoresistors decreases in the presence of light. In the absence of light the resistance of this photocell becomes very high.
3. Photovoltaic cells, also made of semiconductor material, convert light directly into electrical energy. They produce a voltage across their terminals in the presence of light.
4. Photocells are used as transducers (sensors) of light in industrial control systems.
5. The small currents generated by photoconductive cells must usually be amplified in order to actuate control mechanisms such as relays.
6. The voltage output of a photovoltaic cell made of silicon is about 0.6 V in sunlight. The current that it can deliver to a load depends on the area of the light-sensitive material which makes up the cell. The larger the area is, the more current it can deliver. Photovoltaic cells may be connected in series and in parallel to produce higher voltage and power sources.
7. Solar cells are photovoltaic cells.
8. The unique spectral sensitivities (response to frequencies of light) of light-sensitive devices depend on the material from which they are made.

SELF-TEST

Check your understanding by answering these questions.

1. ___________ cells are light-sensitive devices.
2. ___________ and ___________ cells are made of semiconductor material.
3. The ___________ of a photoconductive cell varies inversely with the intensity of light falling on its ___________ surface.
4. Wavelengths of light are sometimes measured in ___________ units.
5. ___________ cells can be used as independent power sources.
6. Photocells can be used as ___________ in industrial control systems.

MATERIALS REQUIRED

- Power supply: Variable dc; variable ac (autotransformer)
- Equipment: 0- to 100-mA meter; 0- to 50-μA meter; resistor decade box, variable in 1-Ω steps, 2 W, from 0 to 100 kΩ
- Resistors: ½-W 220-, 1000-, 1200-, 1500-, 15,000-, 220,000-Ω
- Semiconductors: Photocell CL703 (Clairex) or equivalent; silicon solar cells (0.5 V) two required; transistor D40D4 or the equivalent
- Miscellaneous: 60-W light source with reflector; SPST switch

PROCEDURE

Photoresistive Device

1. See Fig. 8-7. Measure the resistance of the photoconductive cell when there is no light on it (i.e., when the cell is in absolute darkness) and record your reading in Table 8-1. Also measure and record the resistance of the photocell under ordinary light conditions in the laboratory, that is, under conditions of ambient light. Orient the window of the photocell away from any light source in the room.

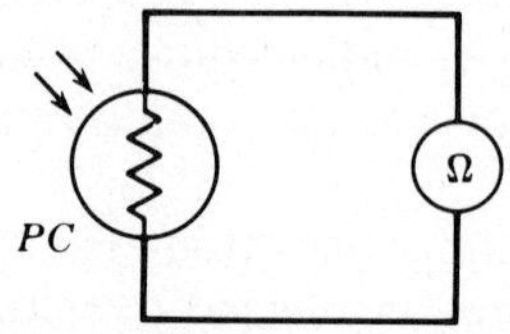

Fig. 8-7. Ohmmeter connected to measure resistance of a photoconductive cell.

2. Plug a 60-W light source into the output of variable autotransformer and the variable autotransformer into a 120-V 60-Hz source. Power ON. Light ON. Adjust the variable autotransformer for 120-V ac output.
3. Position the light source so that it is about 5 in from the photocell and is oriented so that the light enters the window of the photocell. *Do not* change the light orientation or distance from the photocell window.
4. Again measure the resistance of the photocell for this light level and record your reading in Table 8-1.

TABLE 8-1. Photoconductive Cell Characteristics

Step	*Illumination*	*Resistance of Photocell, Ω*
1	Absolute darkness	
	Ambient light	
	Voltage, V	
4	120	
5	100	
	80	
	60	
	40	
	20	
	0	

5. Reduce the output of the autotransformer, in turn, to each of the voltage levels shown in Table 8-1, and repeat step 4. Note that for 0 V, the ambient light of the room illuminates the photocell.

PC in Forward Connection

6. Connect the circuit shown in Fig. 8-5, using the same photocell as in the preceding steps. $R_1 = 1200\Omega$, $R_2 = 1000\Omega$, $R_3 = 220\Omega$, and $R_L = 1500\Omega$. Set V_{CC} at +20 V. A_2 is a microammeter set on the 0- to 300-μA range. A_1 is a 0- to 25-mA milliammeter.

7. Close switch S_1. Be certain that the window of the photocell is not directed at any light source in the room. With the photocell under conditions of room light, measure and record in Table 8-2 the base bias current. Also measure and record collector current.

8. With the light lamp oriented as in step 3, repeat step 7 for each of the light-voltage levels indicated in Table 8-2. Power OFF. Open switch S_1.

PC in Reverse Connection

9. Connect the circuit of Fig. 8-6 using the same photocell. $R_1 = 220{,}000\Omega$, $R_2 = 15{,}000\Omega$. $R_3 = 220\Omega$, and $R_L = 1500\Omega$. V_{CC} is still set at +20 V.

10. Repeat steps 7 and 8, and record your results in Table 8-3.

TABLE 8-2. Photoconductive Cell and Amplifier—Forward Connection

Step	*Illumination*	*Base Current,* μA	*Collector Current,* mA
7	Photocell in ambient room light		
	Voltage, V		
8	120		
	90		
	60		
	30		
	0		

TABLE 8-3. Photoconductive Cell and Amplifier—Reverse Connection

Step	*Illumination*	*Base Current,* μA	*Collector Current,* mA
10	Photocell in ambient room light		
	Voltage, V		
	120		
	100		
	80		
	60		
	40		
	20		
	0		

Characteristics of a Solar (Photovoltaic) Cell

NOTE: In the experimental procedure which follows you will be required to determine the output voltage of a silicon solar cell. These cells are sensitive to heat, and the heat generated by the light source will affect their output. They will therefore need a brief warmup period [about 2 to 3 minutes (min)] to stabilize after the light source has been brought close to their surface. Note also that the output voltage of the solar cells will drop as the current they deliver to a load increases beyond a certain level.

11. Connect the circuit of Fig. 8-8. *PC* is a silicon solar cell 1, rated to produce about 0.6 V in bright sunlight. *V* is a voltmeter set on the 2-V (approx-

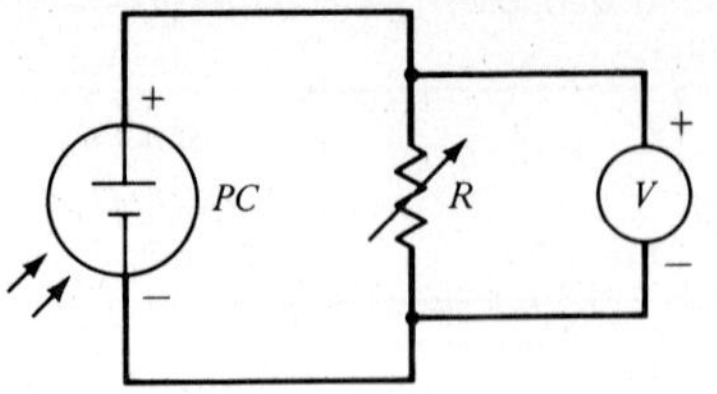

Fig. 8-8. Circuit for determining sensitivity (output) of a photovoltaic cell.

imate) range. *R* is a resistance decade box set at 1000 Ω.

12. With the cell (*PC*) illuminated only by the ambient light in the lab, measure the dc voltage across *PC* and record in Table 8-4.

13. Plug the 60-W light source into the 120-V 60-Hz outlet and observe the effect on the output voltage of the cell as the light is brought close to and is made to illuminate the cell.

14. Orient and set the light source at the distance which produces maximum voltage across *PC*. *Now permit the output of the cell to stabilize.* Measure and record in Table 8-4 the voltage across *PC* after stabilization. NOTE: Keep the light source at the same distance and in the same orientation which produced maximum stabilized voltage, for the remainder of the experiment. Do not move the *PC*.

15. Adjust the decade box so that *R* = 500 Ω. Measure and record the *PC* voltage.

16. Set *R* = 100 Ω. Measure and record the *PC* voltage.

17. Experimentally determine the lowest value of *R* for which the *PC* voltage remains within 10 percent of the voltage measured in step 14. In Table 8-4 record this resistance and the voltage delivered by the *PC*.

18. Reduce *R* to one-half (½) the value found in step 17. Measure and record the voltage delivered by the cell for this value of load resistance. Also record the value of *R*.

19. Compute and record in Table 8-4 the current delivered by the solar cell for each value of load resistance.

20. Replace the first solar cell with the second solar cell. Repeat steps 14 through 19 and record in Table 8-4.

21. Determine, experimentally, the maximum voltage after stabilization produced by the two solar cells connected in series aiding, using the 60-W light source (let *R* = 100 Ω): __________ V.

22. Experimentally determine the maximum voltage after stabilization produced by the two cells connected in parallel (let *R* = 100 Ω): __________ V.

TABLE 8-4. Characteristics of a Solar Cell

Cell 1				*Cell 2*			
Step	*Resistance, Ω*	*Voltage across PC, V*	*Current (Calculated) mA*	*Step*	*Resistance, Ω*	*Voltage across PC, V*	*Current (Calculated) mA*
12	1000				1000		
14	1000				1000		
15	500			20	500		
16	100				100		
17							
18							

QUESTIONS

1. What are two types of photoelectric devices, and what is the principal characteristic of each?
2. From the data in Table 8-1, how does the resistance of a photoconductive cell vary with the light intensity?
3. From the data in Table 8-2 and 8-3 explain the effects on current flow in a photoamplifier of a photoconductive cell as the light which illuminates the photocell is turned ON and OFF if the cell is in the (*a*) forward connection, (*b*) reverse connection.
4. What is meant by spectral sensitivity?
5. From the results in step 21, what appears to be the relationship between the total voltage and the individual voltages of similarly rated photovoltaic cells connected in series aiding? maximum current?
6. From the results in step 22, what appears to be the relationship between the total voltage and the individual voltages of similarly rated photovoltaic cells connected in parallel? maximum current?

Answers to Self-Test

1. Photoelectric
2. Photoconductive; photovoltaic
3. resistance; light-sensitive
4. angstrom
5. Photovoltaic
6. sensors (transducers)

EXPERIMENT 9

RELAYS

OBJECTIVES

1. To study the terminology, construction, operation, and general features of relays
2. To demonstrate forward and reverse photorelay operation

BASIC INFORMATION

Since the early days of Western Union, when the clicking of a relay transmitted a message in code, to modern "contactors" which control three-phase motors in a large machine tool, the relay has been one of the most widely used components in industrial electronics. In combination with transistors, SCRs, vacuum tubes, and other circuit elements, this device performs countless tasks. Among other applications, relays are used as protective devices, for switching, for indication, and for transmission.

In the past, and still today, relays have been electromechanical devices. But in many cases, a solid-state "relay," the SSR, which is discussed in a later experiment, has special characteristics which make it more useful than the electromechanical relay.

Regardless of the type of construction, relays perform the following functions. Protective relays remove from service an element of an electrical system when that element short-circuits, is overloaded, operates in such a way as to damage other elements of the system, or becomes a safety hazard. A ground fault circuit interrupter is an example of the latter application.

Indicating or signaling relays may be used in conjunction with protective relays to show the location of a component which has failed. They may also be employed in connection with lamps, buzzers, ringers, and clocks.

Transmission relays are used with transformers, inductors, and capacitors in power systems and in telephone and communication systems.

Electromagnetic Relays

An electromagnetic relay may be thought of as a remotely controlled switch having one or more sets of contacts. When energized, the relay operates to open or close its contacts or to open some contacts and close others. Contacts which are open when the relay is not energized are called *normally open* (NO) contacts, or simply open contacts. Contacts which are closed when the relay is not energized are called *normally closed* (NC) contacts. Normally open contacts are sometimes referred to as *a* contacts, and normally closed contacts are known as *b* contacts. Relay contacts are shown on schematic diagrams with the relay deenergized. Figure 9-1 shows several schematic representations for open and closed contacts. Some contact arrangements include contacts which "make" before others "break" or "break" before others "make."

A relay is said to "pick up," "pull in," "set," or "trip" when it is energized, and the pickup value is the smallest value of current through the coil which will close an *a* contact and open a *b* contact. When a relay is deenergized, opening an *a* contact and closing a *b* contact, it is said to reset or drop out. An electromagnetic relay will drop out at a lower current than the pickup current and therefore has a type of hysteresis.

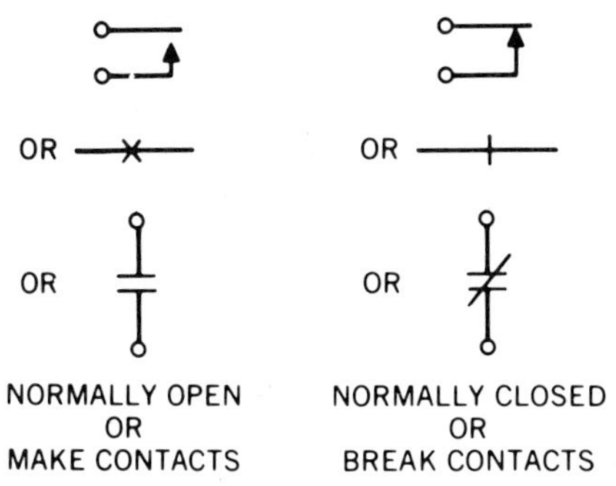

Fig. 9-1. Schematic symbols for relay contacts.

Relay contacts are held in their normal position by either springs or some gravity-actuated mechanism. An adjustment is usually provided to set the restoring force to the value which will cause the relay to operate within specified conditions. In some cases it is a screw and in others it may be a metal tang.

Electromagnetic attraction-type relays, which may be either ac- or dc-actuated, consist of (1) an electromagnet, an armature, and contacts or (2) a solenoid, a plunger, and contacts. The electromagnet consists of a core and a winding around it. The core, armature, and plunger are made of magnetic materials such as iron, silicon steel, or permalloy (an alloy of nickel and steel). The arrangement of parts in an attraction armature-type relay is shown in Fig. 9-2. The movable contact (4) is attached to the armature but insulated from it. A spring with adjustable tension restrains the armature from closing the gap between the stationary and movable contacts. The relay shown has one set of normally open contacts (3 and 4). This construction demonstrates two desirable features of an electromagnetic relay: good isolation between the coil and the contacts, and metal-to-metal contacts which can have a very low resistance when closed and a very high resistance when open.

When terminals 1 and 2 are connected to a source of current, an electromagnet is formed and the armature is attracted to the core. If the resulting magnetic field is sufficiently strong, the restraining force of the spring is overcome and the relay contacts close. The armature will be attracted to the core whether the pole of the electromagnet adjacent to the armature is north or south. Therefore the current energizing the electromagnet may be either ac or dc. However, an ac attraction-type relay differs from a dc relay in that it has a shading ring, whereas the dc type does not. A shading ring is typically a piece of copper or aluminum inserted into a slot cut into the core of the electromagnet at the pole adjacent to the armature. The ring acts as a shorted turn on a transformer and takes a relatively large current which is out of phase with the current in the main winding. The resulting magnetic field minimizes the tendency of the relay contacts to buzz or chatter when alternating current is applied to the coil.

Another contact arrangement is shown in Fig. 9-3. Contacts 4 and 5 are normally closed. When the relay is energized, contact 4 (the "arm") breaks the connection to contact 5 and makes a connection with contact 3. If, at some time, all three contacts are joined, this arrangement is a "make before break" set. As shown, these contacts correspond to those of a single-pole double-throw switch (SPDT). Many different arrangements of contacts are possible and correspond with familiar switch arrangements: SPST, SPDT, DPST, DPDT, etc.

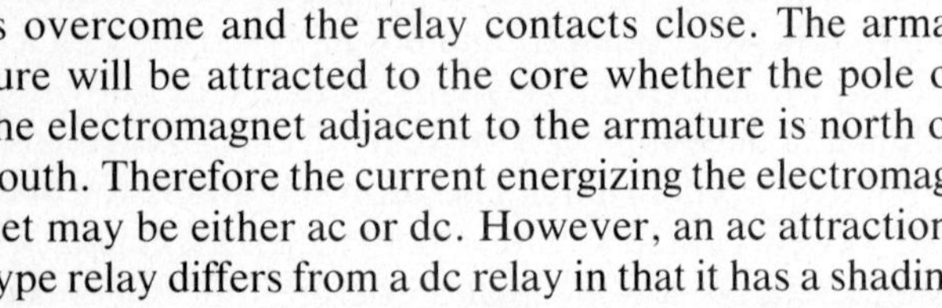

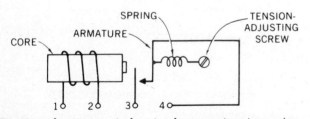

Fig. 9-2. Arrangement of parts of an armature-type relay.

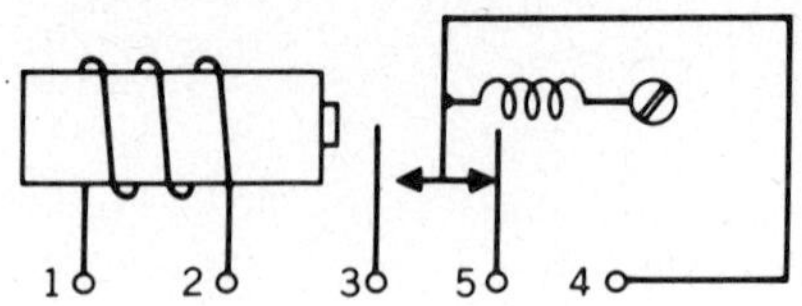

Fig. 9-3. Single-pole double-throw (SPDT) armature relay.

Open- and Closed-circuit Relay System

It is not our intention here to discuss the countless relay circuits used in industry. We will consider the two basic systems which serve as the building blocks for more complex circuits.

The circuit in Fig. 9-4*a* shows a low-voltage, low-power, normally open circuit relay system, used to control a circuit whose load consumes more power at a higher voltage than the relay circuit. When control switch *S* is closed, the relay is energized, closing its contacts. The closed contacts complete the load circuit, permitting current to flow through the load. Figure 9-4*b* is a schematic representation of the circuit in Fig. 9-4*a*.

Relay control of a large-power circuit may also be obtained by the closed-circuit relay system in Fig. 9-5. Here the relay is energized by a switch *S*, which is normally closed. Current flowing in the relay circuit opens the relay contacts, as shown, and keeps the load circuit open, so that no load current is drawn. When it is desired to operate the load, switch *S* is opened, deenergizing the relay circuit, thus closing the relay contacts. This completes the load circuit. Figure 9-5*b* is a schematic representation of the circuit shown in Fig. 9-5*a*. We have used the conventional symbol for a coil in designating the relay coil. Other symbols, as in Fig. 9-5*c*, are also used to designate the relay coil.

Closed-circuit systems have one advantage over open-circuit systems which makes them more desirable in certain applications. Thus, in the circuit shown in Fig. 9-5, any defect in the control system, such as an open circuit or dead battery V_1, will immediately become apparent, because the relay will become deenergized and the load circuit will be made operative. If the load is an alarm bell, the bell will sound continuously, signaling a defect in the control system. However, a defect in the open-circuit system of Fig. 9-4 will not become known until an attempt is made to operate the system.

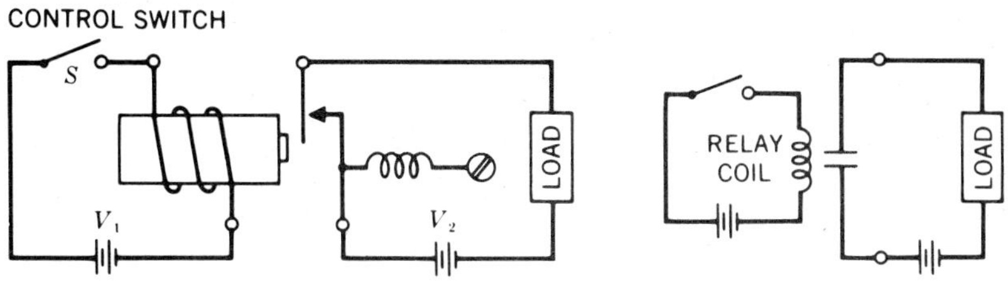

Fig. 9-4. *(a)* Normally open relay circuit; *(b)* schematic of normally open relay circuit.

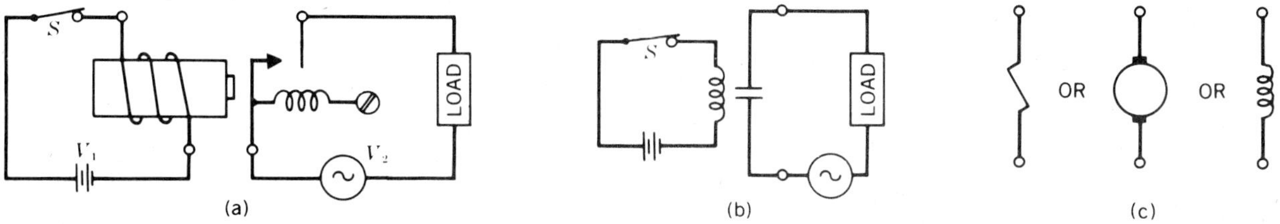

Fig. 9-5. *(a)* Normally closed relay circuit; *(b)* schematic of normally closed relay circuit; *(c)* relay coil symbols.

DC Light-sensitive Relay

We find industrial applications where photocells are used in conjunction with amplifiers and relays in circuits, requiring light-actuated ON-OFF switching arrangements. Thus a photocell amplifier and relay arrangement is used to turn street lights ON automatically as night falls and turn them OFF at sunrise. Let us see how this type of control can be accomplished.

Figure 9-6 is the circuit diagram of a photo relay, which can be used as a switch to apply power to a load when the actuating light which is focused on the photocell is turned OFF. The load can be a motor for opening a garage door, a street light, a buzzer, and so on. In this circuit, Q_1 is the relay amplifier. The relay coil constitutes the load in the collector of Q_1. The relay contacts to which the load is connected are open until the relay is energized. Then they close and complete the ac circuit to the load.

The resistance of the photocell *PC* and R_1 constitutes a voltage divider which forward-biases the emitter-base of Q_1. When enough light falls on the photocell, its resistance is low. The voltage developed across *PC* is inadequate to drive Q_1 sufficiently to energize the relay. Therefore, when light shines on *PC*, the relay is cut off, and no ac power is applied to the load. However, when the illumination on the photocell is turned off, the resistance of *PC* increases dramatically. The voltage at the junction of *PC* and R_1 rises. If the constants of the circuit are properly chosen, enough forward bias is developed on the base-emitter junction to increase collector current appreciably. The increased collector current energizes the relay and the normally open relay contacts close, applying power to the load. Thus, if the load is a street light, it will be turned ON automatically at nightfall and will be turned OFF at daybreak. R_2 is a sensitivity control which is adjusted for the level of light desired to turn OFF load power.

Figure 9-7 is a variation of the circuit shown in Fig. 9-6. The relay in Fig. 9-7 is comparable to an SPDT switch. Two loads can be accommodated in this arrangement. In the position shown, the relay is not energized. Contacts 1 and 2 are normally closed and load 1 receives power from the 120-V line. When the relay

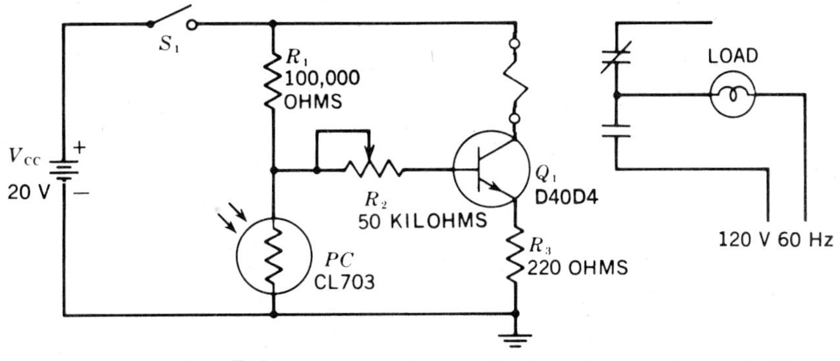

Fig. 9-6. Photo relay—reverse connection. Relay is actuated when light shining on photocell *PC* is turned OFF.

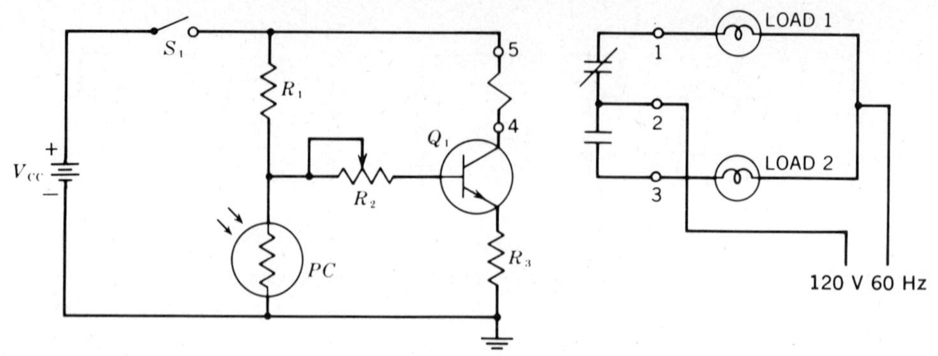

Fig. 9-7. Photo relay—SPDT.

is energized, the movable contact 2 short circuits to 3. Power is now applied to the circuit of load 2 and removed from load 1.

Figure 9-8 shows a photo relay circuit, where light control of a load is accomplished in a manner opposite to that of the circuit of Fig. 9-6. Just as in Fig. 9-6, the resistance of the photocell *PC* and R_1 constitutes a voltage divider which forward-biases the emitter-base of Q_1. Here, however, when no light falls on *PC*, its resistance is very high, and insufficient forward bias is developed on the base-emitter diode, so that the relay is not energized. Hence, when no light, or insufficient light, falls on the photocell, the relay switch remains open and no power is applied to the load. When light falls on the photocell, its resistance decreases, thus increasing the forward bias on the base-emitter section of Q_1. If the load constants are properly selected, the relay is energized and the relay switch closes, applying power to the load. Here again, R_2 acts as a sensitivity control. Its setting determines the light level which will be required to energize the relay.

Relay Specification and Identification of Relay Contacts

Relay manufacturers supply a specification sheet with each of their relays. This ''spec'' sheet contains relay ratings, designates whether the relay is dc or ac, and specifies the location and the ratings of the contacts. For example, the relay which you will use in this experiment is a dc SPDT relay. The resistance of the relay coil is 400 Ω. The contacts of the relay are designed to carry a 5-A current.

The relay coil and contact terminals can usually be located by inspection. If a relay is enclosed in a sealed unit, this may not be possible. An ohmmeter may then be used to identify the terminals. The ohmmeter is used to measure the resistance between any two terminals. There are only a limited number of two-terminal combinations possible. The resistance of the coil will be measured between the two terminals which are connected to the relay coil. Relay coils will vary from very low to very high values of resistance. Your relay coil should read approximately 400 Ω. In the process of finding the coil terminals, you will also locate contact terminals. Normally closed contacts will measure zero resistance. Normally open contacts will read infinite resistance. For example, in the relay in Fig. 9-7, the ohmmeter will read the coil resistance between terminals 4 and 5. There will be zero resistance between terminals 2 and 1 and infinite resistance between terminals 2 and 3. The resistance between terminals 1 and 3 will also be infinite. If you cannot determine by in-

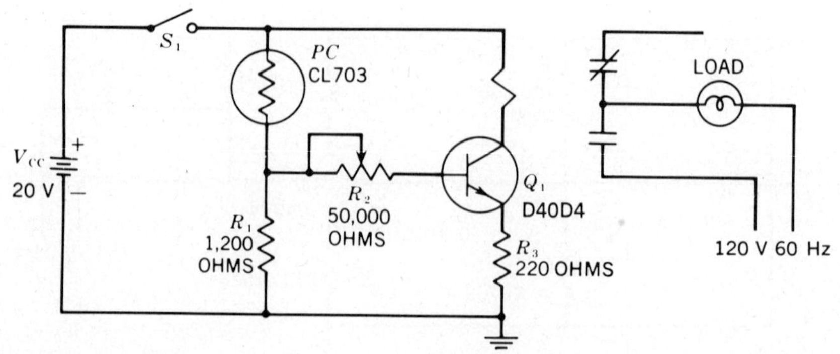

Fig. 9-8. Photo relay—forward connection.

spection whether contact 2 or 1 is the movable contact, the relay should be tripped. A resistance check will then show, as in the case of Fig. 9-7, that there is now zero resistance between contacts 2 and 3, and infinite resistance between contacts 2 and 1 and contacts 3 and 1. It is therefore evident that contact 2 is the movable one. CAUTION: In making resistance checks of relay contacts, be certain the power (to the load) is OFF; otherwise you may damage the ohmmeter.

SUMMARY

1. Relays are remotely operated switches used as protective, indicating and transmitting devices.
2. Protective relays protect good components from the effects of circuit components that have failed.
3. Transmission relays are used in communication systems.
4. Indicating relays may be used to identify a component which has failed, or they may be used with attention-getting devices such as bells and buzzers.
5. Relays may be simple SPST switch-type devices, or they may have complex switching arrangements.
6. The switch contacts of relays may be normally open (NO) or normally closed (NC). The contacts are held in their normal positions by springs or by some gravity-actuated mechanism.
7. Attraction-type relays are either ac- or dc-operated and consist of either (*a*) an electromagnet, a movable armature, and contacts or (*b*) a solenoid, a plunger, and contacts.
8. Terminals are provided on a relay for the winding of the electromagnet and for the relay switch contacts.
9. An advantage of a relay over an ordinary switch is that a low-power source may be used to turn a relay ON and OFF, while in turn, heavy-duty relay contacts open and close the circuit for a high-power load.
10. A relay system may be designed so that its load circuit is open when the relay control switch is open, called an *open-circuit system*. Or a relay may be designed with a *closed-circuit system* in which the load circuit is open when the control switch is closed.
11. Light-sensitive relays whose circuit is controlled by the action of a photocell may be used to turn street lights on automatically at nightfall and off at daybreak.

SELF-TEST

Check your understanding by answering these questions.

1. A dc relay is one in which __________ current in the relay coil actuates the relay mechanism.
2. The movable arm of an attraction-type relay is called the __________ .
3. A relay which "picks up" at 10 mA is one which is turned __________ (ON/OFF) when 10 mA flows in the __________ __________ .
4. Some important electrical specifications of an attraction-type relay are (*a*) the __________ of the coil, (*b*) the __________ of the coil, (*c*) the pickup __________ of the relay, and (*d*) the current-handling capacity of the __________ .
5. In Fig. 9-3, switch contacts __________ and __________ complete the load circuit when the relay is turned ON.
6. In Fig. 9-6, when S_1 is closed and a bright light shines on the photocell, the forward bias of the transistor amplifier is __________ and the relay __________ (is/is not) actuated.
7. In Fig. 9-7, when S_1 is closed and there is no light on the photocell, load light __________ (1 or 2) is ON.
8. In Fig. 9-8, when S_1 is closed and a bright light shines on the photocell, the load light is turned ON. __________ (true/false)

MATERIALS REQUIRED

- Power supply: 120-V 60-Hz source; variable dc source
- Equipment: VOM (20,000 Ω/V); EVM
- Resistors: ½-W 220-, 1200-, 1800-, 100,000-Ω
- Semiconductors: Photocell CL703 or equivalent; D40D4 transistor or equivalent
- Relay: DC; SPDT; 400-Ω field (approximate); 1-A contacts; 7.0-mA pickup current (approximate); RBM, type 10730-8 or equivalent
- Miscellaneous: 60-W light source with reflector; SPST switch; 10-W wired test lamp and socket; 50,000-Ω 2-W potentiometer

PROCEDURE

1. You will receive a relay whose terminals will be numbered. Determine by inspection, if possible, the terminal connections of the coil, the number of contacts, the movable contacts, and whether the contacts are open or closed. If necessary, use an ohmmeter to identify the terminals. Record the relay coil and contact terminal numbers in Table 9-1.

2. Measure and record in Table 9-1 the coil resistance.

3. Connect the circuit of Fig. 9-9. V_B is a source of variable direct current. A_1 is a 0- to 20-mA milliammeter or a VOM set on an equivalent milliampere range. An ohmmeter is connected across normally open terminals 2 and 3. Set V_B for zero output. What will happen when switch S is closed? Will the relay be tripped? Why?

4. Close switch S and slowly increase the dc voltage output of V_B, observing the ohmmeter reading. The ohmmeter will read zero resistance (you may also be able to hear the relay trip). Observe and record in Table 9-1 the minimum current, read on A_1, required to trip the relay. This is the pickup-current value.

Do you think that the relay will reset when the current through the coil is reduced just below the pickup value?

5. Slowly reduce the voltage on V_B, observing the reading on the ohmmeter. When the relay resets, the ohmmeter will read infinite resistance. Observe and record in Table 9-1 the value of current at which the relay resets.

6. Repeat steps 4 and 5, checking your previous pickup and reset readings. Repeat if necessary until your readings are constant. Power OFF.

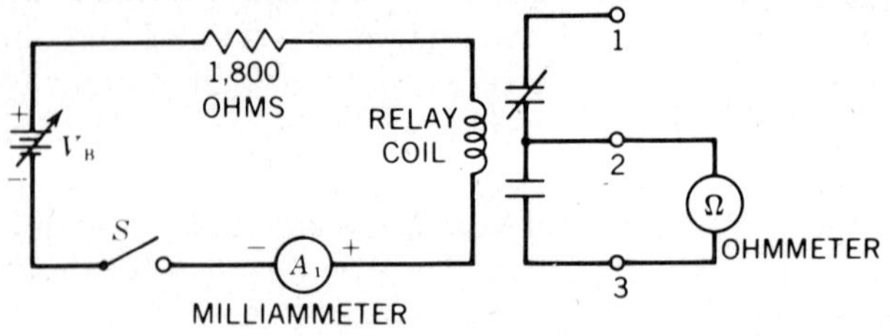

Fig. 9-9. Circuit for determining pickup value of a dc relay.

7. Connect the circuit shown in Fig. 9-6. Switch S_1 is open. The load is a 10-W lamp. The light source is a 60-W bulb and reflector plugged into a 120-V ac source. Position the light source about 5 in from the photocell and orient it so that the light will fall on the photowindow when the 60-W bulb is lit. The photocell is oriented so that it does not face any other light source in the room, just the 60-W light source. NOTE: Set R_2 at 25,000 Ω. During the course of the experiment, it may be necessary to reset R_2 to make the circuit operate properly.

8. Plug the load-power cord into the 120-V ac source, and turn ON the light source. Does the 10-W load lamp light? Why?

9. Close switch S_1, applying power to the photocell and its amplifier. Does the 10-W load lamp light? Why? Record your results in Table 9-2.

10. With an EVM, measure and record in Table 9-2 the transistor voltages V_{BE} (base-emitter), V_{CE} (collector-emitter), V_{CB} (collector-base), and across the relay coil.

11. Do you think the relay will pick up when the light source is turned OFF? Why? Will the 10-W lamp light?

TABLE 9-1. Relay Characteristics

Function	*Terminal Connection*	*Relay Characteristics*	
Relay coil		Coil resistance, Ω	
Normally open contacts		Pickup current, mA	
Normally closed contacts		Reset current, mA	

TABLE 9-2. DC Light-Sensitive Relay Measurements—Reverse Connection

Condition	*Is* 10-W *Load ON or OFF?*	Q_1 V_{BE}	V_{CE}	V_{CB}	*Voltage Across Relay Coil*
Light source ON, steps 9, 10					
Light source OFF, steps 12, 13					

12. Turn OFF the light-source voltage. Observe and record in Table 9-2 the effect on the 10-W load lamp.

13. Again measure and record in Table 9-2 the voltages indicated in the table.

14. Power OFF.

15. What do you think would be the effect on the 10-W load lamp if the photocell were connected in the forward connection, as in Fig. 9-8, when the light source is turned ON? When the light source is turned OFF?

16. Reconnect the photocell circuit as in Fig. 9-8. Q_1 and the relay circuit are identically connected in Figs. 9-6 and 9-8. Set $R_2 = 25{,}000\ \Omega$. Close switch S_1. Plug the load and relay circuits into the 120-V ac source as in Fig. 9-8. The light source is still positioned 5 in from the photocell and is turned ON.

17. Record in Table 9-3 if the 10-W load lamp lights.

18. Repeat step 10 and record in Table 9-3.

19. Repeat step 12 and record in Table 9-3.

20. Repeat step 13 and record in Table 9-3.

21. Power OFF.

TABLE 9-3. DC Light-Sensitive Relay Measurements—Forward Connection

Condition	*Is* 10-W *Load ON or OFF?*	Q_1 V_{BE}	V_{CE}	V_{CB}	*Voltage Across Relay Coil*
Light source ON, steps 17, 18					
Light source OFF, steps 19, 20					

QUESTIONS

1. Define pickup value and reset value.

2. Is it possible to change the pickup value of a relay? How?

3. Explain how you can determine how much current is required to trip a relay.

4. Would it be possible to employ the relay you used in this experiment to turn on a 120-V 2000-W load? Why?

5. From the data in Table 9-1, compute the minimum value of voltage which must be placed across the relay coil to trip the relay. How does this compare with the measured values in Tables 9-2 and 9-3?

6. Explain the results in Tables 9-2 and 9-3 for light source ON and OFF.

7. Which circuit would you use to turn a street light on automatically at night?

Answers to Self-Test

1. direct
2. armature
3. ON; relay coil
4. resistance; voltage; current; contacts
5. 4; 3
6. low; is not
7. 2
8. true

Items for Additional Study

1. Investigate the electrical starting circuit of a modern automobile. Explain how it works and why it is convenient.
2. Ground fault circuit interrupters are used (required by law in some cities) in homes as well as in industrial environments. How do they promote safety?
3. Explain the operation of the circuits given in Fig. 9-10.

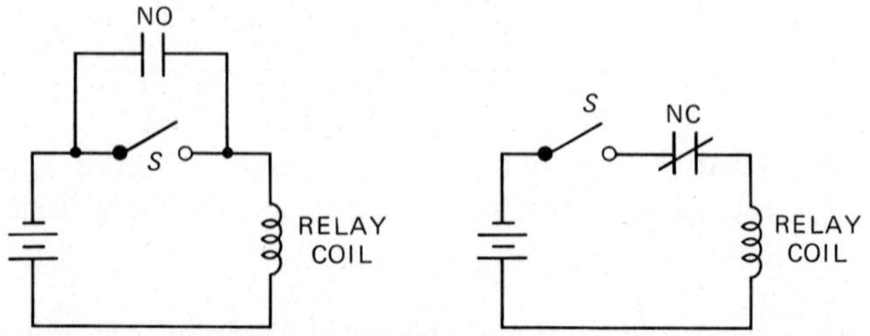

Fig. 9-10. Operation of circuits.

EXPERIMENT

10

OPTOELECTRONICS: THE PHOTOTRANSISTOR

OBJECTIVES

1. To study the characteristics of photodiodes and phototransistors
2. To demonstrate the operation of a light-controlled phototransistor relay circuit

BASIC INFORMATION

A variety of semiconductor devices fall under the heading of optoelectronics: light-emitting diodes (LEDs), photodiodes, phototransistors, light-activated SCRs (LASCRs), optocouplers, optoisolators, and solid-state relays (SSRs). These devices are being used in control applications as diverse as automatic light level controls in photocopying machines and interfaces between digital computers and the machine tools they control. LEDs are used extensively as status indicators, and the seven-segment LED display (as well as other types of seven-segment display) has replaced the analog, meter-type indicator in many applications.

Silicon photodiodes, or photodetectors, are light-sensitive devices which convert light signals into electrical signals. When light shines through a window or lens on the reverse-biased PN junction (Fig. 10-1), hole-electron pairs are created. The movement of these hole-electron pairs in a properly connected circuit results in current flow through the diode. The amount of current is determined by the intensity of the light and is also affected by the frequency (color) of the light falling on the photojunction.

In Experiment 8 it was noted that the sensitivity of the human eye is not uniform over the visible spectrum. As Fig. 10-2 shows, the eye is most sensitive to light whose wavelength is 550 nm and that sensitivity falls off for both shorter and longer wavelengths. The spectral response of the eye, then, is from 400 to 700 nm, with a peak at 550 nm. Outside this range, light which may exist is not perceived by the human eye.

The spectral response of a silicon photodiode is shown in Fig. 10-3. Maximum sensitivity is to radiation at 900 nm, and the total range is from about 400 to 1100 nm, which includes radiation in the visible range and outside it. The total range is called "light" even though light usually refers to the wavelengths within the human visual spectrum. The spectral response of a particular photodiode depends on the materials of which it is made and the type of construction. For maximum

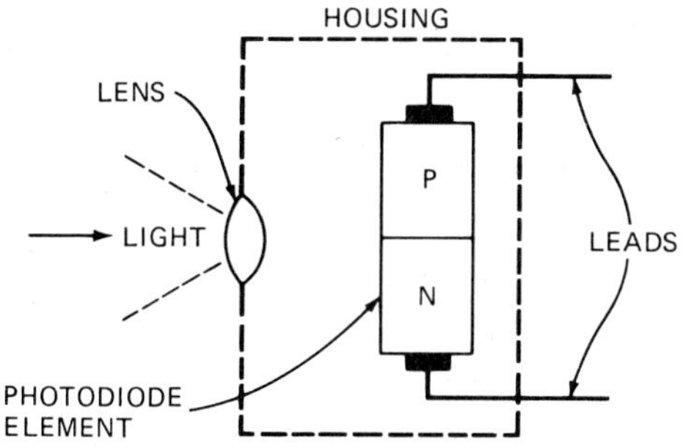

Fig. 10-1. Light falls on the PN junction of a photodiode.

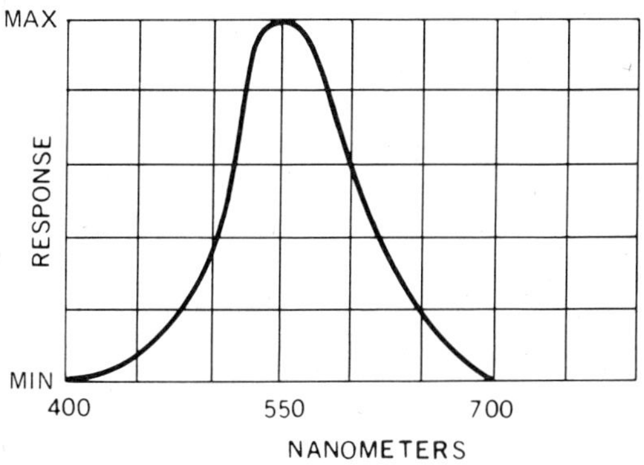

Fig. 10-2. Spectral response of the human eye. (To convert wavelength to angstrom units, multiply above values by 10.)

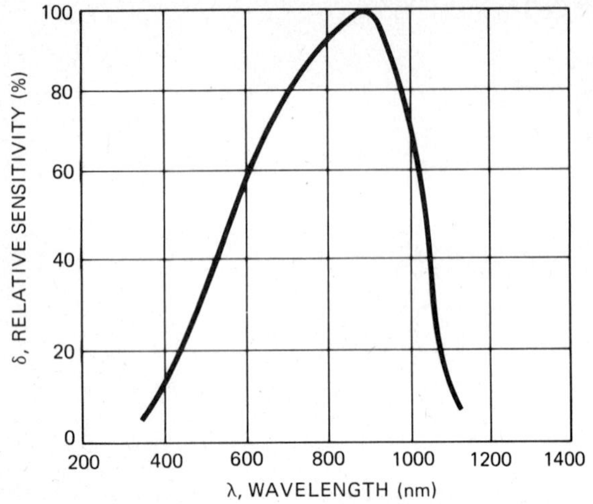

Fig. 10-3. Spectral response of a silicon photodiode. *(Motorola)*

sensitivity the spectral characteristics of the light source (light emitter) used with a photodiode should match the response of the photodiode. For example, for the diode whose response is given in Fig. 10-3, the light source should have a wavelength close to 900 nm.

Phototransistors

The current developed by a photodiode is small and cannot be used directly in most control applications. But with amplification the photocurrent may be great enough to be used in a control system: to energize a relay, for example. The phototransistor is a light detector which is more sensitive than a photodiode because amplification of the photocurrent takes place. Figure 10-4 shows the construction of an NPN phototransistor. Here a lens concentrates the light on the very thin P-type base region between the N-type collector and emitter. In this device, base current is supplied by the current created by the light falling on the base-collector photodiode junction and a larger current flows in the collector-emitter circuit. Either two or three leads may be brought out of the phototransistor. If the base lead is brought out, current from an external source may be used to set the operating point of the phototransistor.

The amount of current through a phototransistor depends mainly on the intensity of light falling on the photo junction and is not strongly affected by the voltage across the phototransistor. Figure 10-5 is a graph of collector current I_C as a function of collector-emitter voltage V_{CE} and of illumination H. It is apparent that the phototransistor acts as a constant-current source and that the current through it depends almost entirely on the illumination. The collector-emitter voltage has little effect.

Figure 10-6 compares the response of a phototransistor with the response of the human eye. It is apparent that the phototransistor can "see" wavelengths that the human eye cannot, particularly in the infrared region (longer wavelengths and lower frequencies).

The angular alignment of a phototransistor and its source of illumination affects the sensitivity of the system. The reason is that the illumination of the photojunction is proportional to the cosine of the angle between the direction of radiation of the light source and the perpendicular to the surface of the photojunction. In addition, the optical lens, or window, and its size further affect the response of the phototransistor. The response characteristic of a typical device is shown by the polar-coordinate graph in Fig. 10-7, where the relative response is the distance from the center to the curve at a particular angle. The inner curve is the response with a lens, and the outer curve is the response with a flat glass window. At the 50 percent point the lens response is about 15° on either side of perpendicular (head on), for a total of 30°.

Phototransistor Relay Operation

Figure 10-8 is a circuit diagram showing the direct operation of a sensitive relay by an NPN phototransistor.

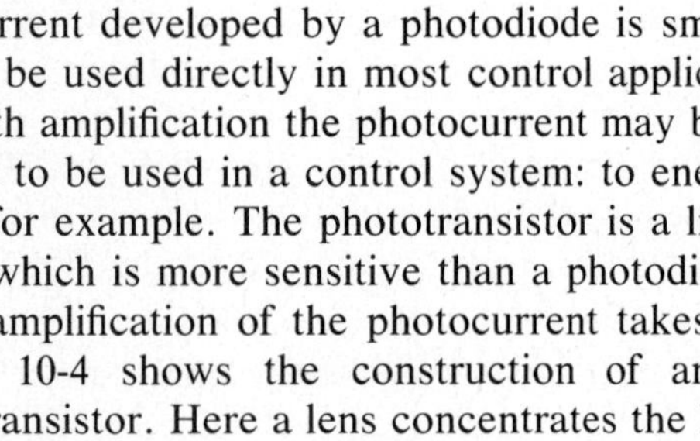

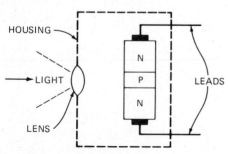

Fig. 10-4. NPN phototransistor.

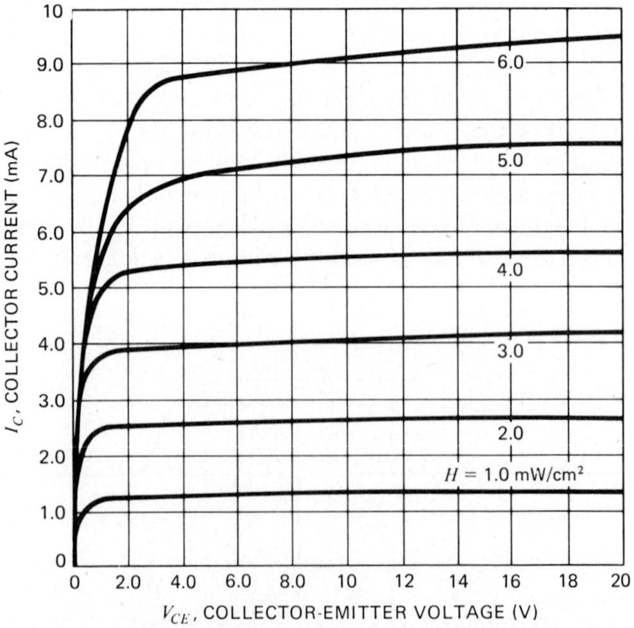

Fig. 10-5. Collector characteristic for the MRD300. *(Motorola)*

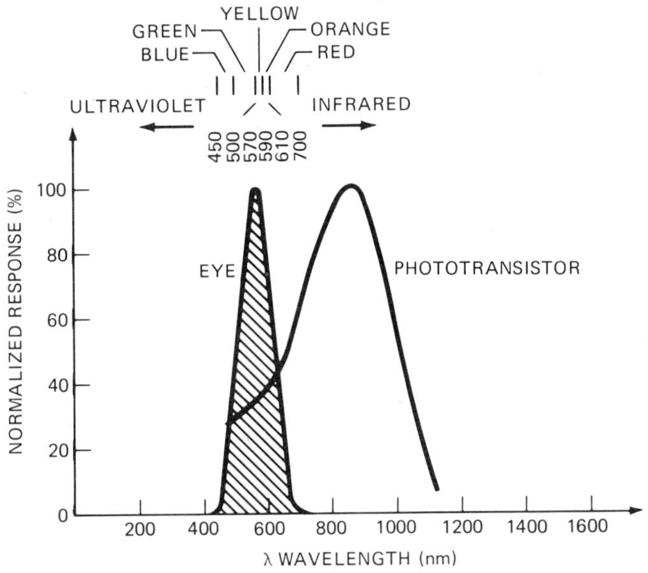

Fig. 10-6. Spectral responses of a phototransistor and the human eye. *(General Electric)*

When a sufficiently strong light falls on the transistor, the relay pulls in, and when the light is removed, the relay drops out. Milliammeter A measures the collector current. If an ordinary 60-W light bulb is used as the light source, it must be fairly close to the photowindow to actuate the relay. The SPDT contacts of the relay may be used to control any other device: a motor, a light, a horn, etc.

Less light is required to actuate the same relay in the circuit shown in Fig. 10-9, where an additional transistor Q_2 amplifies the current through the phototransistor Q_1. The voltage drop across R_1 when Q_1 conducts forward-biases Q_2 and causes a larger current in Q_2, thus energizing the relay. When there is no photocurrent, there is no drop across R_1 and thus Q_2 is biased to cutoff since its base and emitter are at the same potential. The relay is deenergized.

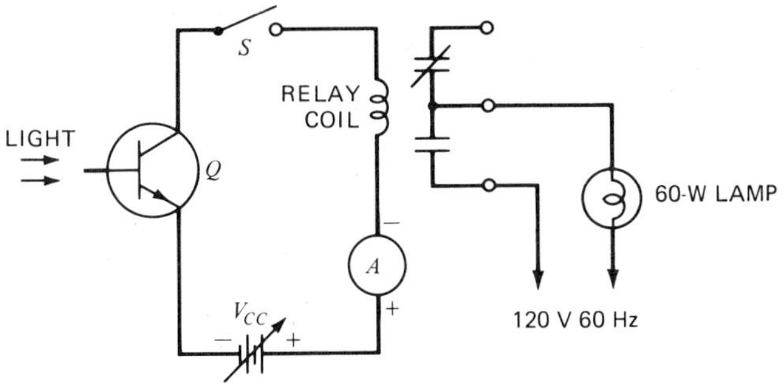

Fig. 10-8. One-stage phototransistor-operated relay.

SUMMARY

1. Optoelectronics deals with light-sensitive semiconductor junction devices used for control of industrial processes.
2. In the optoelectronic family of devices are LEDs, photodiodes, phototransistors, LASCRs, optocouplers (optoisolators), and SSRs.
3. Silicon photodiodes convert light into electrical signals. They may be classified as photodetectors.
4. The current generated by a photodiode, when light shines on its photosensitive PN junction, is proportional to the intensity of the incident light.
5. Photodiode current is very low. If this current is to be used effectively in control applications, it

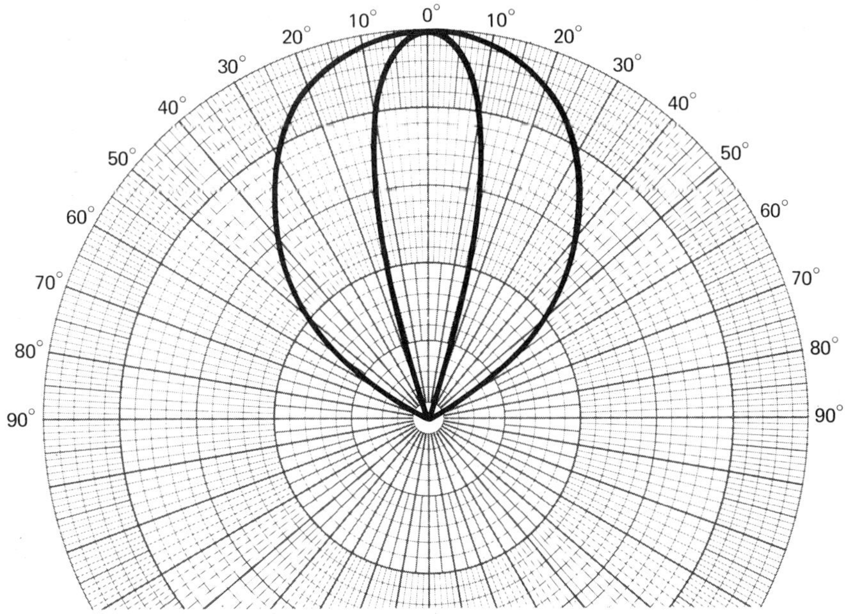

Fig. 10-7. Polar response of the MRD300. Inner curve with lens; outer curve with flat glass. *(Motorola)*

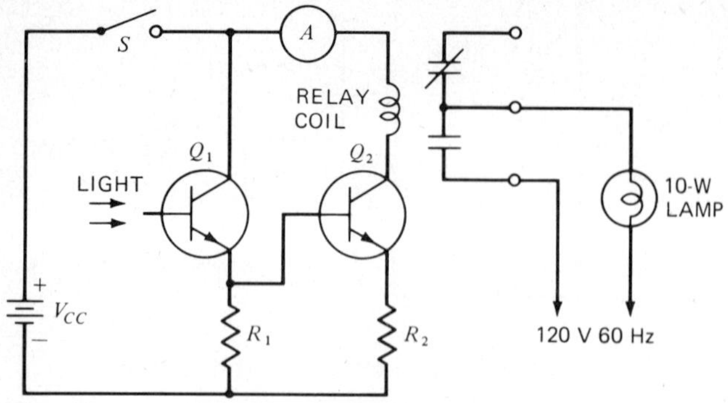

Fig. 10-9. Two-stage phototransistor and amplifier.

must be amplified. This may be accomplished by an external current amplifier or by a device called a *phototransistor.*

6. A phototransistor is a two-junction device, similar to the transistor. The collector-base junction of the phototransistor acts as a photodiode. The current created by this photodiode is the base current for the phototransistor, which then amplifies the photocurrent.
7. Both photodiodes and phototransistors contain a window or optical lens, which is located so that light entering the window falls on the light-sensitive PN junction.
8. Photodiodes and phototransistors exhibit unique spectral characteristics. That is, they are each sensitive to a specific range of ''light'' frequencies. What these frequencies are is determined by the material from which they are constructed and by the extent of doping of the junctions.
9. Photodevices may be sensitive to frequencies at, above, or below the frequencies of the visual spectrum. However, the electromagnetic radiation to which they respond is still called ''light.''
10. Current developed by a phototransistor is dependent mainly on the intensity of incident light—and very little on the applied voltage.

SELF-TEST

Check your understanding by answering these questions.

1. __________ deals with light-sensitive devices.
2. A photodiode is a __________ junction on which light is focused by a __________ or a __________.
3. The human eye is most sensitive to light at __________ nm.
4. The silicon photodiode in Fig. 10-3 has a spectral response which peaks at __________ nm.
5. Phototransistors generate __________ (more/less) current than do photodiodes when the same intensity of light falls on their light-sensitive junctions.
6. Current developed by a phototransistor is practically independent of the external __________ applied to the transistor. It depends almost entirely on the __________ entering the photowindow.

MATERIALS REQUIRED

- Power supply: Variable low-voltage dc source
- Equipment: 0- to 10-mA milliammeter (or 20,000 Ω/V VOM); EVM
- Resistors: ½-W 56-Ω
- Transistor: BPX 38-1 (Litronix) or the equivalent
- Relay: dc SPDT 400-Ω field, 7-mA pickup (approximately) type RBM 10730-8 or the equivalent
- Miscellaneous: Variable autotransformer; isolation transformer; 10-W light bulb and socket; 60-W light bulb and reflector; SPST switch; fused line cord

PROCEDURE

NOTE: *The background lighting in the room (ambient light) should be kept low during this experiment.*

Phototransistor Characteristics

1. Connect the circuit shown in Fig. 10-10. Power ON. *A* is the meter set to measure photocurrent.
2. Connect the 60-W lamp to the autotransformer. Adjust the transformer for 20-V output. Orient the light source so that it is about 26 cm [10 in] away from the phototransistor window. Align the light source with the phototransistor so that the light shines directly into the photowindow and produces maximum photocurrent as measured by *A*. *Do not vary the orientation of the light source or its distance from the phototransistor.*
3. Adjust the output of the dc supply so that the voltage V_{CE} measured from collector to emitter equals 5 V.
4. Measure the photocurrent and record it in Table 10-1.

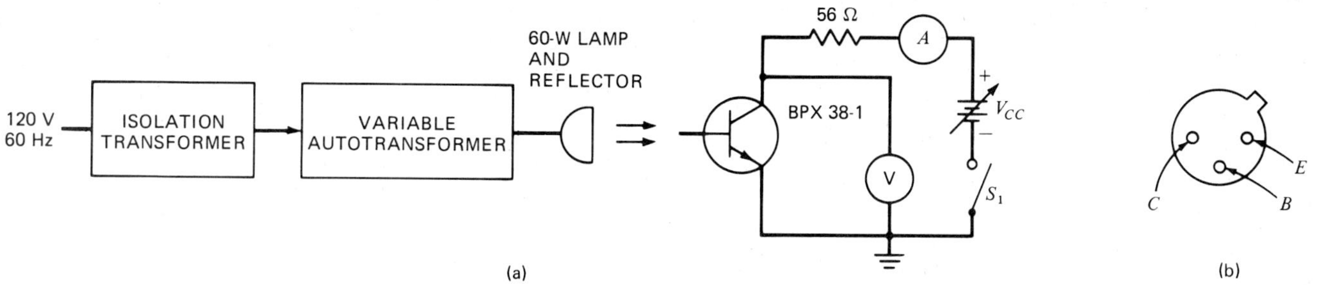

Fig. 10-10. *(a)* Test circuit to determine characteristics of a phototransistor; *(b)* bottom view, BPX 38-1.

5. Adjust the output of the power supply so that V_{CE} equals in turn, 10, 15, 20, and 25 V. For each value of V_{CE}, measure the photocurrent and record it in Table 10-1.

TABLE 10-1. Phototransistor—Current versus Voltage

V_{CE}, V	*Photocurrent* I_C, mA
5	
10	
15	
20	
25	

6. Set V_{CE} again to 10 V. Measure and record, in Table 10-2, photocurrent with the voltage applied to the light source still set at 120 V. Do not change orientation of the light source or its distance from the phototransistor.
7. Adjust the autotransformer so that the isolated voltage for the light source is 100 V.
8. Measure and record in Table 10-2 the photocurrent.
9. Repeat steps 7 and 8 for each value of voltage for the light source shown in Table 10-2.

TABLE 10-2. Phototransistor—Current versus Light Level

Light Source, V	*Photocurrent* I_C, mA
120	
100	
80	
70	
60	

Photo Relay

10. Connect the circuit shown in Fig. 10-11. Set V_{CC} = 10 V. The 60-W light bulb and reflector, plugged into a variable autotransformer—which, in turn, is plugged into an isolation transformer—again serve as the light source.
11. Adjust the variable autotransformer so that the output is 70 V. Place the light source so that it is about 26 cm [10 in] away from the phototransistor window, oriented to produce maximum photocurrent as measured by *A*. *Do not change the distance or the orientation of the light source*. A 10-W light bulb is the load to be controlled by the action of the relay. Indicate in Table 10-3 if the relay is ON or OFF. Also record photocurrent (relay coil current).

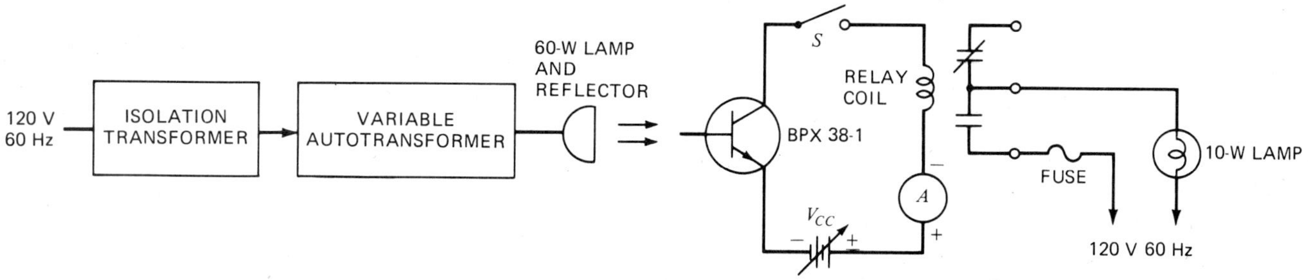

Fig. 10-11. One-stage photoresistor-operated relay test circuit

12. Turn the 60-W light source OFF. Measure and record in Table 10-3 the relay coil current. Indicate if the relay is ON or OFF.

13. Turn the 60-W light source ON. The output of the isolation transformer is still 70 V. Measure and record in Table 10-3 the relay coil current. Indicate if the relay is ON or OFF.

14. Gradually increase the output voltage of the variable autotransformer just to the point where the relay pulls in and the 10-W load light goes ON. Measure and record in Table 10-3 the voltage output of the transformer at this point and the current at which the relay turned ON.

TABLE 10-3. Photo Relay Measurements

Step	*Voltage for Light Source,* V	*Relay Coil Current (Photocurrent I_C),* mA	*Relay*	
			ON	OFF
11	70			
12	0			
13	70			
14			ON	

QUESTIONS

1. What do the results in Table 10-1 indicate about the dependence of photocurrent on the voltage V_{CE}?
2. At what voltage does the 60-W light source deliver more light, at 70 V or at 120 V?
3. What do the results in Table 10-2 indicate about the dependence of photocurrent on the intensity of light falling on the phototransistor?
4. What would be the effect on relay current of bringing the light source closer to the phototransistor? further away?
5. On what is photocurrent more dependent, the intensity of light shining on the phototransistor, or the voltage V_{CE}?

Answers to Self-Test

1. Optoelectronics
2. PN; window; lens
3. 550
4. 900
5. more
6. voltage; light

EXPERIMENT 11

OPTOELECTRONICS: THE LASCR, THE LED, AND THE OPTOISOLATOR

OBJECTIVES

1. To study the characteristics of LASCRs, LEDs, and optoisolators/optocouplers
2. To experimentally observe the operation of LEDs and optoisolators

BASIC INFORMATION

The light-activated SCR (LASCR) is an optoelectronic device used as a light sensor. Like the conventional SCR, it has three leads: the anode, cathode, and gate, as shown in Fig. 11-1. Unlike the ordinary SCR, the LASCR has a window through which light may fall on a light-sensitive silicon pellet (Fig. 11-2). It is the action of the light on this pellet which provides the gate current to trigger the LASCR. External bias may be provided for the gate to set the operating point on which the photo-derived gate current is superimposed.

Since the photocurrent supplied to the gate is very small, the LASCR is constructed so that it is very sensitive and can be triggered by this small current. To accomplish this a thin silicon pellet with small dimensions is used. These small dimensions limit the current carrying capacity of typical LASCRs to a few amperes.

In addition to its light sensitivity, the LASCR is also sensitive to high junction temperature: the amount of light required to trigger it decreases as junction temperature increases. It is more sensitive than the SCR to the rate of change of the anode-to-cathode voltage (dv/dt) and has a longer turn-off time.

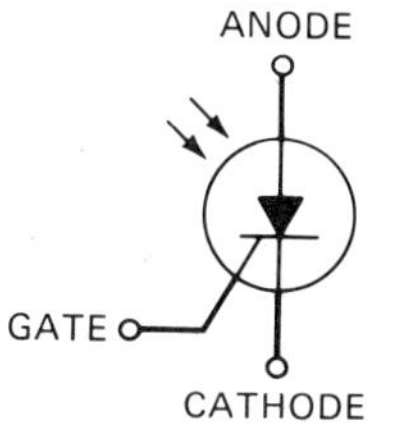

Fig. 11-1. LASCR symbol.

Figure 11-3*a* shows the light-junction temperature triggering characteristics of the GE L8/L9 LASCR. Figure 11-3*b* is a polar plot which shows the sensitivity of these LASCRs to the angle of the incident light from a tungsten filament light source. The spectral response, or sensitivity to light of different colors, of these LASCRs is seen in Fig. 11-3*c*. It is evident that these devices will respond to light which is not visible to the human eye. Their maximum sensitivity is at 1000 nm, well outside the range of visible wavelengths.

The light which triggers the LASCR into conduction loses control once the LASCR has fired, and to turn the LASCR OFF, the anode voltage must be reduced to zero or to a negative voltage relative to the cathode.

Triggering Higher-Power SCRs

LASCRs are relatively low power devices, but it is possible to increase their power-controlling capacity by using them as gate-triggering sources for high-power SCRs, as in Fig. 11-4*a* and *b*. In Fig. 11-4*a* the driven SCR is normally OFF; in the absence of light, the LASCR is nonconducting and no gate current is delivered to the SCR. When light falls on the LASCR, the

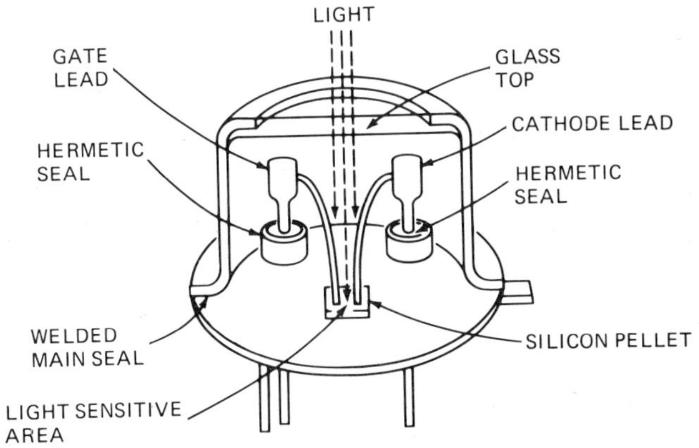

Fig. 11-2. LASCR construction. *(General Electric)*

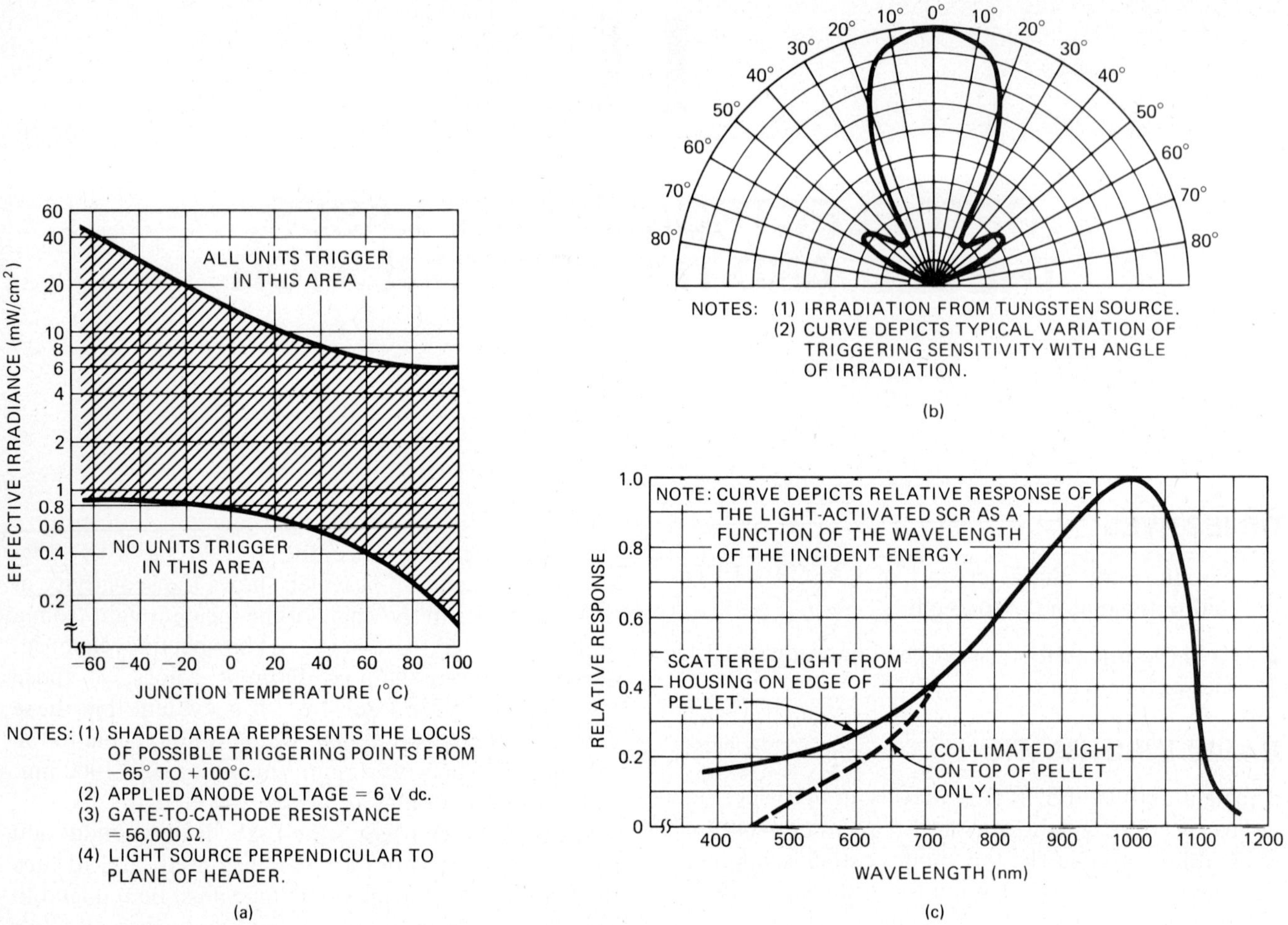

Fig. 11-3. Characteristics of the L8/L9 LASCR. *(General Electric)*

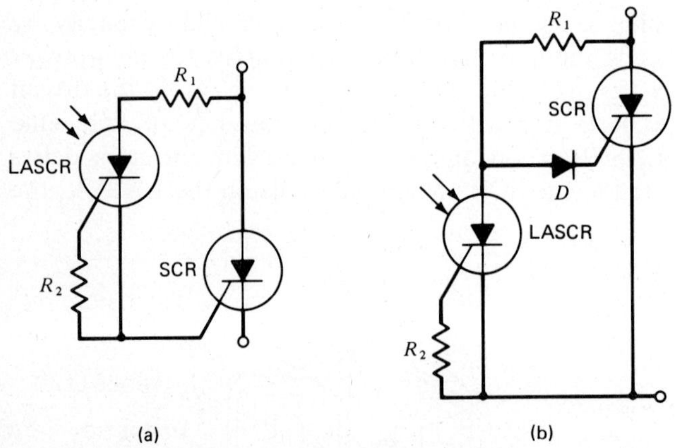

Fig. 11-4. Triggering high-current SCRs by light. *(a)* SCR normally OFF; *(b)* SCR normally ON.

SCR is triggered ON. Therefore, in this application, the SCR acts like the normally open contacts of a relay which close when light falls on the LASCR.

In Fig. 11-4*b*, in the absence of light, the SCR is normally ON because the gate is forward-biased by current flowing through the diode D and R_1. When sufficient light falls on the LASCR, it is turned ON and effectively shorts the gate voltage of the SCR to its cathode, thus turning the SCR OFF. In this arrangement the SCR acts like a set of normally closed contacts on a relay in the absence of light.

A practical circuit in which a LASCR is connected to trigger a high-power SCR is shown in Fig. 11-5. In the absence of light, the LASCR is turned OFF, the SCR is OFF, and no current flows through the load. When light falls on the LASCR, it turns ON and supplies gate current to the SCR. The SCR now acts as a closed switch and applies power to the load.

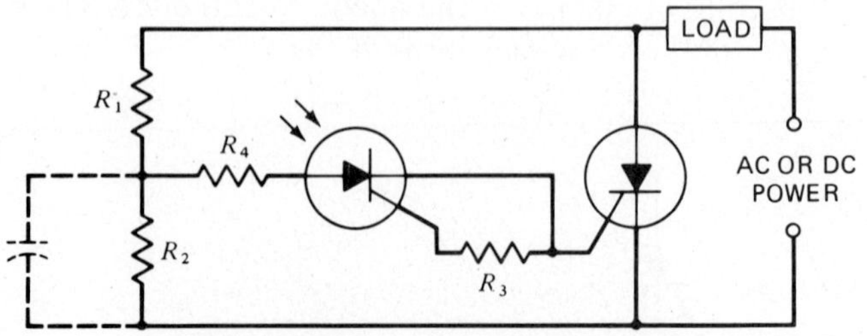

Fig. 11-5. Use of LASCR to trigger high-power SCR. *(General Electric)*

LASCR Applications

A LASCR, together with other solid-state components, can serve as a solid-state relay (SSR). This application is studied in Experiment 12. We have already seen that a LASCR can be used to trigger a high-power SCR. There are many other applications of LASCRs, such as:

Light-activated motor controls

Light-activated remote (photographic) flash units

Automatic sun-tracking systems

Automatic security and safety lighting

Light-Emitting Diodes

The photodiode, the phototransistor, and the LASCR are the light detector members of the optoelectronics family. They permit current flow when their light-sensitive junctions are exposed to light, and their useful characteristics are due to this property.

The LED (Fig. 11-6) is a PN junction device which emits light when forward-biased, a phenomenon called *electroluminescence*. The light emitted from the diode may be visible or invisible, from infrared (IR) to ultraviolet (UV). For many applications, LEDs are constructed as IR-emitting diodes (IREDs) and emit light which is invisible. IREDs are used as the ''light'' source for the photodetector in the component called the optoisolator, or optocoupler, the heart of optoelectronic solid-state relays. We will explore the characteristics of the optoisolator with LASCR output in this experiment.

The LEDs which emit visible light (red, green, yellow, white, orange, and blue) serve as indicators or readouts in a variety of applications in computers, calculators, status displays, and so on. They require less power than the older tungsten filament light bulbs, are smaller, and have a longer life. Their disadvantages include less intensity and a smaller viewing angle than filamentary lamps.

LED OPERATION

Recombinations of holes and electrons occur at the junction in every semiconductor diode and, in some cases, generate heat. But in LEDs and IREDs some of these recombinations release light. The spectral content of the emitted ''light'' depends on the type of semiconductor material from which the diode is constructed and on the concentration of dopants or impurities at the PN junction. Thus gallium phosphide can be used to produce either green or red LEDs. The circuit schematic symbol for a LED is shown in Fig. 11-7. Note that the arrows point away from the diode, showing that light is being emitted rather than absorbed.

The small physical size of the LED wafer and, therefore, its limited ability to dissipate heat determine the maximum current which an LED can safely pass. A typical LED, such as the one used in this experiment, operates well with 1.6 V across its terminals when passing a current of 20 mA. Figure 11-8 shows an experimental circuit which we will use to study the characteristics of a visible LED. Note that the LED, considered a conventional diode, is forward-biased. We will observe that increasing the current through the diode causes it to emit brighter light. With insufficient current, or reverse bias, no light is emitted.

The Optocoupler (Optoisolator)

An optocoupler (or optoisolator) is a solid-state component in which a light source, light path, and light detector are all enclosed within a plastic package which resembles other ICs. Figure 11-9*a* shows a dual-in-line package (DIP) optocoupler. The schematic symbol for this device, Fig. 11-9*b*, shows its internal components, an IRED light source, and a phototransistor. Its internal physical structure is shown in Fig. 11-9*c*. Figure 11-9*d* shows that the peak light output from the IRED is near the peak sensitivity of the phototransistor, which is a desirable feature.

When current flows through the IRED, the light it emits causes the phototransistor to conduct. The ratio of the detector current to the emitter current is called the *current transfer ratio* (CTR) and is a measure of the device's effectiveness. The input and output circuits are said to be ''isolated'' because there is no electrical path between them. (An ohmmeter test between any input terminal and any output terminal should show an ''open'' circuit.) An electrical signal is only transmitted in one direction, that is, through the light path.

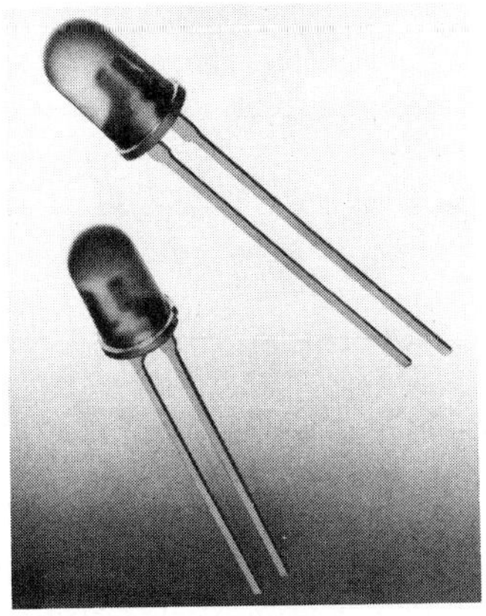

Fig. 11-6. LED with translucent housing permits light emitted by the diode to be visible. *(Litronix)*

Fig. 11-7. LED symbol.

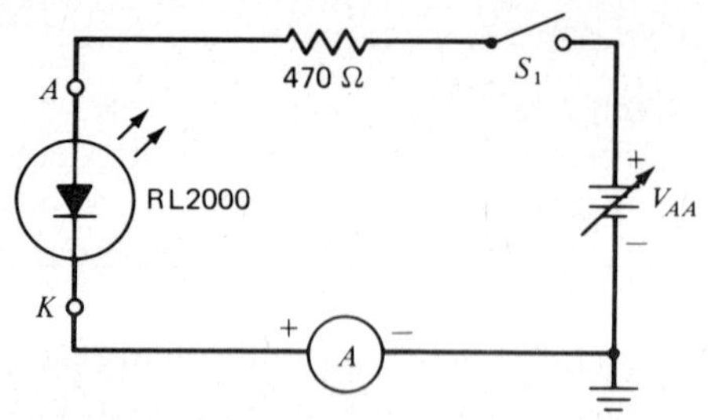

Fig. 11-8. Experimental circuit to determine LED characteristics.

The degree of isolation between input and output depends on the kind of material in the light path and the distance between the light emitter and the light detector. The greater this distance is, the better the isolation is. But greater distance also reduces the current transfer ratio. Therefore, a thin layer of glass is used to produce an acceptable compromise between good isolation and acceptable CTR (other dielectrics are also used).

Optoisolators which consist of an IRED and a LASCR (Fig. 11-10) are also made. As was learned from a previous discussion, the photodetector part of the LASCR develops a small current which is used as the gate drive. For this reason, the construction of the LASCR in the optoisolator is such that the SCR portion must be very sensitive to gate current so that it will be triggered by the light emitted by the IRED and directed at the photojunction of the LASCR. This device provides excellent isolation between the components of a microprocessor or microcomputer (which are very susceptible to damage by narrow, high-amplitude voltage "spikes") and the external relays, motors, or other inductive devices the LASCR may control.

SUMMARY

1. A LASCR is a light-activated SCR which acts as both a photodetector and an SCR.
2. A LASCR has an optical window through which light may pass to trigger it into conduction.
3. As with other optoelectronic devices, LASCRs may respond to light of various wavelengths (colors).
4. Photodetectors and photoemitters are members of the optoelectric family of electronic components.
5. Photodetectors produce an electrical output in response to a light input. Photoemitters produce a light output in response to an electrical input.
6. A LED is a diode which emits light when it is forward-biased and current flows through it.
7. An IRED emits energy in the IR spectrum when forward-biased. IR radiation is not visible to the human eye.
8. Visible LEDs are used as readouts in alphanumeric displays in calculators, machine tools, etc., and as pilot lights and indicators in a wide variety of applications.
9. Two circuits are said to be "isolated" if there is no direct electrical path between them. Energy is transferred in an optoisolator by means of a light beam. In a transformer, energy is transferred by a magnetic field.
10. There is almost perfect isolation between the input and output terminals of an optoisolator or optocoupler.

SELF-TEST

Check your understanding by answering these questions.

1. The LASCR described in Fig. 11-3*c* is less sensitive to light whose wavelength is 800 nm than light whose wavelength is 1000 nm. ________ (true/false)
2. In Fig. 11-5, until light falls on the LASCR the SCR is ________ (conducting/nonconducting)?
3. An IRED emits red visible light. ________ (true/false)
4. LEDs constructed of different materials emit light of different ________ .
5. Optoisolators include within their housings both a light ________ and a light ________ .
6. To emit light LEDs must be ________ -biased.
7. As the current through an LED increases, the intensity of the emitted light ________ .

MATERIALS REQUIRED

- Power supply: Variable low-voltage dc source; 50-V dc source
- Equipment: Two 0- to 100-mA dc milliammeters; EVM or 20,000 Ω/V VOM
- Resistors: ½-W 470-, 10,000-Ω; 20-W 1250-Ω
- Solid-state devices: RL 2000 LED (Litronix) or equivalent; H11C6 optoisolator (GE), or equivalent
- Miscellaneous: Fused line cord; two SPST switches

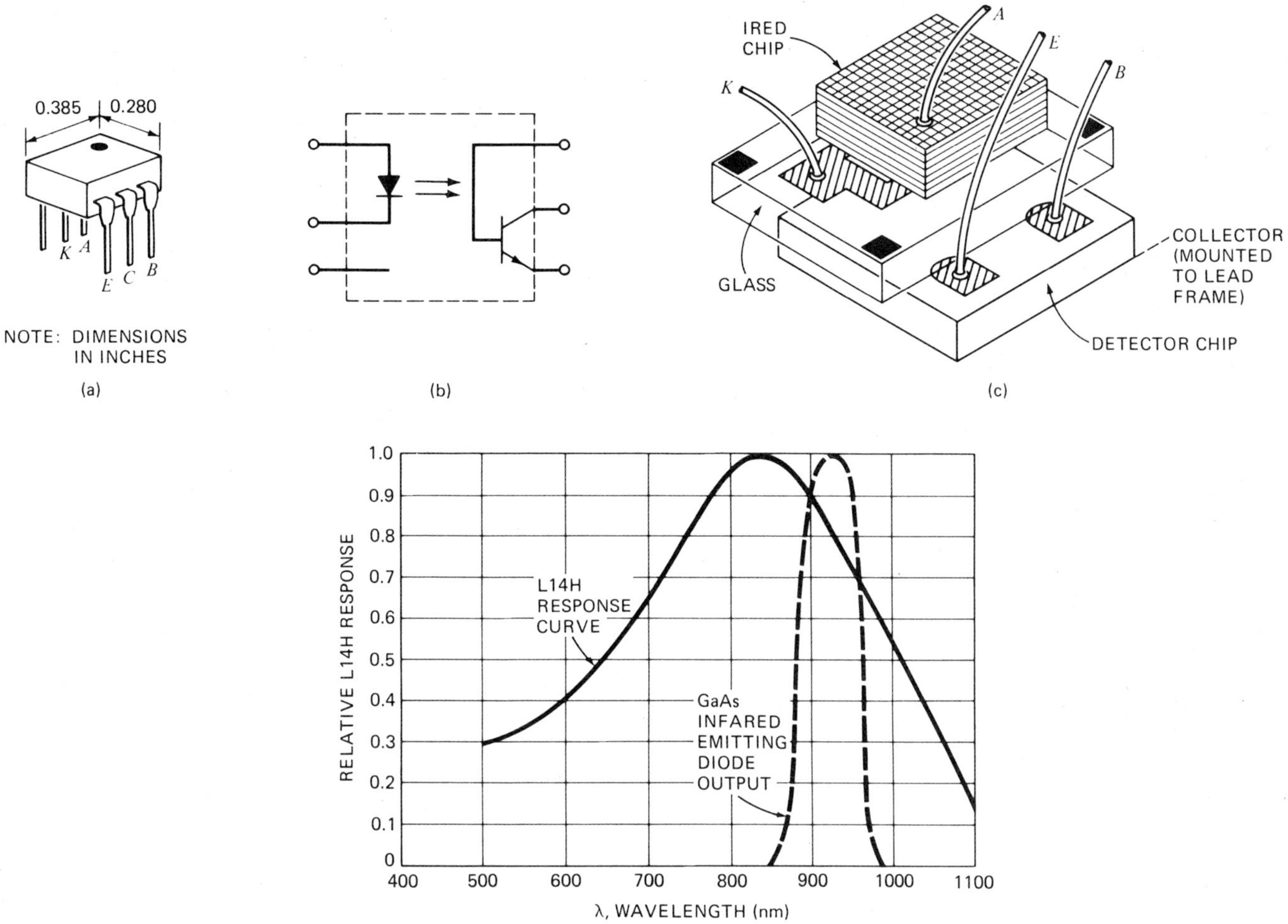

Fig. 11-9. Optoisolator. *(a)* Dual-in-line package; *(b)* symbol showing photodiode and phototransistor; *(c)* structural features; *(d)* spectral curves of the IRED and phototransistor. *(General Electric)*

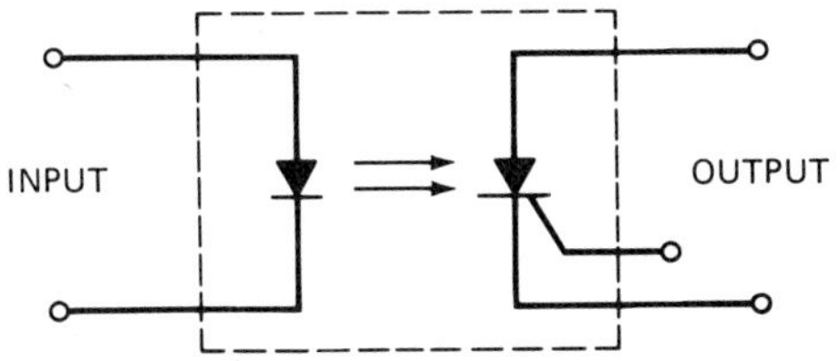

Fig. 11-10. Optoisolator made of LED with LASCR output.

PROCEDURE

LED Characteristics

NOTE: *The longer lead of the RL2000 is the anode; the shorter is the cathode.*

CAUTION: *Do not* exceed 20 mA of LED current.

1. Connect the circuit shown in Fig. 11-8, S_1 is open. Set $V_{AA} = 0$ V.
2. Close S_1. Power ON. Gradually increase the output of V_{AA}, while observing the LED, until the LED just glows. Measure and record in Table 11-1 the current in the LED (I_{LED}) and the voltage V_{AK} across the LED.
3. Adjust the output of V_{AA} until $I_{LED} = 10$ mA. Indicate in Table 11-1 whether the LED is brighter

TABLE 11-1. LED Characteristics

Step	*Light Intensity of LED*	I_{LED}, mA	V_{AK}, V
2	Just glows		
3		10	
4		20	
6			

or less bright than in step 2. Measure and record the voltage V_{AK} across the LED.

4. Repeat step 3 for I_{LED} = 20 mA.

5. Open S_1. Reverse the leads of the LED in the circuit. Set V_{AA} = 10 V.

6. Close S_1. Measure and record in Table 11-1 the current I_{LED}, if any, and V_{AK}. Indicate also whether the LED is or is not lit.

Optoisolator Operation

7. Connect the circuit shown in Fig. 11-11. S_1 and S_2 are open. A_1 and A_2 are milliammeters to measure the optoisolator's input and output current, respectively. Set V_{AA1} = 0 V, V_{AA2} = 50 V.

8. Close S_1 and S_2. Observe, measure, and record in Table 11-2 the input current I_{in} and the output current I_{out}, if any, and V_{AB}, the voltage across the LASCR.

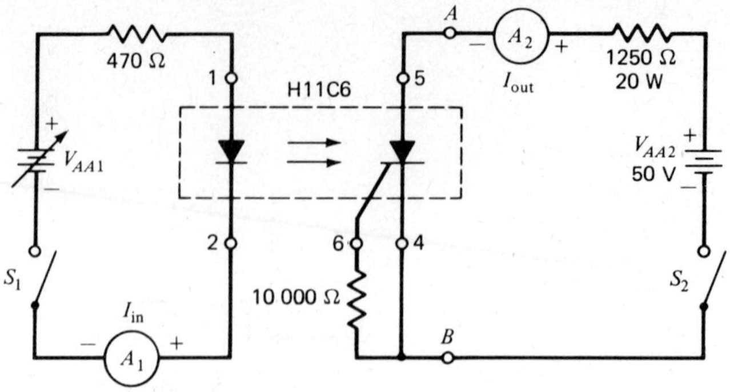

Fig. 11-11. Experimental circuit to determine how an optoisolator with LASCR output operates.

TABLE 11-2. Optoisolator Operation

Step	I_{in}, mA	I_{out}, mA	V_{AB}, V
8			
9			
10			

9. Gradually increase the output of V_{AA1} just to the point where there is measurable current I_{out}. Measure and record in Table 11-2 I_{in}, I_{out}, and the voltage V_{AB} across the LASCR.

10. Increase the output of V_{AA1} until I_{in} is twice the value measured in step 9. Measure and record in Table 11-2, I_{in}, I_{out}, and V_{AB}.

11. Open S_1. Does this affect load current?

QUESTIONS

NOTE: *Refer to your experimental results.*

1. What is the result of current flow through a LED?

2. What determines the intensity of light emitted by a LED?

3. In the optoisolator shown in Fig. 11-11, what is the minimum value of input current which causes the LASCR to turn ON?

4. In Fig. 11-11, once the optoisolator is turned ON, does increasing the input current have any effect on the output current? Why?

5. In Fig. 11-11, explain the result on output current of opening S_1 once the optoisolator is turned ON.

6. In step 8, was the LASCR conducting? Explain.

Answers to Self-Test

1. true (0.6)
2. nonconducting
3. false
(infrared; invisible to the human eye)
4. colors
5. emitter; detector
6. forward
7. increases

EXPERIMENT 12

OPTOELECTRONICS: SOLID-STATE RELAY (SSR)

OBJECTIVES

1. To study the characteristics of SSR
2. To demonstrate the operation of an SSR

BASIC INFORMATION

Solid-State Relay

In Experiment 11 the optocoupler/optoisolator was studied, and it was noted that the optocoupler can be used by itself as an SSR (the solid-state equivalent of an electromechanical "relay").

Two desirable characteristics of the optocoupler which make it useful as an all-electronic, nonmechanical "relay" are (1) its high degree of isolation between input and output, and (2) the fact that a small input current can be used to control a much larger output current.

A comparison of the SSR with the conventional electromagnetic relay (EMR) shows that the input of an SSR (IRED diode) corresponds to the coil of an EMR and the output of the SSR (the anode-cathode circuit of the photodetector) corresponds to the contacts of the EMR. These contacts are either closed or open, depending on whether the output device is conducting or nonconducting. As with EMRs, commercially available SSRs can have either ac or dc inputs.

In the previous discussion it was assumed that the devices shown in Fig. 12-1 were complete SSRs. In actual practice the internal circuitry is more complex and includes other active or passive components, depending on the performance characteristics desired. Figure 12-2 shows the additional components of a fairly complex input circuit to an optocoupler. Transistors Q_1 and Q_2 and related components are connected as a Schmitt trigger circuit which will change a sine-wave input voltage to a rectangular wave at the collector of Q_2. This waveform is suitable for turning the IRED ON by driving Q_3 fully into saturation, or turning it OFF when Q_2 conducts and insufficient drive is available to turn Q_3 ON. For the values given, the upper trigger level of the input circuit is about 4 V. If the ac input voltage is positive-going, the Schmitt trigger circuit changes states at that voltage. When the input voltage is negative-going (positive but decreasing), the trigger circuit goes back to the untriggered state at about 1 V. Figure 12-3 shows that the IRED conducts from *A* to *B* and is OFF from *B* to *E*. Depending on the amplitude of the ac input waveform, the IRED will be ON, and the

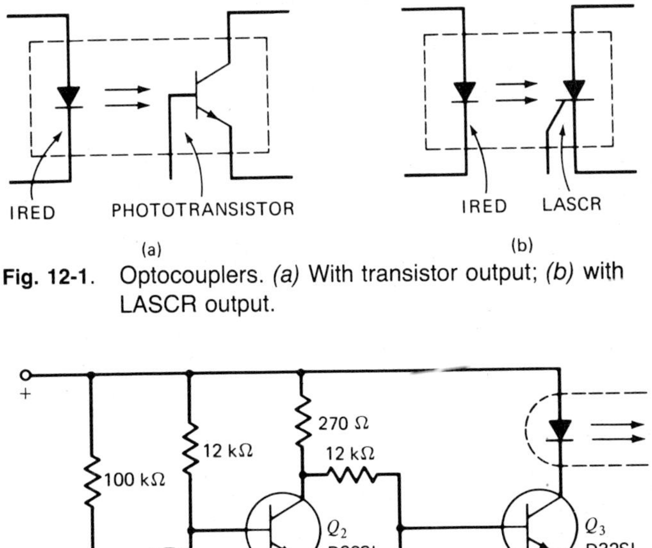

Fig. 12-1. Optocouplers. *(a)* With transistor output; *(b)* with LASCR output.

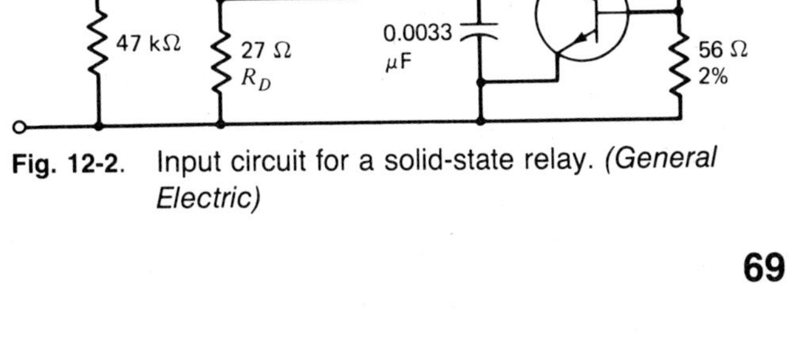

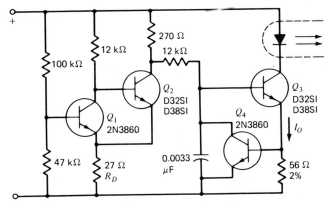

Fig. 12-2. Input circuit for a solid-state relay. *(General Electric)*

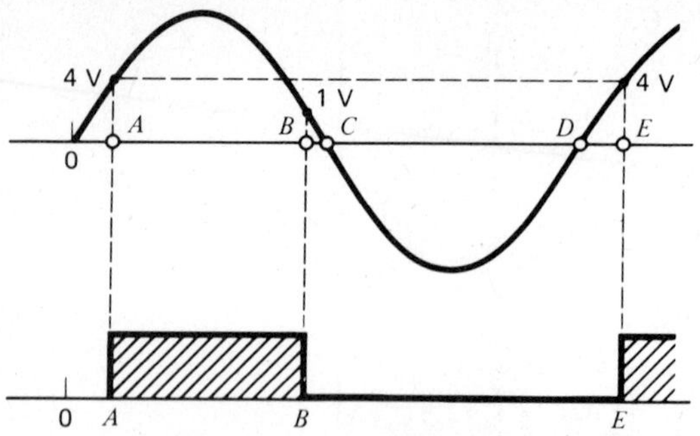

Fig. 12-3. In Fig. 12-2 the SSR is ON during the period *AB* and OFF from *B* to *E*.

output device will also be ON, for nearly one half-cycle and OFF for the remainder of the cycle.

The relay circuit just described is that of an ac relay. SSRs are available for either ac or dc input and for various standard voltages.

Phototransistor and LASCR output relays are relatively low current devices, but they can be used to trigger higher-powered SCRs and TRIACs so that larger loads may be controlled. For example, an SSR together with a full-wave bridge rectifier can be used with a triac to provide controllable, full-wave ac power to a load. For low-power full-wave ac loads, two photocouplers with SCR outputs can be connected back to back to act as an ac SSR, as shown in Fig. 12-4. In this circuit the two IREDs are connected in series to a dc control source and the two output SCRs are connected in parallel. The SCRs are connected in series with the source and conduct alternately, providing full-wave ac power to the load.

SSR Advantages

The SSR has several advantages over the electromechanical relay: (1) no moving parts to wear or break; (2) no contact "bounce," wear, or arcing; (3) no electromagnetic interference (EMI) when the input is deenergized; (4) fast operation ("contacts" close in microseconds rather than milliseconds); (5) resistance to shock, vibration, and other harsh environmental conditions; (6) high sensitivity (small actuating power); (7) good isolation and low coupling capacitance between input and output.

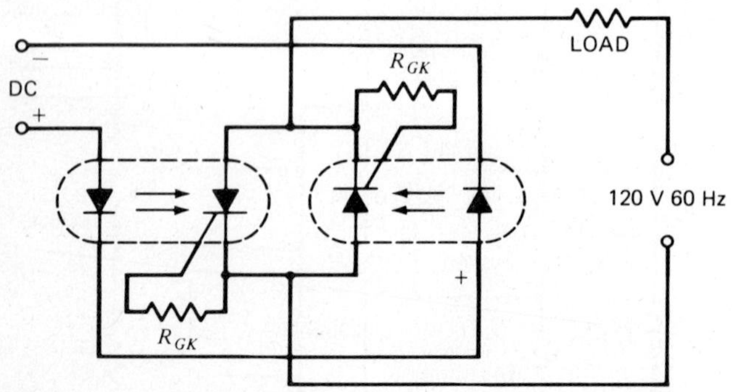

Fig. 12-4. Use of two photocouplers, SCR output, as an ac relay to provide full-wave ac power to a load. *(General Electric)*

SSR Disadvantages

Some of the disadvantages of SSRs are (1) excessive voltage transients ("spikes") on the input voltage may damage the solid-state components in the SSR; (2) the output "contacts" are not "metal to metal," as with an EMR, so there is a voltage drop (about 1 V) across the SSR when it conducts; (3) there is a small leakage current through the SSR even when it is OFF so the resistance between "open" contacts is not as high as with an EMR; (4) they may be damaged by nuclear radiation; and (5) they are expensive.

SUMMARY

1. Optocouplers are useful in SSR applications because they have good isolation between input and output and because they can control a large output current with a small input current.
2. An optocoupler by itself can act as an SSR.
3. Practical SSRs usually include an optocoupler and other components.
4. Since optocouplers conduct only in one direction, for full-wave ac power to a load, two optocouplers are connected back to back.
5. For high current applications, an optocoupler may be used to trigger a high-power SCR or triac.
6. Because they have several advantages, SSRs are used instead of electromechanical relays in many applications. In some cases, however, the EMR is preferred because of its low contact resistance when ON and high open circuit resistance when OFF.

SELF-TEST

Check your understanding by answering these questions.

1. For the same current through the contacts, an SSR has the same voltage drop as an EMR. __________ (true/false)
2. A circuit useful for converting a sine wave into a rectangular wave is the __________ __________.
3. In Fig. 12-2, if Q_3 is ON, the IRED is __________ (ON/OFF).
4. __________ optocouplers are required for a simple SSR which can pass full-wave ac to a load.
5. The 60-W lamp receives __________ current if the IRED in Fig. 12-5 is reverse-biased.

6. The current through the 60-W lamp in Fig. 12-5 will be __________ [dc (in one direction)/ac (in two directions)].

MATERIALS REQUIRED

- Power supply: Variable low-voltage dc source; 120-V 60-Hz source
- Equipment: EVM or 20,000 Ω/V VOM; oscilloscope
- Resistors: ½-W 470-, two 10,000-Ω
- Solid-state devices: Two H11C6 optoisolators or equivalent
- Miscellaneous: Isolation transformer; 60-W light bulb and socket; two SPST switches; fused line cord

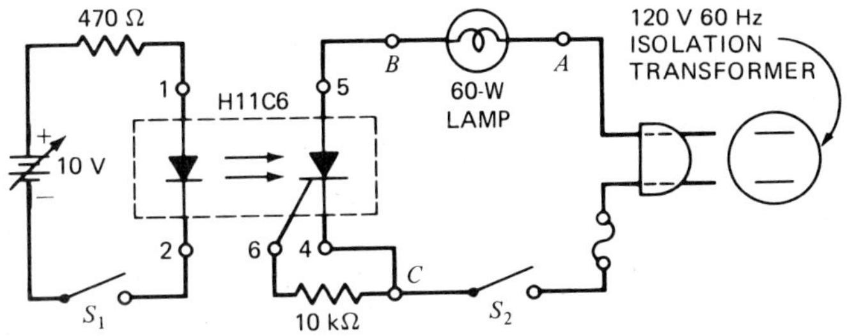

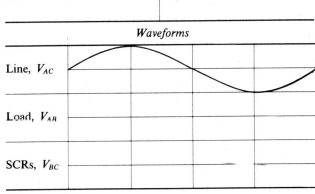

Fig. 12-5. SSR turns ON 60-W load lamp, half-wave operation.

PROCEDURE

CAUTION: *Hazardous voltage. Observe safety precautions.*

Solid-state Relay: Half-Wave Operation

1. Connect the circuit shown in Fig. 12-5. S_1 and S_2 are open. Plug the line cord into the 120-V 60-Hz line-isolated outlet of the isolation transformer.
2. Power ON. Adjust dc source to 10 V. Close S_1 and S_2.
3. Record in Table 12-1 if the 60-W load lamp goes ON. If it does, indicate if it is bright or dim.
4. With an oscilloscope which is either line-synchronized (triggered) or externally synchronized to the ac power line, observe and record, in proper time phase, the waveform across the line V_{AC} (hot lead on A), the waveform across the load V_{AB}, and the waveform across the output contacts of the relay V_{BC}. Record the waveforms in Table 12-1.
5. Open S_1. Does the load lamp remain lit? __________ .
6. Remove the load lamp from the circuit and plug it into the 120-V 60-Hz line. Is the lamp brighter than in steps 3 and 4? __________ .

Solid-state Relay: Full-Wave Operation

7. Power OFF. Connect the circuit shown in Fig. 12-6. S_1 and S_2 are open.

TABLE 12-1. Solid-State Relay, AC Half-Wave Operation

60-W Lamp	
ON or OFF	*Brightness*

Waveforms				
Line, V_{AC}				
Load, V_{AB}				
SCRs, V_{BC}				

8. Close S_1 and S_2. Power ON.
9. Record in Table 12-2 if the load lamp goes ON. If it does, indicate if it is brighter or less bright than in step 3.
10. With an oscilloscope which is line-synchronized or externally synchronized, observe and record in Table 12-2, in proper time phase, the waveforms across the line V_{AC}, the load V_{AB}, and across the SCRs V_{BC}. Power OFF.

TABLE 12-2. Solid-State Relay, AC Full-Wave Operation

60-W Lamp	
ON or OFF	*Brightness*

Waveforms

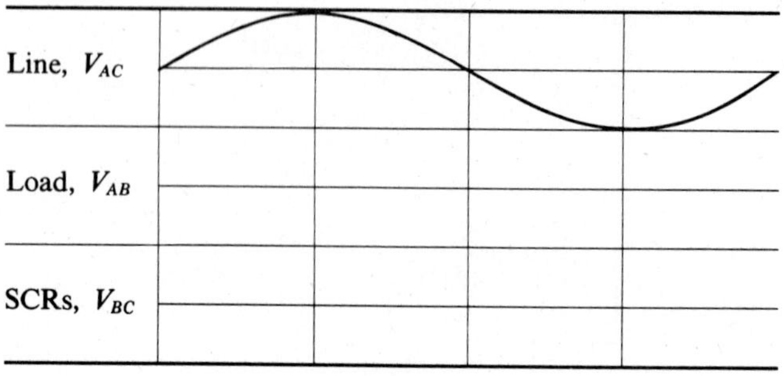

QUESTIONS

1. The intensity of the light emitted by the 60-W lamp in Fig. 12-5 should be less than the intensity when it is plugged directly into the 120-V 60-Hz line. Why?
2. Will the 60-W lamp in Fig. 12-5 receive a *full* half-cycle of the ac source voltage? Explain.
3. What is the effect of opening S_1 in Fig. 12-5? Why?
4. Why are the waveforms in Table 12-1 different from those in Table 12-2?
5. Compare the power *controlled* (approximately 60 W) with the *controlling power* taken from the 10-V dc source. Assume 0 V dropped across the IRED and the LASCR (Fig. 12-6).
6. Why are back-to-back photoisolators used in Fig. 12-6?

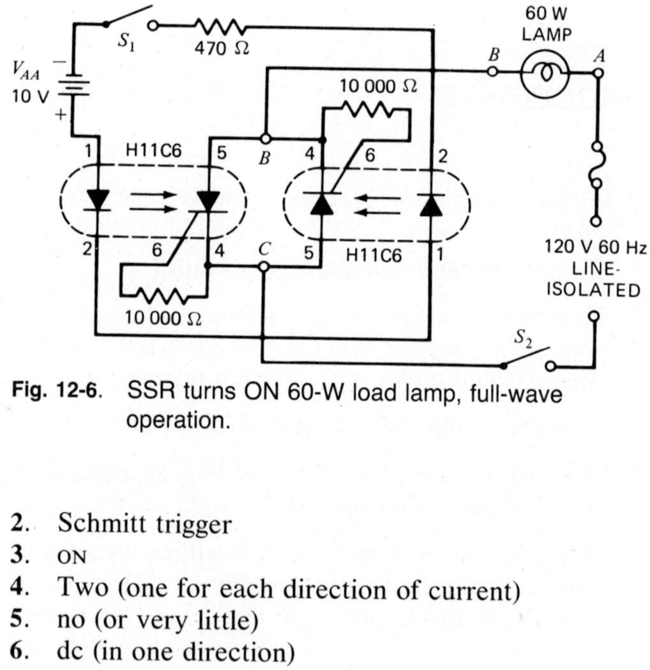

Fig. 12-6. SSR turns ON 60-W load lamp, full-wave operation.

Answers to Self-Test

1. false (greater across the SSR)
2. Schmitt trigger
3. ON
4. Two (one for each direction of current)
5. no (or very little)
6. dc (in one direction)

EXPERIMENT

13

PHOTOELECTRIC CONTROL OF AN SCR

OBJECTIVE

To study and demonstrate how a photoconductive cell may be used to trigger an SCR

BASIC INFORMATION

In the preceding experiment we found that we could use a photoconductive cell and an amplifier to energize and deenergize a relay. Thus, with a low-power input to the photocell and its amplifier, we were able to control much higher power delivered through the relay contacts to an external load. Another method involves the use of a photocell and an SCR. The photocell triggers the SCR under proper conditions of light. The SCR in turn acts as a switch to turn the load ON and OFF.

In Fig. 13-1, when a sensitive SCR replaces the transistor amplifier in a photoelectric circuit, SCR anode current energizes the relay coil connected in the anode circuit of the SCR, when the SCR is turned ON. Since the SCR can fire only on the positive alternations of the applied ac voltage, capacitor C across the relay coil is used to keep the relay from becoming deenergized during the negative ac alternations. The charge on C is delivered to the relay coil during the negative alternation, maintaining sufficient current in the relay coil to keep the relay ON. This prevents relay chatter.

The photocell PC, in series with R_5, R_4, and R_3, constitutes an ac voltage divider, which is designed to trigger the SCR for a predetermined light level. When no light falls on PC, its resistance is very high, and the

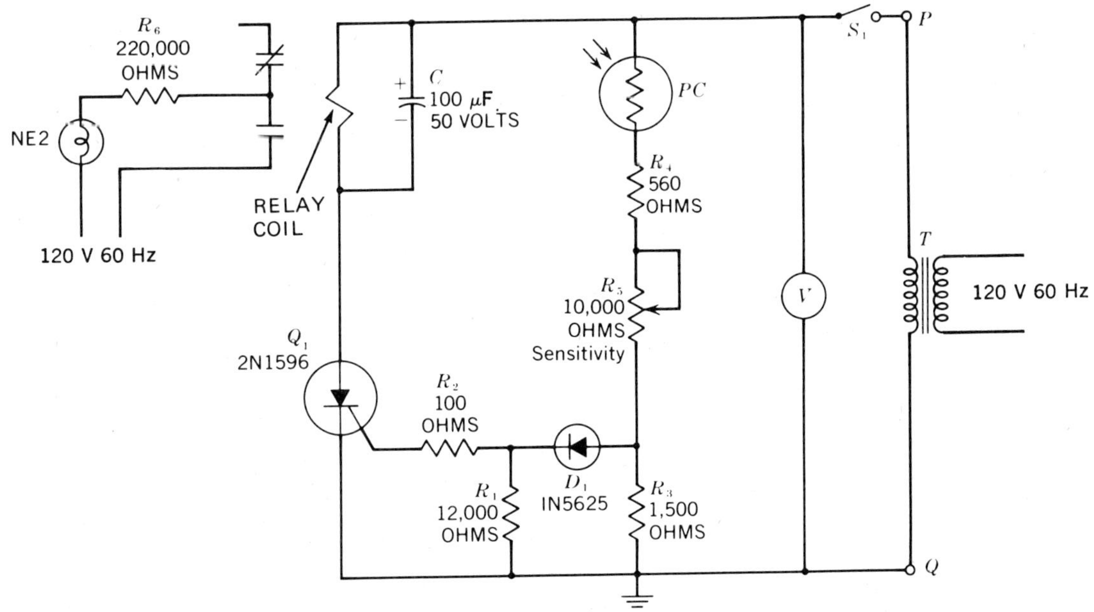

Fig. 13-1. SSR triggered by light falling on photoconductive cell.

voltage across R_3 is very low. Hence there is insufficient gate current I_G to trigger the SCR. As the light level increases to a point determined by the setting of R_5, the *PC* resistance decreases, the voltage across R_3 increases, and there is enough gate current to trigger the SCR. D_1 acts as a rectifier which passes only the positive ac alternation to the gate circuit. R_2 is a gate-current-limiting resistor. The peak level of gate current I_G may be determined by measuring the peak ac voltage V_2 across R_2 and dividing this value by R_2. Thus,

$$I_G\ (peak) = \frac{V_2\ (peak)}{R_2}$$

SUMMARY

1. An SCR may be used in conjunction with a photocell and a relay to control power to a high-power load.
2. The "dark" resistance of a photoconductive photocell is high and decreases when light falls on its active surface.
3. The combination of photocell, SCR, and relay produces a circuit having high power amplification, or "sensitivity."
4. The circuit of Fig. 13-1 does not "latch" but is energized only when light falls on the photocell.

SELF-TEST

Check your understanding by answering these questions.

1. In the circuit of Fig. 13-1 the gate current to the SCR is __________ in the absence of light falling on the photocell because the resistance of the PC is __________ .
2. If the sensitivity control R_5 is set at maximum resistance, the amount of light required to trigger this circuit will be __________ . (greatest, least)
3. Diode D_1 allows the gate to be only __________ (forward/reverse)-biased
4. The component in Fig. 13-1 which helps to smooth out the voltage applied to the relay coil is the __________ *C*.
5. The size of the load which this circuit can control is determined by the ratings of the __________ __________ .

MATERIALS REQUIRED

- Power supply: 120-V 60-Hz source; variable autotransformer; isolation transformer
- Equipment: EVM; oscilloscope
- Resistors: ½-W 100-, 560-, 1500-, 12,000-, 220,000-Ω
- Capacitors: 100-μF 50-V
- Solid-state devices: 2N1596 SCR or equivalent; 1N5625 diode or equivalent; photocell CL703 or equivalent
- Relay: SPDT, 1-A contacts, 400-Ω armature coil, 7.0-mA pickup (approximate)
- Miscellaneous: 60-W 120-V light source, reflector and housing; SPST switch; 10,000-Ω 2-W potentiometer; neon lamp, NE-2; power transformer, 120-V primary, 25-V secondary

PROCEDURE

1. Connect the circuit shown in Fig. 13-1. The step-down transformer is rated at 120-V primary, 25-V secondary. The 60-W light source is plugged into a variable autotransformer, so that the light level produced may be varied by adjusting the voltage applied to the light source.

   Orient the window of the photocell so that it does not face any light source in the room other than the 60-W lamp. Position the 60-W light source so that it is about 5 in from the photocell. *Light source is* OFF.
2. Close S_1. Vary R_5 over its entire range. Is the relay energized at any setting of R_5 with the light source OFF? Record your answer (yes or no) in Table 13-1.
3. Turn the oscilloscope ON. The scope must be "floating" off line ground. Calibrate the oscilloscope for 3 V/cm and 30 V/cm peak-to-peak.
4. *PC* light source is still OFF. Connect your oscilloscope vertical-input leads across *PQ*, the secondary of *T*, ground lead to point *Q*, the return of the circuit (Fig. 13-1). Set the oscilloscope on line sync or external sync. If external sync is used, connect the external-sync jack to point *P* through a 0.002-μF capacitor. Adjust sync, sweep, and centering controls until the reference waveform across *PQ* appears as in Table 13-1. Measure and record in Table 13-1 the voltage V p-p of this waveform.

TABLE 13-1. Circuit Measurements—No Light on Photocell

Step 2	*Relay Energized as R_5 Varied?*		
	Test Point	*Waveform*	*Voltage, V p-p*
Step 4,5	*PQ*	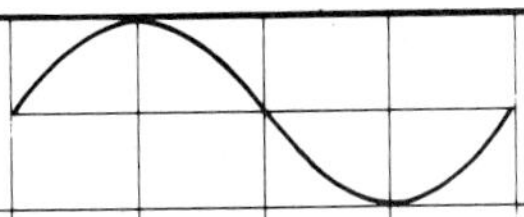	
	SCR anode-to-cathode		
	Across R_3		
	Across R_2		
	Across relay		
	Across photocell		

5. Measure also and record, in proper time phase with the reference, the following waveforms: across the SCR (anode-cathode), across R_3, across R_2, across the relay coil, and across the photocell. In all cases, connect the ground lead of the scope closest to the ground return of the circuit.

6. Set R_5 for minimum resistance. Apply power to the variable autotransformer and set its output to 0 V. Plug the 60-W light source into the autotransformer. Slowly increase the output of the variable autotransformer just to the point where the relay sets. Measure and record in Table 13-2 the voltage output of the variable autotransformer applied to the 60-W light source which just turns ON the relay.

7. Open S_1. Set $R_5 = 500\ \Omega$. Close S_1. Gradually increase the variable voltage source to the point where relay sets. Measure and record in Table 13-2 the *PC* light-source voltage.

8. Repeat step 7 for $R_5 = 1000\ \Omega$; $R_5 = 500\ \Omega$. Open S_1. Set light-source voltage at 120 V. Determine the maximum resistance of R_5 which will permit the relay to set. Record in Table 13-2.

TABLE 13-2. Effect of Sensitivity Control on Light Level Required to Turn On Relay

Step	*Resistance of R_5, Ω*	*Light-Source Voltage, V*
6	0	
7	500	
	1000	
8	1500	
		120

9. With $R_5 = 1000\ \Omega$ and the relay energized (light source is at 120 V), repeat steps 4 and 5 and record your results in Table 13-3.

10. With relay ON, disconnect *C*. Observe effect and record in Table 13-3.

QUESTIONS

1. Using your data in Table 13-1, calculate the "dark" resistance of the photocell. From the data in Table 13-1, determine the photocell resistance in step 9.

2. What effect does the setting of the sensitivity control R_5 have on the light level required to energize the relay? Explain.

3. Determine the peak gate current by dividing the peak voltage drop across R_2 by 100.

TABLE 13-3. Circuit Measurements—Light on Photocell: Relay Energized

Step	*Test Point*	*Waveform*	*Voltage,* V *p-p*
9	*PQ*	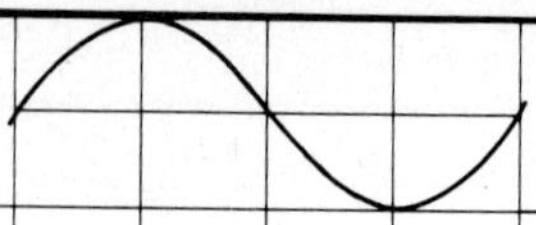	
	SCR anode-to-cathode		
	Across R_3		
	Across R_2		
	Across relay		
	Across photocell		
10	Effect on circuit operation of disconnecting *C*:		

4. Which of the following best describes the contact arrangement of the relay: SPST, SPDT, DPST, or DPDT?

5. Would it matter if the connections to the photocell were reversed? Explain.

Answers to Self-Test

1. very small (zero); very high
2. greatest
3. forward
4. capacitor (filter capacitor)
5. relay contacts

EXPERIMENT

14

TIMING CIRCUITS

OBJECTIVES

1. To study the timing characteristics of *RC* and *RL* circuits
2. To determine the input resistance of an electronic voltmeter (EVM)

BASIC INFORMATION

Time in Electronics

In electronics we are often concerned with "time" and circuits which measure or determine time intervals. Remember that, for periodic waveforms, the "period," or time for one complete cycle of the waveform, is the reciprocal of the frequency:

$$t = \frac{1}{f} \qquad f = \frac{1}{t}$$

The period of a 60-Hz sine wave is one-sixtieth of a second. An electromechanical clock keeps exact time because the line frequency is maintained precisely at 60 Hz by the power company and the clock uses a synchronous motor whose rotational speed is exactly related to the line frequency. A digital clock is basically a "counter" which counts the number of line-derived pulses in a given time interval.

As we have seen in a previous experiment, if the voltage supplied to the gate of an SCR is delayed with respect to the anode-to-cathode voltage, the current to the load in the anode circuit can be varied.

In television systems, synchronized oscillators in the TV studio and in the TV receiver cause the scanning of the same picture elements to occur at exactly the same time at both ends of the system.

In calculators, computers, TV sets, radars, and countless other familiar devices, *RC* and *RL* networks generate timing waveforms and timing intervals that are essential to the operation of these systems.

Capacitors and inductors are energy storage devices. Since the flow of an electrical current is similar to the flow of a fluid, it seems obvious that it takes time to "fill" a capacitor or inductor with electricity in the same way that it takes time to fill a gasoline tank with gas. It takes more time to fill a large capacitor than it does to fill a small capacitor. The charge Q stored in a capacitor is expressed by the equation:

$$Q = C \times V \qquad (14\text{-}1)$$

where C = size of the capacitor, F
V = voltage across the capacitor, V
Q = electrical charge on the capacitor, C

Since a coulomb is a quantity of electrons and since these electrons must flow one after the other into the capacitor, the rate of flow of these electrons with time, dQ/dT (current), determines the rate at which the voltage across a capacitor changes with time, dV/dT.

RC Time Constant

If we know the value of the current flowing into a capacitor of given size C, the voltage change in a given time is $V = I \times t/C$. To generate a constantly increasing voltage (a "ramp") across a capacitor, it is only necessary to charge it from a constant-current source.

Consider the circuit shown in Fig. 14-1. If the capacitor has no initial charge on it and S_1 is now closed, the waveforms at important points in the circuit are

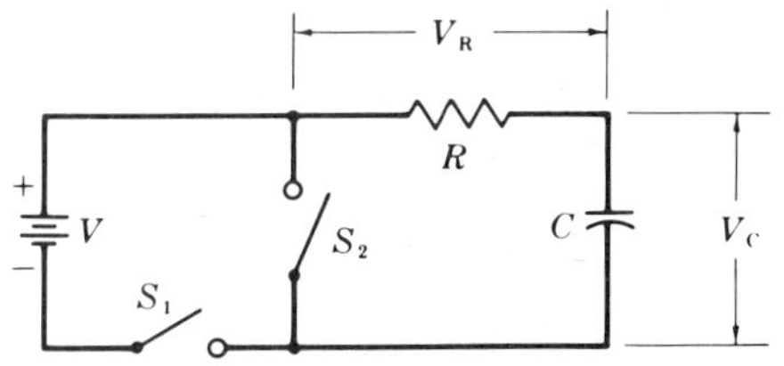

Fig. 14-1. *RC* circuit for charging a capacitor.

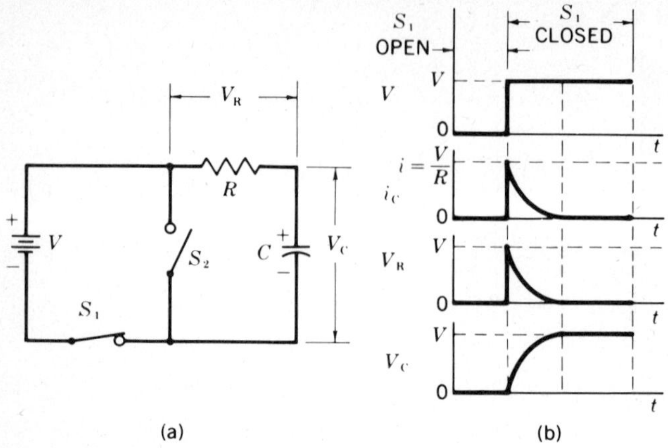

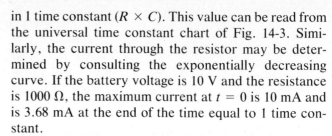

Fig. 14-2. Capacitor C charging an RC circuit.

given in Fig. 14-2. At the instant S_1 is closed, the battery sees only the resistance R because the capacitor has no initial charge ($V_c = 0$) and acts as a short circuit. Therefore, the initial current is $I = V/R$.

This current is now charging C, which develops a voltage across it with the same polarity as the battery. Since the capacitor voltage subtracts from the battery voltage, the current through R is less. When the capacitor is fully charged to the source voltage, the current falls to zero. In Fig. 14-2, the current and the voltage drop across R are exponentially decreasing while the voltage across the capacitor is exponentially increasing.

An important property of this circuit is its time constant:

$$t = R \times C \tag{14-2}$$

where t = time, s
R = resistance, Ω
C = capacitance, F

For the circuit of Fig. 14-1, the voltage across the capacitor will reach 63.2 percent of the battery voltage in 1 time constant ($R \times C$). This value can be read from the universal time constant chart of Fig. 14-3. Similarly, the current through the resistor may be determined by consulting the exponentially decreasing curve. If the battery voltage is 10 V and the resistance is 1000 Ω, the maximum current at $t = 0$ is 10 mA and is 3.68 mA at the end of the time equal to 1 time constant.

Note that the first part of the charge curve is nearly a straight line. For some purposes, this part of the charge curve is sufficiently linear that "constant-current" charging is assumed. For timing applications this is the best portion of the curve to use. Beyond 2 time constants, however, the rate of change is very slow and definitely nonlinear. In fact, the current in Fig. 14-2 never reaches exactly zero, and the voltage across the capacitor never reaches 100 percent of the battery voltage. For most purposes, after 5 time constants the capacitor is fully charged (99 percent) and the current is less than 1 percent of the maximum value.

How does the capacitor discharge? Assume that switch S_1 in Fig. 14-2 has been closed for a very long time and C has charged to the full battery voltage. If S_1 is now opened and S_2 is closed, the capacitor is connected across R and will discharge. Figure 14-4 shows the important waveforms. Note that the current is now flowing in the opposite direction to the charging current and the waveforms are similar to those of Fig. 14-2, but in the opposite direction. The universal time constant chart can also be used to determine the values of voltage or current at any time t after switch S_2 is closed. The time t must be expressed in time constants. For example, if the time constant of the circuit is ½ s and the elapsed time is 1 s, the entry to the chart is at "2" on the x axis. The important consideration in all cases is not "how much time has elapsed" but "how many *time constants* have elapsed."

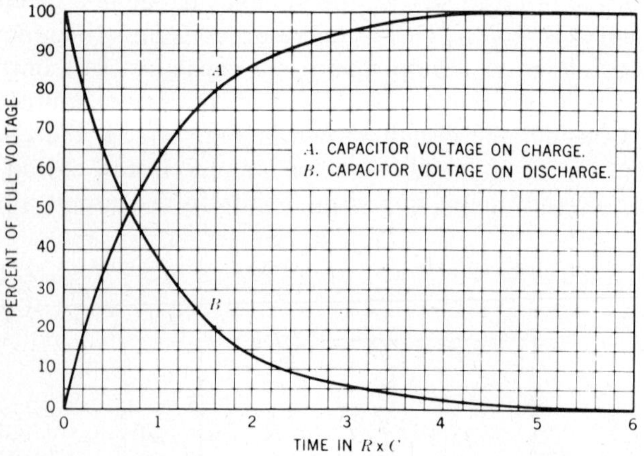

Fig. 14-3. Universal chart for charge and discharge of a capacitor.

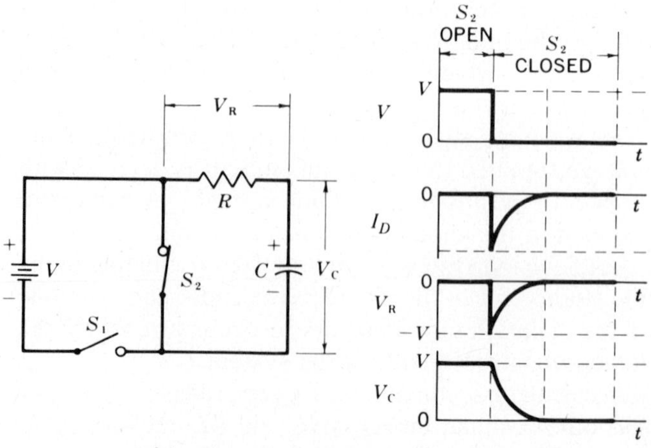

Fig. 14-4. Capacitor discharging in an RC circuit.

RL Time Constant

An important property of an inductor is that it opposes any change of current through it. If a battery is connected to an inductor, the current through the inductor builds up gradually because of the counter electromotive force (emf) generated by the self-inductance of the coil. As current builds up through the inductor, energy is stored in the magnetic field around the coil. Eventually, in some period of time, the inductor is fully charged and the maximum current, which is limited by the resistance of the coil, is reached. If the battery is now disconnected, the magnetic field collapses and again causes a counter emf to be generated, tending to keep the current through the inductor constant. With no discharging resistance across the coil, very large voltages are generated.

Consider the circuit of Fig. 14-5*a*. When S_1 is closed, *current* begins to flow through the circuit and follows an exponential charge curve like the *voltage* across the capacitor in Fig. 14-2. The current starts from zero to the steady-state value V/R. This maximum current flows after approximately 5 time constants. For an *LR* circuit, the time constant is

$$t = \frac{L}{R} \qquad (14\text{-}3)$$

where t = time, s
L = inductance, H
R = resistance, Ω

The current through the inductor in Fig. 14-5 reaches approximately 63 percent of the maximum value in 1 time constant. When S_1 is opened and S_2 is simultaneously closed, the voltages and currents in the circuit are as shown in Fig. 14-5*c* (discharge).

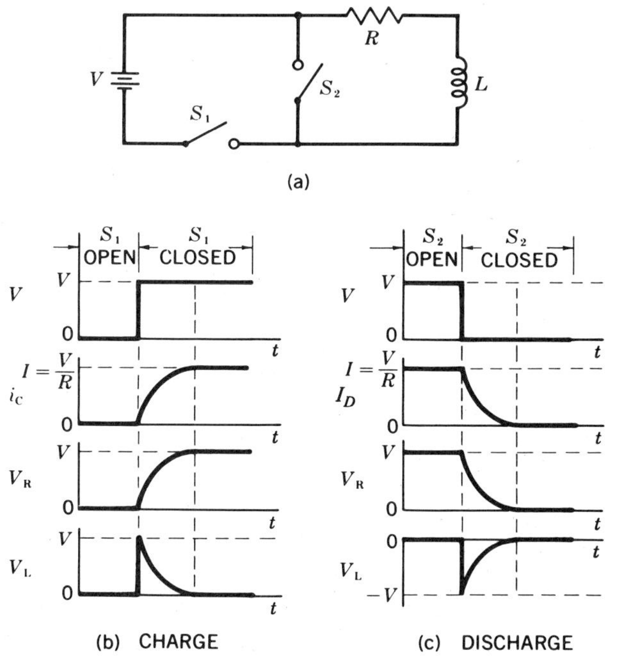

Fig. 14-5. *RL* charge-discharge circuit and waveforms.

When using the charts of Fig. 14-3, note that *current* rise and fall through an inductor corresponds to *voltage* rise and fall across a capacitor.

For precise calculations, the voltage rise across the capacitor in an *RC* circuit is calculated from this equation:

$$V_c = V_b(1 - \epsilon^{-t/RC}) \qquad (14\text{-}4)$$

where V_c = voltage on C at time t
V_b = battery voltage
ϵ = 2.71828 (base of natural logarithms)
RC = time constant, s
t = elapsed time, s

On discharge of a capacitor, use this equation:

$$V_c = V(\epsilon^{-t/RC}) \qquad (14\text{-}5)$$

where V_c = voltage on C at time t
V = voltage on capacitor at t = zero
t = elapsed time, s
RC = time constant, s

For inductive circuits, these equations may be modified by substituting currents for voltages and replacing the time constant RC by L/R.

Experimental Techniques in Observing a Capacitor Charge and Discharge

We can set up an experiment using simple equipment to demonstrate how long it takes a capacitor to charge and discharge. However, our results will be approximate because we will not be using precision laboratory timing or measuring equipment.

Refer to Fig. 14-6*a*. By selecting the values of R and C, we can find the length of time it takes C to charge and discharge, using an ordinary watch with a second hand. In determining the charge time for C, we use a large nonelectrolytic capacitor, say, a 1-μF capacitor, a 1-MΩ resistor, an EVM, a variable dc voltage source, and a watch. When we close switches S_1 and S_2, C will start charging toward the dc supply voltage. The EVM will read V_C, the voltage across C. We will find that in approximately five *RC*s the capacitor will be fully charged.

We should note here that C will not actually charge up to the total applied voltage, but to a value determined by the voltage divider, consisting of the internal resistance of the EVM and the value of R. Thus, if we use an EVM with an input resistance of 11 MΩ and if R is a 1-MΩ resistor, C will charge up to 11/12 of the applied dc voltage. Thus, if we wish to have C charge to 100 V, we should set the output of our dc supply to 109 V (approximately).

To observe the length of time it takes C to discharge, we open S_1. Now C will discharge through the input

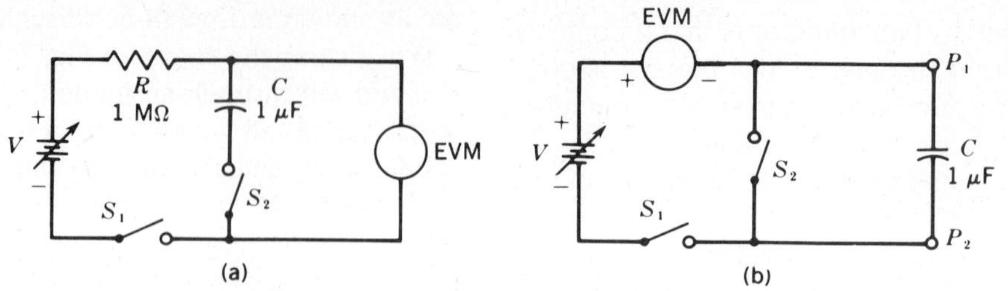

Fig. 14-6. Experimental circuit for the charge and discharge of a capacitor.

resistance of the meter, which we assumed to be 11 MΩ. The discharge time constant is C times the resistance of the meter, and in this case would be 11 s. We can verify the discharge voltage at RC intervals, 11-s intervals in this case. Our results should roughly conform to the capacitor discharge (Fig. 14-3).

If, in the circuit shown in Fig. 14-6*a*, R is replaced by the EVM, as in Fig. 14-6*b*, another means is provided for determining the charge time of C. When switch S_1 is closed, C will charge through the input resistance R_{in} of the EVM. At the moment S_1 is closed, the full battery voltage V appears across R_{in}, and the EVM measures the full voltage V. There is, of course, *zero* voltage across C at that instant. As C charges, the voltage V_C developed across C increases, while the voltage V_R developed across the input resistance of the EVM decreases. At the end of one RC, only 37 percent (100 − 63) of the applied voltage V will be measured on the meter. At the end of two RCs, 14 percent (100 − 86) will be measured by the EVM. At the end of the third, fourth, and fifth RCs, the EVM will show respectively, 5, 2, and 1 percent (approximately) of the applied voltage V. Of course, the charging time constant in this circuit is $R_{in}C$.

Using this method to determine the length of time it takes C to charge, we would connect the circuit of Fig. 14-6*b* and set V to some value, say, 100 V. The start of the timed interval begins when S_1 is closed, and ends when the meter voltage drops to 1 V (approximately). If the meter reading stabilizes at a measurable value of voltage, say, 3 V, it does so because of capacitor leakage. In that event, readings should be discontinued just when the voltage reaches this stable level. NOTE: As the voltage measured on the EVM falls, we can switch the meter to a lower range without disturbing the charging time constant, as long as the input resistance of the EVM remains constant on the various ranges.

By this method, we can measure the charging time of C somewhat more accurately than by the preceding means. Hence it will be used in this experiment.

Visual Method of Observing the Charge and Discharge of a Capacitor

Figure 14-7 illustrates another technique which we can employ to observe directly the charge and discharge of a capacitor. The output of a square-wave generator is applied across an RC circuit, and an oscilloscope is used for viewing the voltage across C. The oscilloscope pattern will give us a direct graph of the exponential rise and fall of voltage across C, as the capacitor charges during the positive alternation and discharges during the negative.

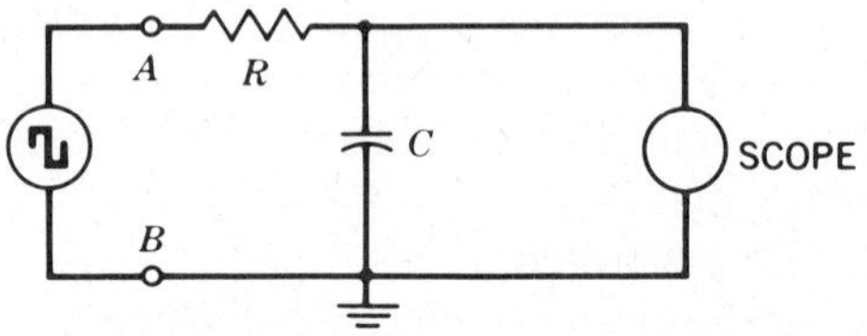

Fig. 14-7. Experimental circuit for showing visually the charge and discharge of a capacitor.

To understand why this is so, refer to Fig. 14-8. This is a graph of a square wave, showing voltage plotted as a function of time. Two cycles are given. Each cycle is made up of a positive and a negative alternation. A square wave is a periodic waveform, each of whose periods is equal. Moreover, the time of each alternation is equal.

Refer again to Fig. 14-8. Consider the first positive alternation. We can describe this part of the waveform as a voltage which rises rapidly from zero voltage to a value V. The voltage remains at this level V during the time interval t_1 seconds.

During the second alternation, that is, during the next t_1 seconds, the voltage drops to zero and remains at zero. The second cycle is simply a repetition of the first.

The square wave may be compared to the voltage supplied by the battery V in Fig. 14-2*b*. While S_1 is

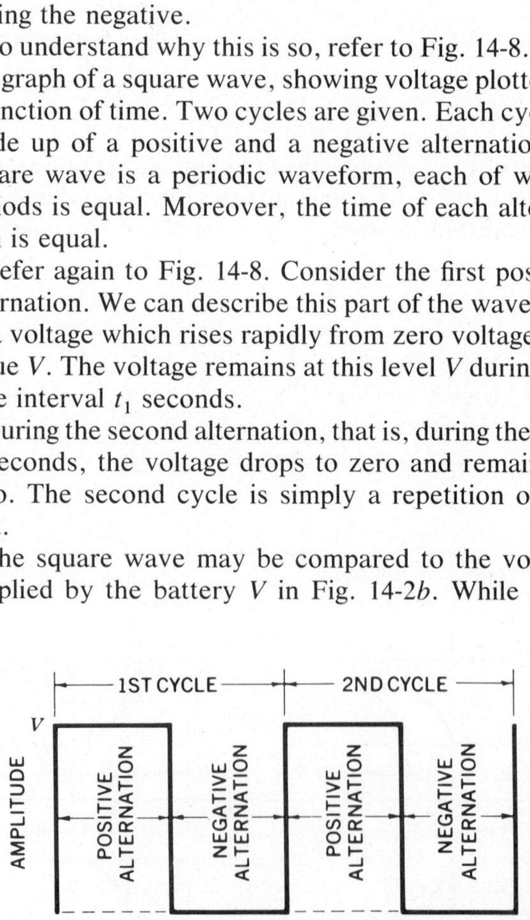

Fig. 14-8. Graph of a square wave.

open, there is zero voltage. If we close S_1 and keep it closed for t_1 seconds, voltage V is applied to the circuit. As soon as we open S_1, the applied voltage drops to zero. If S_1 is kept open for t_1 seconds, the voltage across the circuit remains zero, and so on. We can conclude that a square wave acts as a switch which closes a battery circuit during the positive alternation and opens the battery circuit during the negative.

Let us refer again to the circuit shown in Fig. 14-7. During the positive alternation, C charges through R (and through the resistance of the generator). It attempts to charge to the voltage level V represented by the flat top of the square wave. Whether C will charge all the way to V depends on the time t_1 of one alternation. If t_1 is equal to or greater than five RCs, then C will charge fully. If t_1 is less than five RCs, C will not charge fully, and the level to which it will charge will be determined by the ratio of t_1/RC. Thus, if $t_1/RC = 1$, that is, if $t_1 = 1RC$, the capacitor will charge up 63 percent, following the universal time constant chart.

During the negative alternation, C discharges through R, and the same considerations apply as during the charge interval.

You will recall that the period T of a cycle of a periodic ac waveform is

$$T = \frac{1}{f}$$

where T is the time in seconds and f is the frequency in hertz. We can set the frequency of a variable-frequency square-wave generator with relation to a particular RC circuit so that capacitor C will or will not have sufficient time to charge.

Will an oscilloscope pattern indicate whether C has or has not charged fully? The answer is "yes." But the operator must know how to interpret properly the oscilloscope presentation.

Consider the waveforms in Fig. 14-9. Waveform a is a square wave, as it would appear on an oscilloscope screen. Assume that this waveform is applied to an RC circuit as in Fig. 14-7. The waveforms b, c, and d are oscilloscope presentations, which appear across C for different values of R and C, that is, for different time constants. What does each of these three waveforms indicate?

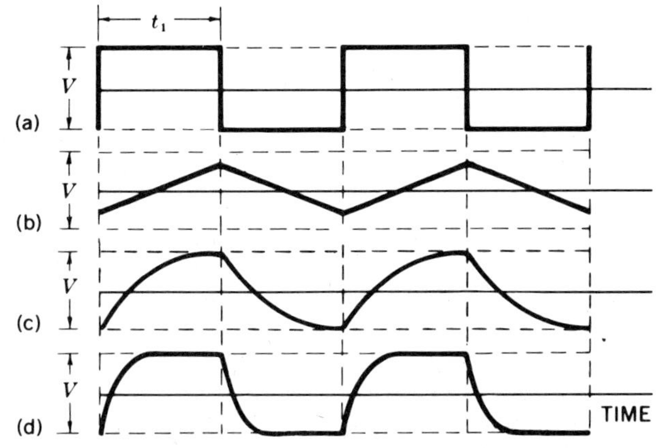

Fig. 14-9. Oscilloscope waveform interpretation.

The p-p amplitude of waveform b is less than the p-p amplitude V of the applied square wave. This means that capacitor C has not charged fully to V. Our previous discussion must lead us to conclude that the charging (and discharging) time t_1 is less than five RCs. The sawtooth waveform b, therefore, shows that the capacitor C charges only part of the way, then discharges, charges part of the way again, and so on.

The p-p amplitude of waveforms c and d is V. This means that capacitor C charges to the full value of the applied voltage V, in both cases. However, in the case of waveform c, the capacitor just about reaches the value V at the end of the positive alternation, that is, in t_1 seconds. We can conclude, therefore, that in the case of waveform c, $t_1 = 5RC$ (approximately).

In the case of waveform d, the capacitor C reaches the value V long before the end of the positive alternation. We can conclude that in this case t_1 is greater than five RCs.

In our description of the waveforms which appear across C, we have assumed a fixed square-wave frequency and variable RC circuit. We can also demonstrate the effects of the charge and the discharge of a capacitor by using a fixed RC circuit and varying the frequency of the square-wave generator. We should note, in this connection, that an increase in frequency of the square wave reduces the period of each cycle and hence reduces the time during which capacitor C can charge and discharge. On the other hand, reducing the frequency of the square wave increases the time during which C may alternately charge and discharge.

SUMMARY

1. The charge Q on a capacitor $= C \times V$.
2. The time constant of an RC circuit $= R \times C$.
3. The capacitor in a simple RC circuit charges to 63 percent of the applied voltage in 1 time constant.
4. It takes about 5 time constants for a capacitor to fully charge or fully discharge.
5. The universal time constant charts are useful for quick calculations related to the charge and discharge of RC and RL circuits
6. In most cases the important consideration is the amount of elapsed time measured in time constants
7. In an LR circuit, the time constant $= L/R$.
8. The current through an inductor builds up exponentially like the voltage across a capacitor.
9. Capacitors and inductors are energy storage devices.

SELF-TEST

Check your understanding by answering these questions.

1. The time constant of a 0.25-μF capacitor and a 2.2-MΩ resistor is __________ s.
2. If the capacitor in question 1 is charging through the resistor toward 50 V dc, it will reach how much voltage in 1 time constant.
3. A 0.05-μF capacitor is charged to 100 V dc and begins discharging through a 220-kΩ resistor. It will discharge to 37 V in __________ ms.
4. The universal time constant curves (Fig. 14-3) are also called __________ curves.
5. If the values in Fig. 14-5 are $R = 100$ kΩ and $L = 8$ H, the time constant is __________ μs.
6. In Fig. 14-5, if $V = 10$ and $R = 100$ kΩ, the maximum current through the inductor is __________ μA. It occurs at __________ .
7. For the values of V, L, and R above, the current through the inductor will equal 55 μA __________ .

MATERIALS REQUIRED

- Power supply: Variable dc source
- Equipment: Square-wave generator; oscilloscope; EVM
- Resistors: ½-W 1-MΩ, 12-MΩ (5 percent)
- Capacitors: 0.01-μF, 1-μF
- Miscellaneous: 1-MΩ 2-W potentiometer; two SPST switches

PROCEDURE

Input Resistance R_{in} of EVM

1. Connect the circuit shown in Fig. 14-10. R is a 12-MΩ 5 percent resistor. Set $V = 25$ V as measured by the EVM.
2. Close S_1. The applied voltage V will be distributed proportionately across the input resistance of the meter R_{in} and across the 12-MΩ resistor R, according to the formula

$$\frac{V_{meter}}{V} = \frac{R_{in}}{R_{in} + R} \qquad (14\text{-}6)$$

Solving for R_{in} in Eq. 14-6, we find that

$$R_{in} = \frac{V_{meter}}{V - V_{meter}} \times R \qquad (14\text{-}7)$$

Substituting 25 V for V and 12 MΩ for R, we find that

$$R_{in} = \frac{V_{meter}}{25 - V_{meter}} \times 12\ \text{M}\Omega \qquad (14\text{-}8)$$

3. Measure and record in Table 14-1 the voltage V_{meter}.

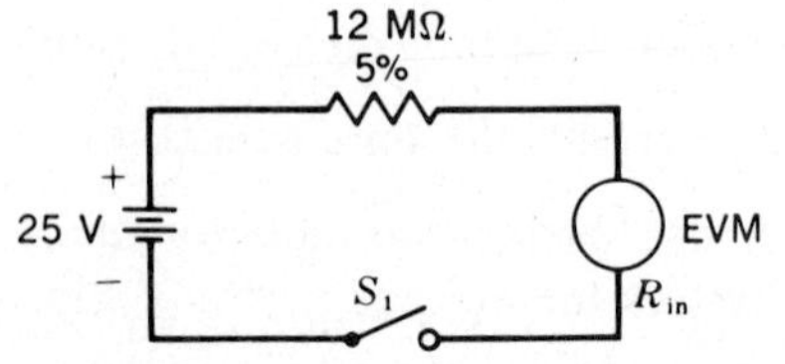

Fig. 14-10. Circuit for determining input resistance R_{in} of an EVM.

4. Substitute this measured value V_{meter} in Eq. (14-8) and compute R_{in}, the input resistance of the EVM. Record the computed value R_{in} in Table 14-1. Record also the rated value for R_{in}, including the resistance of the EVM probe. The computed and rated values should be approximately the same. Open S_1.
5. Connect the circuit in Fig. 14-6*a*. S_1 and S_2 are open. Close S_1. Adjust the output of the power supply V until the meter reads 30 V (V_{meter}). Maintain V_{meter} at 30 V, because this is the value to which the capacitor will charge when S_2 is closed.
6. Close S_2. Observe that the voltage indicated on the EVM falls to zero the moment S_2 is closed and then rises as C charges. After approximately 10 s, C should be fully charged and the EVM should measure 30 V. NOTE: Some meters are heavily damped and should be avoided in this experiment where instantaneous time is a factor.

Discharge Time Constant

7. Compute the discharge time constant $R_{in}C$ and record in Table 14-1. Use the rated value of R_{in}, or if this is not available, use the measured value.
8. Open S_1. Observe and record V_C, the voltage across C, at every discharge time-constant interval shown in Table 14-1. For example, if $R_{in}C = 11$ s, readings should be made at 11-s intervals and recorded in column labeled "V_C, 1st Trial," until C is completely discharged. NOTE: When the voltage read on the meter stabilizes at some level, you are reading a voltage produced

TABLE 14-1. Discharge Time Constant of a Capacitor

Step	Applied Voltage V, V	V_{meter}, V	R_{in}, Ω Computed	R_{in}, Ω Rated
1–4				
7	Discharge time constant, $R_{in}C$ = _____ s			

Step	Discharge time, RC	V_C, V 1st Trial	2d Trial	3d Trial	Average	Computed
8–13	1					
	2					
	3					
	4					
	5					
	10					

by the dielectric strain within the capacitor. This may occur at about the tenth time constant.

9. Close S_1 and permit C to charge fully (30 V).

10. Repeat step 8. Record your readings in column labeled "V_C, 2d Trial."

11. Repeat steps 9 and 10, recording your readings in column labeled "V_C, 3d Trial."

12. Average the three values of V_C for one RC and record in column "V_C, Average." Average also and record the value of V_C for two RCs, three and so on. Draw a graph (in red pencil) of V_C (average) versus RC. Let RC be the horizontal axis.

RC Time Constants

13. Compute and record in Table 14-1 the voltage V_C which should appear across C after one RC, two RCs, and so on. For comparison with step 11, draw a graph (in green) of the computed values of V_C versus RC.

Charge Time Constant

14. Connect the circuit shown in Fig. 14-6*b*. Set the EVM on the 50-V or next-higher range. Close S_1 and S_2 and set V at 30 V, as measured on the EVM.

15. Open S_2. Observe and record V_R, the voltage across the EVM, at every charge time constant

TABLE 14-2. Charge Time Constant of a Capacitor

Step	Charge Time Constant $R_{in}C$ = _____ s								
	Charge time RC	V_R, V 1st Trial	V_R, V 2d Trial	V_R, V 3d Trial	V_C, V 1st Trial	V_C, V 2d Trial	V_C, V 3d Trial	Average V_C	Computed V_C
14–19	1								
	2								
	3								
	4								
	5								
	10								

interval shown in Table 14-2. NOTE: The charge time constant is the same as the discharge time constant, $R_{in}C$, recorded in Table 14-1. Record readings in the column labeled "V_R, 1st Trial." Compute the voltage V_C across the capacitor at every time constant interval and record in the column "V_C, 1st Trial." Use the equation

$$V_C = V - V_R = 30 - V_R$$

16. Close S_2, thus discharging capacitor C. V should be 30 V. Open S_2 and repeat step 15. Record the measured voltage in the column "V_R, 2d Trial," and the computed values in the column labeled "V_C, 2d Trial."

17. Repeat step 16. Record the measured voltage in the column "V_R, 3d Trial," and the computed values in the column "V_C, 3d Trial.

18. Average the three values of V_C at every time constant interval and record in column "V_C, Average."

19. Compute and record in Table 14-2 the voltage V_C which should appear across C at the end of one RC, two RCs, and so on. For comparison with the graph in step 18, draw a graph of the computed values of V_C versus RC (in ink).

Visual Method for Observing the Charge and Discharge of a Capacitor

20. Power OFF. Discharge C.

21. Connect the circuit of Fig. 14-11. Set the square-wave generator at 60 Hz. Calibrate the vertical amplifier of your oscilloscope at 1 V/cm, and connect the vertical scope leads across the output of the square-wave generator. Adjust the output attenuator of your square-wave generator at 3 V p-p output.

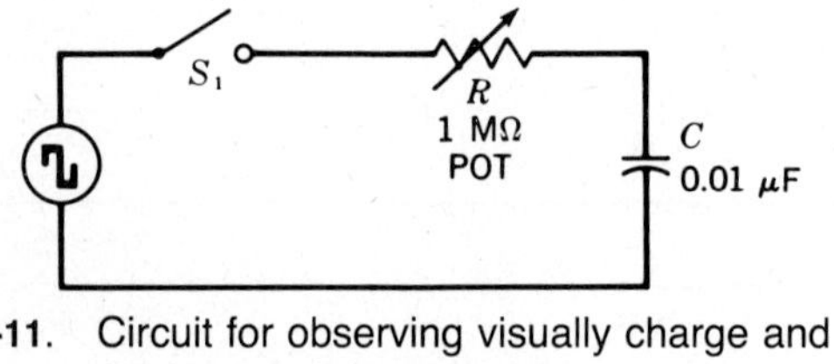

Fig. 14-11. Circuit for observing visually charge and discharge of a capacitor.

22. Adjust the oscilloscope sweep and sync controls so that there are two square waves presented on the screen as in Table 14-3.

23. Adjust the 1-MΩ potentiometer so that R measures 170,000 Ω.

24. Close S_1, applying the output of the square-wave generator across the RC circuit.

25. Connect the oscilloscope across C. Observe, measure, and record in Table 14-3 the waveform and its p-p amplitude.

26. Connect the oscilloscope across R.

27. Observe, measure, and record in Table 14-3 the waveform and its p-p amplitude.

28. Set frequency of the square-wave generator at 300 Hz, maintaining a 3-V p-p output. Readjust oscilloscope sweep and sync controls for two square waves, as in step 22.

29. Connect the oscilloscope across C.

30. Observe, measure, and record in Table 14-3 the waveform and its p-p amplitude.

31. Set frequency of square-wave generator at 30 Hz and repeat step 21.

32. Repeat steps 29 and 30.

33. Set square-wave generator frequency at 100 Hz and repeat step 28.

34. Connect oscilloscope across C.

35. Vary R over its entire range, from minimum to maximum. Observe how waveform across C changes. Record in Table 14-3 the waveforms across C for minimum and maximum values of R.

QUESTIONS

1. What is meant by the term "time constant" in RL and RC circuits?
2. Why is it important to use *nonelectrolytic* type capacitors in long time constant, critical circuits?
3. Calculate the period of a rectangular waveform whose frequency is 1000 Hz.
4. What value of C should be used with a 100-k resistor for a time constant of 1 ms?
5. What is the maximum current that will flow through a 10-H inductor having a 100-Ω resistance if 100 V dc is applied to the terminals of the inductor?
6. If the frequency of the square-wave generator in Fig. 14-7 were increased, would the p-p output voltage increase or decrease? Why?

TABLE 14-3. Waveforms in *RC* Circuit

Frequency, Hz	*Oscilloscope across*	*Waveform*	*Peak-to-Peak Amplitude,* V
60	Generator	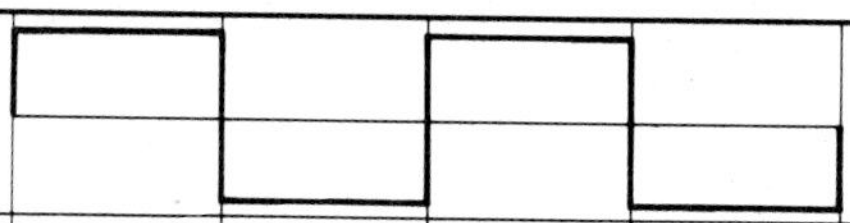	3
60	*C*		
60	*R*		
300	*C*		
30	*C*		
100	*C* *R* minimum		
100	*C* *R* maximum		

Answers to Self-Test

1. 0.55
2. 31.5
3. 11
4. exponential
5. 80
6. 100; $t = 400\ \mu s$, approximately
7. 64 μs after S_1 is closed

EXPERIMENT

15

TRANSISTOR TIME-DELAY RELAY

OBJECTIVE

To study and demonstrate the operation of a transistor time-delay relay

BASIC INFORMATION

In many industrial control situations there is not sufficient voltage, current, or power available from the sensors or transducers to directly operate a relay or other control device. The current-amplifying capability of transistors can be utilized to increase the small control signal to a level which will operate a relay. The relay, in turn, can control large motors, heaters, or other industrial machinery. In Figs. 15-1 and 15-2, the biasing resistors R_1 and R_2 are chosen to provide sufficient forward bias to the base-emitter circuit of the transistor so that the resulting collector current is great enough to energize the relay when switch S is closed. Note that the polarity of the battery voltage V_{aa} is reversed in the PNP circuit compared with the NPN circuit. The base current in these circuits will be much smaller than the collector current.

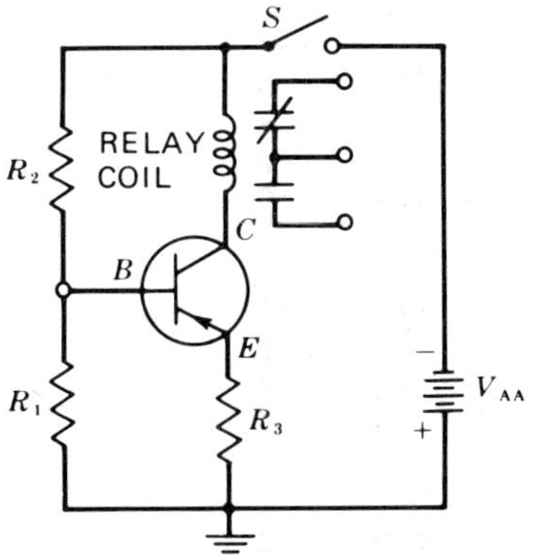

Fig. 15-1. PNP-transistor-operated relay.

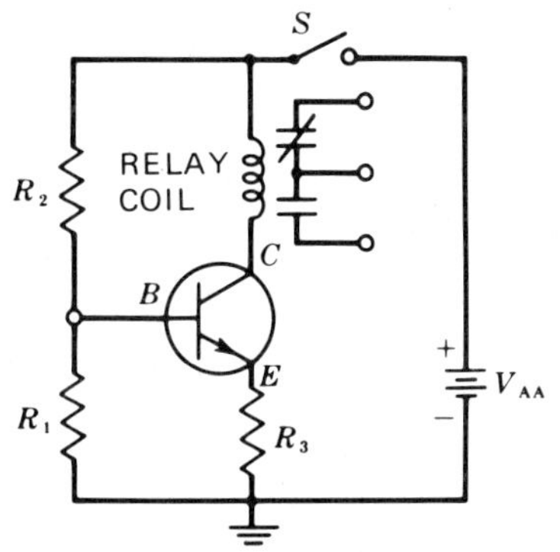

Fig. 15-2. NPN-transistor-operated relay.

Time-Delay Relay

Timers of all kinds are used in the design of industrial machinery. Older equipment still uses mechanical timers based on electrical motors operating cams and microswitches, and other methods which may utilize vacuum tubes. But modern equipment uses solid-state timers based on *RC* charge curves or on digital counting techniques.

The circuit shown in Fig. 15-3 uses the charging time of capacitor C to determine the time the relay is pulled in. To understand the operation of this circuit, assume that capacitor C is discharged and S_1 and S_2 are both open. When S_2 is closed, the capacitor starts to charge toward the supply voltage V_{EE} through R_1, R_2, and the base-emitter circuit. These values are so chosen that the initial charging current supplies sufficient base current, and resulting collector current, that the relay is energized. But as C continues to charge, the amount of charging current decreases, thereby decreasing the base and collector currents until the relay eventually drops out.

The length of time the relay is energized depends on the time constant $C \times (R_1 + R_2 + R_{be})$, the battery voltage, the characteristics of the relay, and the cur-

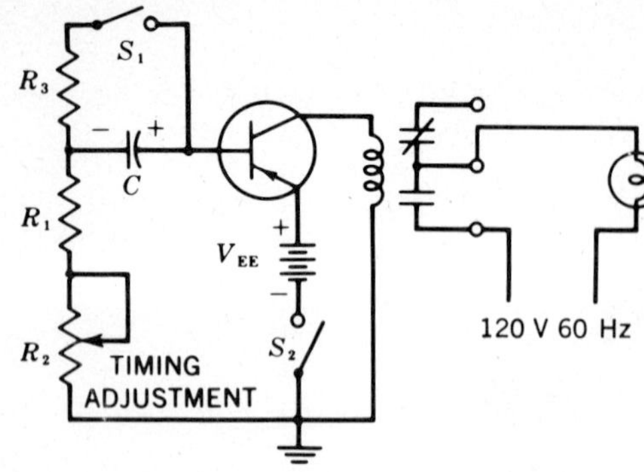

Fig. 15-3. Time-delay relay utilizing capacitor charge time.

rent gain of the transistor. By varying R_2 we vary the time during which the relay is energized.

When C is charged, the relay is deenergized. If S_1 is closed briefly, discharging C, the relay is immediately reenergized and a new cycle is begun [S_1 may be a spring-loaded, normally open (NO), push-button switch].

Figure 15-4 illustrates a circuit in which the discharge time of a capacitor is used to determine the timing interval. To analyze the operation of this circuit, assume that switch S_1 is open and that S_2 is in position B. The capacitor is not charged. If S_1 is now closed, the relay remains deenergized since the base-emitter junction is not forward-biased ($V_{BE} = 0$). But if S_2 is briefly thrown to position A, thereby charging it to 12 V, and then returned to position B, the base-emitter circuit becomes forward-biased and the relay is energized. The capacitor starts to discharge through the parallel paths ($R_1 + R_2$) and (R_4 + the base-emitter circuit), thereby reducing the base and collector currents. At some point the relay drops out because the collector current is too small. The length of time the relay is energized is determined by the discharge time constant $C \times R_{eq}$, where R_{eq} is the effective parallel resistance between point B and ground. This time constant, and the length of time the relay is energized, may be controlled by adjustment of R_1. Here, too, S_2 may be a spring-loaded, SPDT push-button switch. A small 10-W lamp controlled by the NO relay contacts will light during the timed interval.

If a set of normally closed (NC) contacts on the relay are connected between points A and B, as in Fig. 15-5, the controlled light bulb will blink on and off and the circuit becomes a free-running, low-frequency oscillator.

As soon as S_1 is closed, the capacitor charges very rapidly to V_{AA}. This causes the relay to pull in, opening the contacts between A and B. The capacitor then begins to discharge. During the discharge period, the relay is energized, but when the capacitor voltage has sufficiently decreased, the relay drops out. Now the contacts between A and B close, repeating the cycle. This circuit has no stable condition but oscillates between the two possible states due to "feedback" from the output (contact closure) to the input (the base circuit).

SUMMARY

1. The ON and/or OFF time of transistor/relay timers may be determined by RC circuits.
2. The timed interval may be determined either by an RC charge or RC discharge curve.
3. The circuits of time-delay relays include transistors (and/or ICs), resistors, capacitors, relays, a power source, and other components required by a given application.
4. As a part of a timer, electromechanical relays can provide complex contact arrangements, for example, DPDT.
5. The adjustable element of variable time-delay relays is usually a potentiometer or rheostat, which is part of the RC timing network.

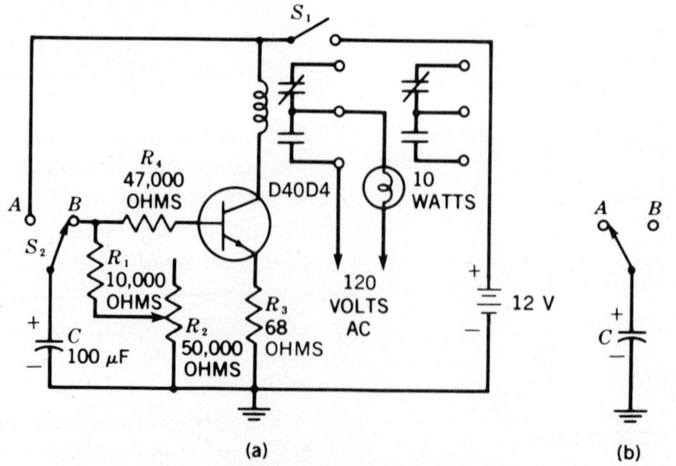

Fig. 15-4. Time-delay relay utilizing discharge time of capacitor C for timing interval.

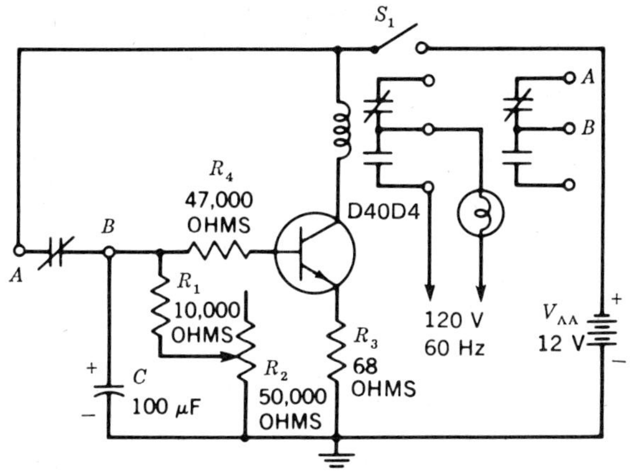

Fig. 15-5. Transistor "blinker."

SELF-TEST

Check your understanding by answering these questions.

1. The timers of this chapter are based on the fact that the voltages and currents in an *RC* circuit change with __________ .
2. The circuit of Fig. 15-3 could be used to control a photographic exposure light. If S_1 and S_2 are both open, __________ should be closed to turn on the light for a timed interval.
3. In any of these timers, a larger capacitor will cause a __________ (longer/shorter) timed period.
4. There should be a small resistor between point *A* and the switch *S* in Figs. 15-4 and 15-5 to __________ .
5. The blinker circuit of Fig. 15-5 is an electromechanical __________ .

MATERIALS REQUIRED

- Power supply: Variable dc source
- Equipment: EVM
- Resistors: ½-W 33-, 68-, 4700-, 10,000-, 47,000-Ω
- Capacitors: 100-μF 15-V
- Relay: Direct current, SPDT, 400-Ω field (approximate), 1-A contacts, 7-mA pickup current (approximate)
- Transistor: D40D4
- Miscellaneous: Associated transistor socket or mount; 10-W wired test lamp; two SPST switches; SPDT switch; 50,000-Ω 2-W potentiometer

PROCEDURE

The connections of the D40D4 transistor which you will use in this experiment are shown in Fig. 15-6.

Relay on Capacitor Charge

1. Connect the circuit of Fig. 15-7. Switches S_1 and S_2 are open. V_{AA} is a 12-V battery or a variable source of dc voltage. With an EVM, monitor the dc output voltage of this supply as you carefully adjust it for 12 V.

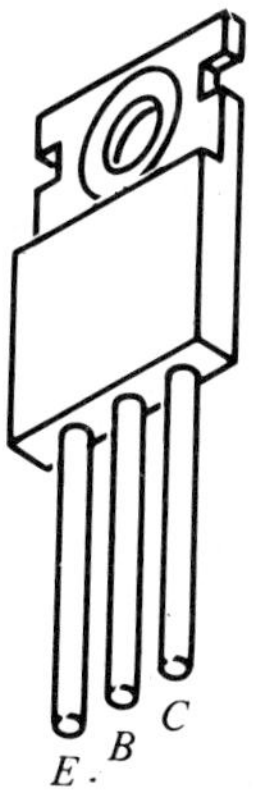

Fig. 15-6. Identification of transistor leads.

2. Close switch S_1, keeping S_2 open. Note that the relay is at once energized. Observe that the light goes on and stays on for a timed interval. Measure the delay time, using a stop watch or digital timer. Record in Table 15-1 the time delay for this random setting of R_2.
3. With an EVM, measure V_{CE}, the voltage from collector to emitter, and record in Table 15-1. Also measure V_{CB}, the voltage from collector to base, and V_{EB}, the voltage from emitter to base. Record in Table 15-1. These voltages should be measured after the relay has reset, that is, after the light has gone out.
4. Set R_2 for minimum resistance. Close S_2 for an instant to discharge *C* and then open it. With an EVM, monitor V_{EB}. Observe and record in Table 15-1 the range of bias V_{EB} during the interval that the relay is energized, as capacitor *C* is charging.
5. Measure and record in Table 15-1 the length of time that the relay remains energized after S_2 is opened.
6. Set R_2 for maximum resistance. Repeat step 4.
7. Repeat step 5, with R_2 still at maximum resistance.

TABLE 15-1. Time-Delay Relay Measurements

Step	DC Voltage V_{CE}	V_{CB}	V_{EB}	Condition of R_2 and S_2	Delay, s
2, 3				R_2 set at random	
4, 5	X	X		R_2 set at minimum resistance	
6, 7	X	X		R_2 set at maximum resistance	
11				S_2 in position *B*	X
13, 14		X	X	S_2 in position *B*, R_2 is minimum	
16, 17		X	X	S_2 in position *B*, R_2 is maximum	

8. Reduce dc supply voltage to 3 V. Does the relay function? Why? Power OFF.

Relay on Capacitor Discharge

9. Connect the circuit shown in Fig. 15-4. Switch S_1 is open, and switch S_2 is in position *B*. Adjust the supply voltage for 12 V as before.
10. Close S_1. Is the relay energized? Why?
11. Measure and record in Table 15-1 V_{CE}, V_{CB}, and V_{EB}.
12. Adjust R_2 for minimum resistance. Throw S_2 to position *A*, thus permitting capacitor *C* to charge.
13. Throw S_2 to position *B*. With an EVM, monitor V_{CE} while the relay is energized and record in Table 15-1.
14. Measure and record in Table 15-1 the length of time the relay remains energized.
15. Throw S_2 to position *A*. Set R_2 for maximum resistance.
16. Repeat step 13.
17. Repeat step 14. Power OFF.

QUESTIONS

1. What is the purpose of R_4 in Fig. 15-7?
2. Make a step-by-step list of instructions for using the time-delay circuit, Fig. 15-7.
3. What would happen in Fig. 15-7 if S_1 and S_2 are both closed?
4. What determines the minimum time delay in Fig. 15-7?
5. How can the maximum time delay be increased in Fig. 15-7? Is there a limit to the maximum delay?

Answers to Self-Test

1. time
2. S_2
3. longer
4. protect the switch and relay contacts by limiting the current charging *C* to a reasonable value (5 or 10 Ω)
5. oscillator

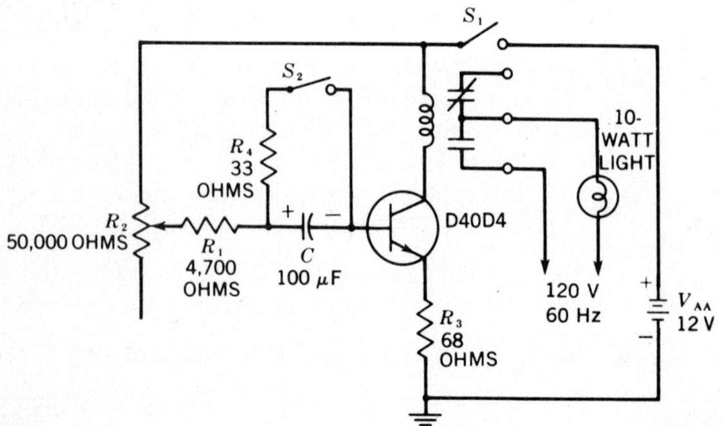

Fig. 15-7. Experimental time-delay circuit.

EXPERIMENT

16

THE UNIJUNCTION TRANSISTOR AS A CONTROL DEVICE

OBJECTIVES

1. To study the characteristics of a UJT
2. To observe and measure the operation of a UJT oscillator circuit.
3. To learn how an SCR may be triggered by a UJT

BASIC INFORMATION

In designing a trigger source for an SCR, several factors must be considered: triggering from a low-power source may cause erratic operation; too much power, while assuring consistent SCR turn-on, may overheat the gate and cause it to burn out. We have previously studied dc and ac triggering methods. While these methods are satisfactory in some cases, a more ideal method involves triggering the SCR with a sharp, high-powered pulse of short duration, whose peak and average power do not exceed the power capabilities of the SCR gate for which they are intended. Because they can generate such pulses, unijunction transistors (UJTs) are frequently used to trigger SCRs.

UJT Construction and Characteristics

A cross-sectional diagram of the physical construction of a UJT is shown in Fig. 16-1. A ceramic disk serves as the base for this device, assuring rigidity and ruggedness. A conductive gold film is deposited on this base, on both sides of a very narrow slit in the center. An N-type silicon bar is symmetrically attached to this film, forming two resistive (nonrectifying) contacts called *bases*, designated B_1 and B_2. A PN "emitter" junction is formed on the bar at a point closer to B_2 than B_1. The finished and encapsulated unit may resemble a conventional transistor, but its electrical characteristics are quite different.

Figure 16-2 shows the schematic symbol for the UJT and a simplified equivalent circuit is shown in Fig. 16-3. Important characteristics of this diagram include the diode formed by the PN junction, and the interbase resistance R_{BB} of the N-type bar. R_{BB} consists of two parts, R_{B1} and R_{B2}. Their sum for a 2N2160 UJT lies in the range 4000 to 12,000 Ω. R_{B1} is shown as a variable resistor because this value decreases sharply when emitter-B_1 current flows.

When there is no emitter current I_E, the voltage from point A to B_1 may be written

$$V_{AB1} = V_{BB} \times \frac{R_{B1}}{(R_{B1} + R_{B2})u} = V_{BB} \times \frac{R_{B1}}{R_{BB}} \quad (16\text{-}1)$$

The ratio R_{B1}/R_{BB} is called the *intrinisic standoff ratio* of the UJT and is designated by η, the Greek letter eta. Thus, Eq. 16-1 may be written

$$V_{AB1} = \eta \times V_{BB} \quad (16\text{-}2)$$

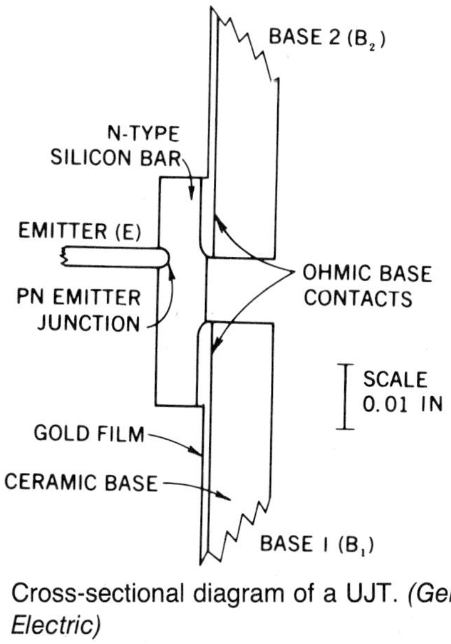

Fig. 16-1. Cross-sectional diagram of a UJT. *(General Electric)*

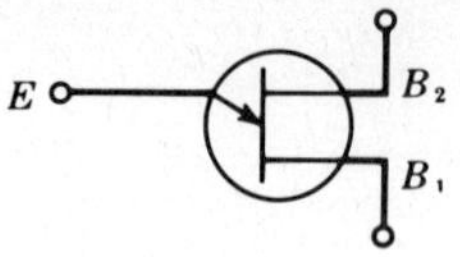

Fig. 16-2. UJT circuit symbol.

Values of η are different for different UJTs and lie in the range 0.51 to 0.81. If the emitter voltage V_E is less than $\eta \times V_{BB}$, the emitter–base 1 junction is reverse-biased and only leakage current flows. But when the emitter voltage is greater than $\eta \times V_{BB}$, emitter current flows and, as the emitter current increases, the emitter–base 1 voltage decreases, as seen in Fig. 16-4. The region to the left of the peak point is called the *cutoff region* and there is little emitter current. The region between the peak point and the valley point has a unique characteristic called *negative resistance* because decreasing emitter voltage corresponds with increasing emitter current. (The slope of the curve is negative.) It is this region which makes possible the UJT relaxation oscillator circuit. To the right of the valley point is the saturation region where an increasing emitter voltage causes increasing emitter current.

The peak point voltage V_P is given by the equation

$$V_P = \eta \times V_{BB} + V(diode) \qquad (16\text{-}3)$$

Because the diode voltage drop decreases with increasing temperature, it is desirable to provide a degree of temperature compensation for this effect, which would otherwise change V_P. The interbase resistance R_{BB} has a positive temperature coefficient (R_{BB} increases as temperature increases). By placing a small resistor in series with B_2, the change in R_{BB} will cause V_{BB} to increase with temperature, thereby increasing the peak point voltage. Since the peak point voltage decreases due to the effect of temperature on D, the result of adding the resistor R_2 is to keep the peak point constant as temperature changes. For the GE 2N2160 UJT, the approximate value of R_2 in Fig. 16-5 is

$$R_2 = \frac{0.7 \times R_{BB}}{\eta \times V_1} + \frac{(1-\eta)R_1}{\eta} \qquad (16\text{-}4)$$

The output pulse generated by the UJT is usually taken across R_1.

UJT Relaxation Oscillator

Connected as a relaxation (RC) oscillator, Fig. 16-6*a*, the UJT generates the waveform V_{B1} across R_1 in Fig. 16-6*b*, which is suitable for triggering the gate of an SCR. When S_1 is first closed, applying power to the circuit, C_T starts charging exponentially through R_T toward the applied voltage V_1. When this voltage, applied to the emitter, reaches V_P, the UJT is turned ON and the resistance of R_{B1} is rapidly decreased. A sharp pulse of current (limited by R_1 and the R_{B1} resistance)

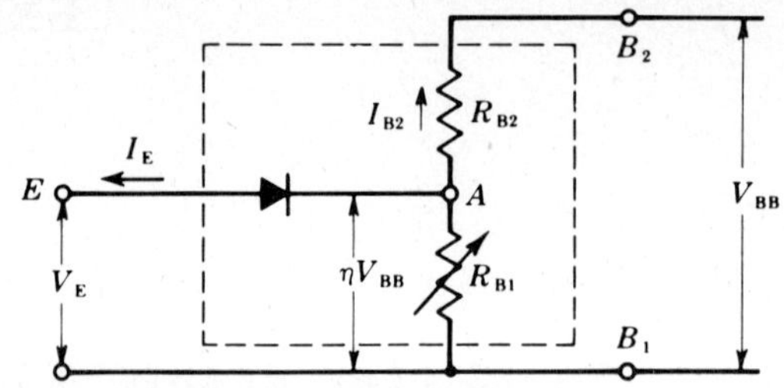

Fig. 16-3. UJT simplified equivalent circuit.

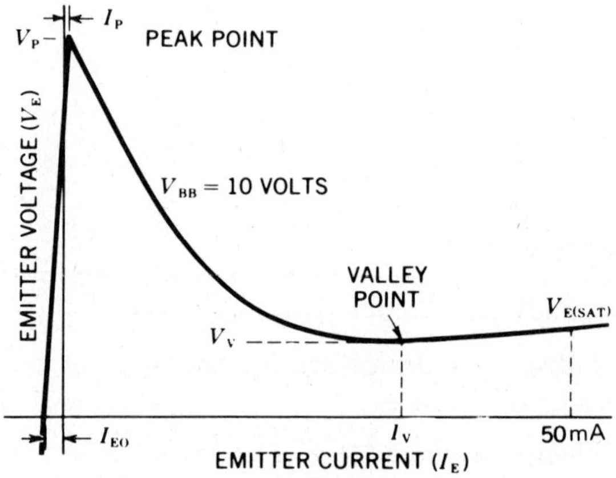

Fig. 16-4. Static emitter characteristic curve showing important parameters. *(General Electric)*

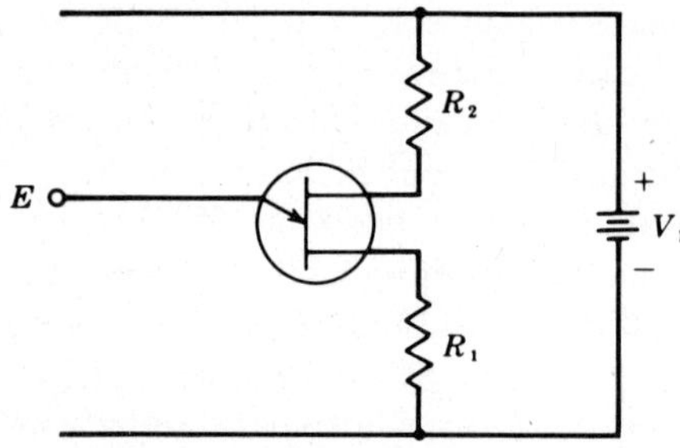

Fig. 16-5. UJT bias-stabilization circuit.

flows in the emitter–base 1 circuit, rapidly discharging C_T and producing a narrow, high-amplitude pulse across R_1. When the voltage across C_T has dropped to about 2 V, the UJT turns OFF and the cycle is repeated.

For some purposes the resulting sawtooth waveform across C_T may be useful (Fig. 16-6*b*), but not for SCR triggering.

The frequency f of the relaxation oscillator circuit depends on the time constant $R_T \times C_T$ and on the intrinsic standoff ratio of the UJT. For small values of R_1 (less than 100 Ω), the period between pulses T is given approximately by

$$T = R_T \times C_T \times \frac{ln(1)}{1-\eta} \qquad (16\text{-}5)$$

where T = period, s
$R_T \times C_T$ = time constant, s
ln = natural logarithm

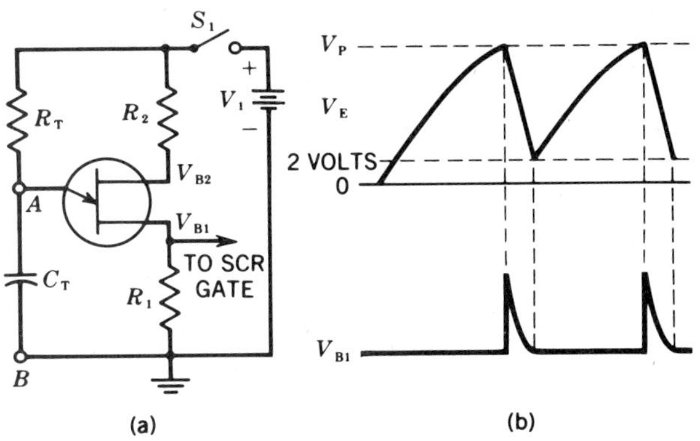

Fig. 16-6. UJT relaxation oscillator.

Since $f = 1/T$, the frequency of oscillations can be calculated from the period. A very coarse approximation for f is $f = 1/R_T \times C_T$. The value of R_T should lie in the range of 3 kΩ to 3 MΩ. For long time periods (low frequencies), the capacitor C_T should be a low-leakage, *nonelectrolytic* type capacitor. V_1 is usually 10 to 35 V.

SCR Triggered by UJT Relaxation Oscillator

In the experimental circuit shown in Fig. 16-7, the pulses developed across R_1, in base 1 of the UJT, are used to trigger the SCR, as in the circuit shown in Fig. 16-8. The UJT is connected as a relaxation oscillator. The frequency of the sawtooth voltage developed across C is determined by the time constant R_4C. R_4 is made variable so that the timing of the trigger pulses, developed across R_1, may be adjusted to control the firing of the SCR at different points on the pulsating input anode wave. The fact that *full-wave* rectified alternating current (pulsating direct current) is employed as a power source for the SCR, rather than a straight sinusoidal input, doubles the load current capabilities of the circuit, because this arrangement eliminates the half-cycle during the negative alternation when the SCR would normally be cut off. The pulsating dc source, however, permits the SCR to turn OFF when the anode voltage is reduced below the holding level of the controlled rectifier.

Of interest is the arrangement of resistor R_3 and zener diode Z_1. The zener clips the tops of the positive alternations and provides a relatively stable voltage level to which capacitor C can charge through resistor R_4. Straight direct current cannot be used as the charging level for capacitor C in this relaxation oscillator, because the frequency of the pulse output of the UJT would then not be synchronized to the frequency of the pulsating direct current applied to the SCR. The flat-topped voltage alternations to which C charges effect synchronization, because the sawtooth waveforms developed across C occur in groups at a *recurrent* frequency which is the same as the frequency of the full-wave rectified ac source. Thus, although the basic frequency of the oscillator is still determined by the time constant R_4C, the pulse-recurrent frequency is fixed by the power source. In this manner, the full-wave rectified signal, obtained from an appropriate rectifying source, supplies both power to the SCR and synchronization to the trigger circuit.

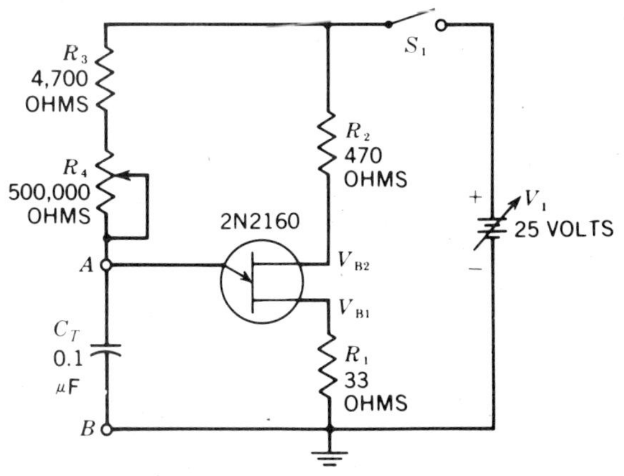

Fig. 16-7. Experimental relaxation oscillator.

SUMMARY

1. A UJT oscillator circuit can provide the short-duration, high-powered pulse which is desirable for reliable triggering of an SCR.
2. A UJT has one PN junction and two "bases."
3. The emitter–base 1 circuit becomes forward-biased when the peak point voltage is reached and the UJT begins to conduct.
4. Because the emitter current increases between V_P and V_V while the emitter voltage decreases, the UJT is said to have "negative resistance."
5. The frequency of oscillation of the UJT oscillator circuit must be synchronized with the anode-cathode voltage of the SCR.

SELF-TEST

Check your understanding by answering these questions.

1. The total resistance between B_1 and B_2 is designated R_{BB} and is usually many megohms. ________ (true/false)
2. The intrinsic standoff ratio of a UJT is always less than 1. ________ (true/false)
3. If the current through a device increases as the voltage across the device increases, the device has ________ resistance.

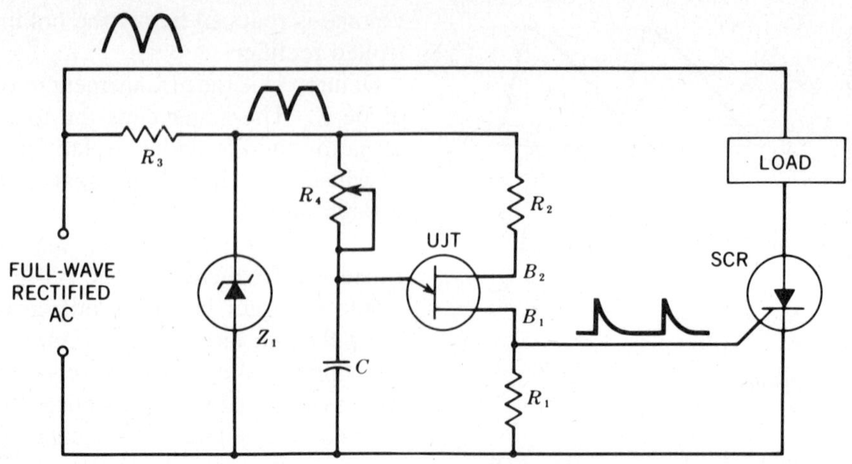

Fig. 16-8. SCR triggered by UJT.

4. A small resistor in the B_2 circuit of a UJT provides __________ compensation.
5. In Fig. 16-7, as the value of R_4 is increased, the frequency of oscillation will __________ .
6. If the emitter–base 1 voltage of a UJT does not reach V_P, the UJT will not __________ .
7. The intrinsic standoff ratio of a UJT is indicated by the Greek letter __________ .

MATERIALS REQUIRED

- Power supply: Variable regulated dc voltage source; variable 60-Hz source (variable autotransformer); isolation transformer
- Equipment: Oscilloscope; dc milliammeter; EVM or VOM; frequency-calibrated AF sine-wave oscillator (as a comparison source for checking frequency)
- Resistors: ½-W 33-, 100-, 220-, 470-, 1200-, 4700-Ω; 1-W 1000-Ω; 5-W 250-, 5000-Ω
- Capacitors: 0.1-μF 400-V
- Semiconductors: SCR 2N1596; UJT 2N2160; zener diode, 18-V 1-W; four silicon rectifiers, type 1N5625 or equivalent
- Miscellaneous: Two SPST switches; power transformer, 120-V primary, 25-V 1-A secondary; 500,000-Ω 2-W potentiometer

PROCEDURE

Emitter Characteristic (V_E versus I_E)

1. Connect the circuit shown in Fig. 16-9. The autotransformer is plugged into an isolation transformer. Switches S_1 and S_2 are open. Power OFF. Output of the autotransformer is set at 0 V. Output of the dc supply is set at 0 V, as measured by voltmeter V.
2. The *horizontal sweep* switch of the oscilloscope is set on external sync.

   Calibrate the vertical amplifier of the oscillo-

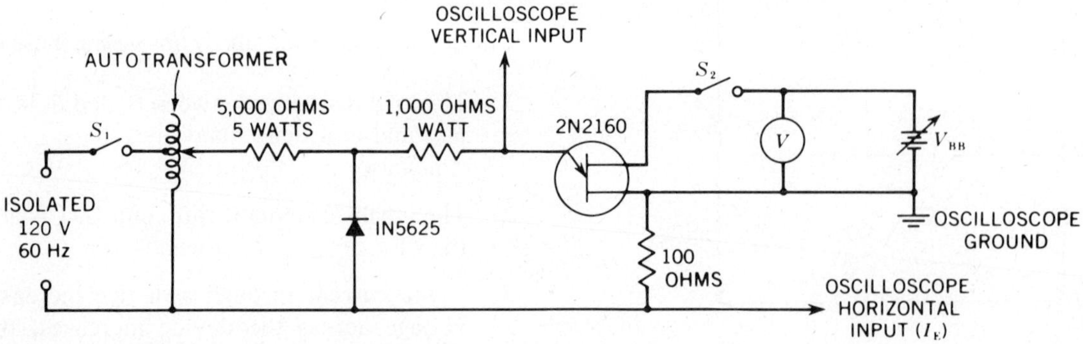

Fig. 16-9. Test setup to display *VE* versus *IE* characteristic of a UJT.

scope at 2.5 V/cm, and the horizontal amplifier at 100 mV/cm. Position the horizontal trace of the scope on the lowest horizontal line of the graticule.

3. Close switch S_2 and set V_{BB} at 5 V dc. Close switch S_1 and adjust output of the autotransformer until the horizontal deflection on the oscilloscope is 10 cm. (NOTE: This limits I_E to 10 mA because each centimeter of horizontal deflection corresponds to 1 mA of emitter current.) Observe the trace on the screen and record on the graph in Table 16-1. Mark this curve $V_{BB} = 5$ V. Identify V_P, the peak voltage point. NOTE: The curve observed on the oscilloscope may appear reversed, as compared to that in Fig. 16-4.

4. Repeat step 3 in turn for (*a*) $V_{BB} = 10$ V, (*b*) $V_{BB} = 15$ V, (*c*) $V_{BB} = 20$ V.
Draw and identify by V_{BB} all the curves in Table 16-1. Identify also the peak voltage point V_P for each curve. Open switches S_1 and S_2. Power OFF.

Relaxation Oscillator

5. Connect the circuit shown in Fig. 16-7. Set R_4 for maximum resistance. S_1 is open. Adjust the output of the power supply V_1 for 25 V. Calibrate the dc vertical amplifiers of the oscilloscope for 5 V/cm. Set the trace on the lowest horizontal line of the graticule. The oscilloscope is set on triggered (or free-running) sweep. Set R_4 in the middle of its range.

6. Close S_1, applying power to the circuit. Connect the vertical input leads of the oscilloscope across capacitor C, hot lead on A, ground lead on B. *Externally* synchronize the oscilloscope with the voltage waveform V_{B2} at base 2. Adjust scope sweep controls for at least two or three complete waveforms. Draw and record the waveform, labeled V_E, in Table 16-2. Measure and record the p-p amplitude of the waveform. Measure and record the voltage level to which the waveform falls (still using the dc vertical amplifiers of the oscilloscope). Also measure and record the frequency of the waveform.

7. Observe the waveform V_{B1} at base 1. Draw it in Table 16-2, in proper time phase with the waveform V_E. Also measure and record its p-p amplitude.

8. Adjust R_4 for the minimum resistance. Observe and record the waveforms V_E and V_{B1}; measure and record in Table 16-2 their p-p amplitude and frequency as in steps 6 and 7.

9. Adjust R_4 for half of its total resistance and repeat step 8. Power OFF.

TABLE 16-1. UJT Emitter Characteristic Curve

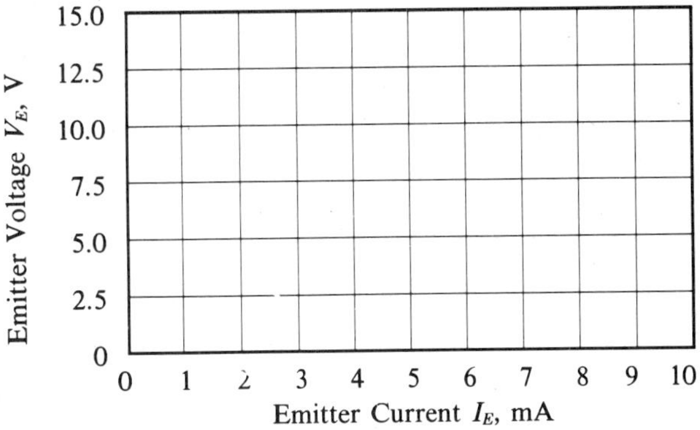

TABLE 16-2. UJT Relaxation Oscillator Waveforms

Step		*Waveform*	*Voltage, V p-p*	*Frequency, Hz*
5	V_E			
6	V_{B1}			
7	V_E			
	V_{B1}			
8	V_E			
	V_{B1}			

SCR Triggered by UJT

10. Connect the circuit of Fig. 16-10. *T* is a step-down transformer (120-V primary, 25-V center-tapped secondary). Close S_1, Power ON.

11. Calibrate the vertical amplifiers of your oscilloscope at 10 V/cm and 3 V/cm. With the oscilloscope set on line sync or externally synchronized by the voltage from point *A* in the secondary of *T*, observe the waveform V_{AB} across the secondary of *T*, vertical lead of the oscilloscope on point *A*, ground lead on point *B*. Adjust the sweep, sync, and centering controls until the reference waveform appears as in Table 16-3 on page 92. Measure and record in Table 16-3 the p-p amplitude of the waveform.

12. Observe, measure, and record in Table 16-3, in proper time phase with the reference, the waveforms V_{CD} and V_{FD}.

13. With R_4 set at minimum (zero resistance), observe, measure, and record in Table 16-3 the waveform V_{JD}, in proper time phase with the reference. Measure on milliammeter M_1 and record the load current I_F.

14. Vary R_4 over its entire range. Observe the effect on load current and on the load waveform.

15. With R_4 set at maximum resistance, observe, measure, and record in Table 16-3 the waveform V_{JD} and the load current.

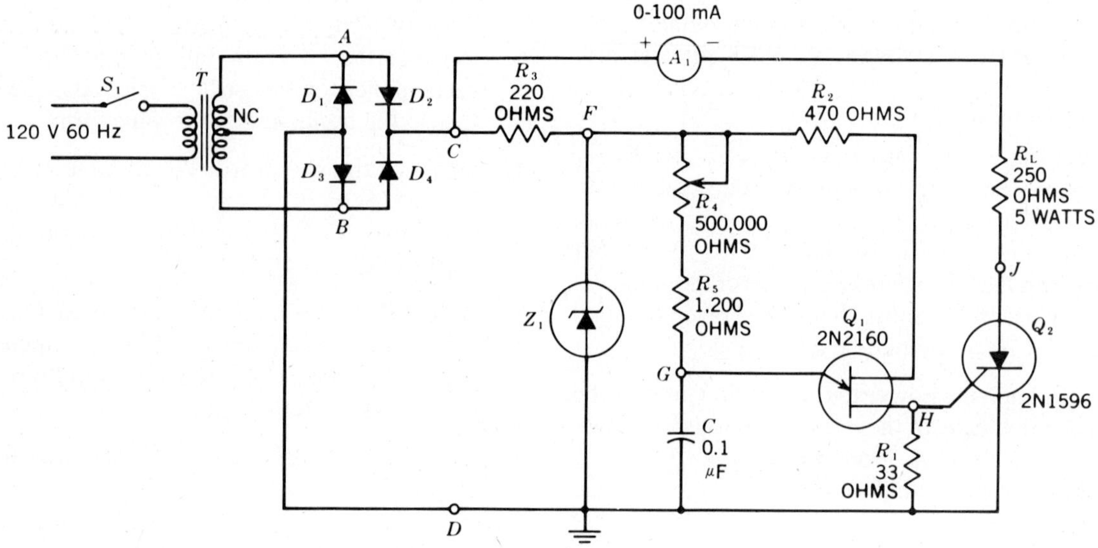

Fig. 16-10. Experimental UJT trigger controls the firing of the SCR.

QUESTIONS

1. What are the desirable characteristics for an SCR gate-triggering source?

2. Does a phase-shift sinusoidal gate-triggering source meet the characteristics you listed in answer to question 1? If not, why not?

3. Draw the circuit diagram of a UJT circuit which can be used as a gate trigger.

4. Explain how the circuit that answers question 3 operates.

5. Assume the circuit shown in Fig. 16-6 acts as the gate-triggering source for an SCR. The anode-cathode circuit of the SCR is powered from a 60-Hz sinusoidal source. What problem, if any, do you see in the use of a dc voltage for the gate trigger and a 60-Hz sine wave for the anode-cathode of the SCR?

6. Would it be possible to achieve 180° (approximately) control of an SCR using a UJT trigger circuit? If yes, explain why.

7. Explain in detail, with waveforms, the operation of the experimental circuit shown in Fig. 16-7.

8. What relationship, if any, is there between the resistance of R_4 in Fig. 16-7 and the frequency of the output waveform? Refer to your experimental data in Table 16-2 to confirm your answer.

TABLE 16-3. Measurements in UJT-Triggered SCR Circuit

Step	*Condition*	*Test Points*	*Waveform*	*Voltage,* V *p-p*	*Load Current,* mA
11	X	*AB*	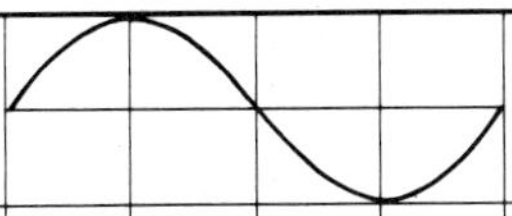		X
12	X	*CD*			X
		FD			
13	$R_4 = 0$	*JD*			
15	$R_4 = 500{,}000\ \Omega$	*JD*			

9. What relationship, if any, is there between the resistance of R_4 in Fig. 16-10 and the conduction angle of the SCR? Refer to your experimental data in Table 16-3 to confirm your answer.

10. From your measurements in Table 16-3, what is the zener voltage (approximately) for Z_1?

Answers to Self-Test

1. false
2. false
3. negative
4. temperature
5. decrease (period becomes longer)
6. fire/trigger
7. η (eta)

EXPERIMENT

17

UJT-CONTROLLED SCR TIME-DELAY CIRCUITS

OBJECTIVE

To study the operation of ac- and dc-powered UJT/SCR time-delay circuits

BASIC INFORMATION

DC-operated Time-Delay Circuits

Time-delay circuits are frequently used in industrial control systems to apply or remove power from a load at a specified time or for a specified period after a starting signal is given. In some cases, to prevent damage to a piece of equipment, a "warm-up" period must precede application of full operating power. In other situations, a preset amount of time is required to perform a desired function. The applications for time delays and the methods of achieving them are many and varied. While digital techniques are increasingly being used, many timing circuits are based on the charge or discharge time of a capacitor in an *RC* circuit together with other electronic components such as UJTs, which will trigger at a known point on the charge or discharge curve.

The time-delay circuit of Fig. 17-1 utilizes previously studied ideas and operates in the following way: When power is first applied by closing S_1, the relay R_{L1} is energized, load 2 is turned ON, and the timing capacitor C_1 is unshorted, thereby starting a timing interval.

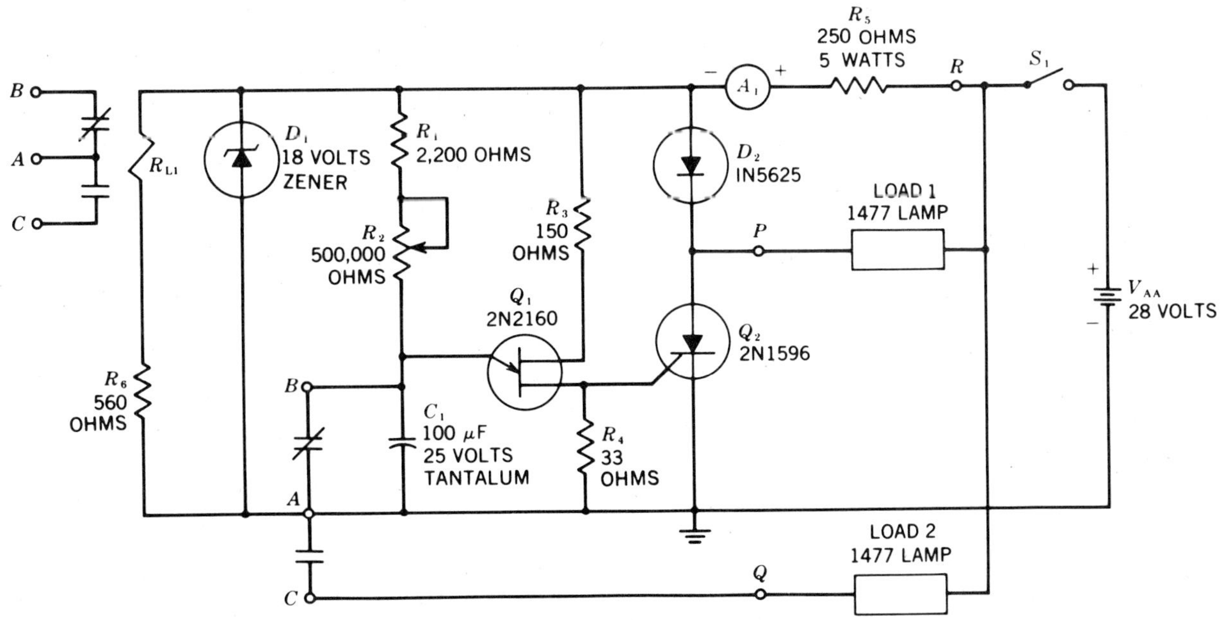

Fig. 17-1. Experimental dc-powered UJT-controlled SCR.

Load 1 is OFF because the SCR has not been triggered. After the timed interval, the UJT fires, the SCR is turned ON, and load 1 is energized. Because the voltage drop across diode D_2 and the SCR Q_2 is small, the relay drops out and load 2 is turned OFF when the contacts A-C open. Load 1 stays ON until S_1 is opened.

In this circuit the UJT fires only once per closure of S_1 because the voltage supplied to Q_2 and its RC circuit drops to about 2 V due to current flow through R_5 after the SCR fires. This current also acts as a "holding current" for the SCR and keeps it conducting if load 1 should be disconnected. The zener diode in this example tends to keep the voltage applied to the UJT timing circuit constant regardless of changes in the input voltage V_{AA}, thereby stabilizing the timing interval. For long-term, reliable operation of this timer, C_1 should be a low-leakage capacitor. Because of its capacitance value, it would be a tantalum capacitor rather than an aluminum foil type.

AC-powered Time-Delay Relay

If the ac bridge rectifier circuit and associated components of Fig. 17-2 are substituted for the dc source of Fig. 17-1, we have an ac-powered time-delay relay. A more sophisticated, precision, ac-powered time-delay relay is shown in Fig. 17-3. This timer includes coarse and fine selection of the timed interval (S_1 and R_{10}), operation controlled by push-button "start" and "stop" switches, and provision to accurately calibrate the timed interval by R_5.

Details of circuit operation are as follows: When ac power is first applied, C_4 has no charge. Charging current through it provides gate current to SCR_1, turning it ON and energizing the relay. Contacts S_{1B} are now closed, P_{L1} is ON, S_{1C} is open, and P_{L2} is OFF. The timing capacitor is shorted to ground through 10 Ω by contacts S_{1A}. These conditions continue until the start switch is pressed.

When the start switch is pressed briefly, the SCR is shorted anode-to-cathode through contacts S_{1B} and the SCR is turned OFF. This deenergizes the relay, P_{L_2} comes ON, P_{L1} goes OFF, and the timing capacitor C_3 is unshorted, thus beginning the delay period. At the end of the delay period, the UJT fires, SCR_1 turns ON, and the initial conditions are reestablished.

At any time before completion of the timed interval, pressing the stop switch will provide gate current to SCR_1, and the initial conditions will be reestablished (S_1 ON, P_{L1} ON, P_{L2} OFF). A new cycle may be started by briefly pressing the start switch SW_2.

R_{10} is a precision, 10-turn potentiometer (helipot) which is used to adjust the delay period in fine increments. SW_1 provides coarse adjustment in three time ranges. The fourth position is the timing OFF control for the circuit. This timer provides a range of adjustable time delays from 0.25 to 20 s. Unused contacts on the relay are used to control external loads.

SUMMARY

1. Time-delay circuits are very common in industrial control applications. Electronic timers with no moving parts are preferred over older mechanical timers.
2. Time-delay timers may be either digital types or use RC networks to determine the delay.
3. Loads may be turned either ON or OFF by timers for a preset period of time.
4. Time-delay timers use a combination of circuits which may be studied separately and individually understood.

SELF-TEST

Check your understanding by answering these questions.

1. In Fig. 17-1, when S_1 is open, the relay is __________ and the loads are both __________ .

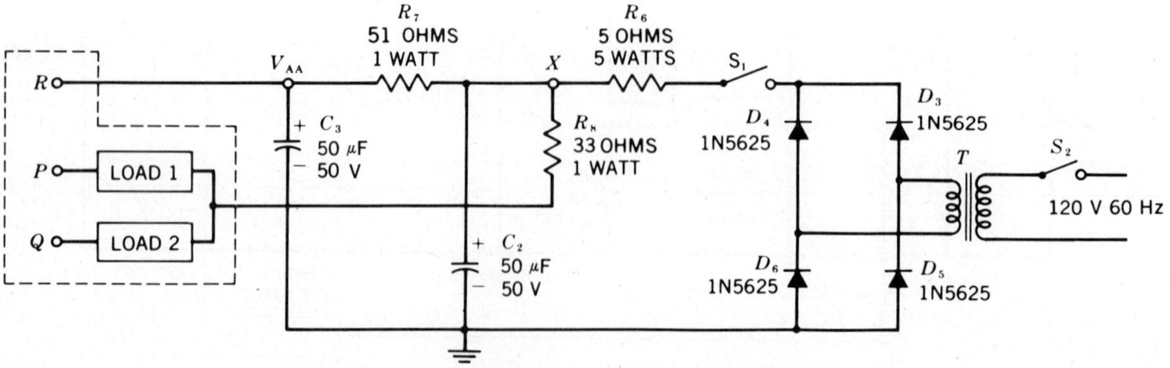

Fig. 17-2. AC bridge rectifier supplies filtered dc voltage to time-delay circuit of Fig. 17-1.

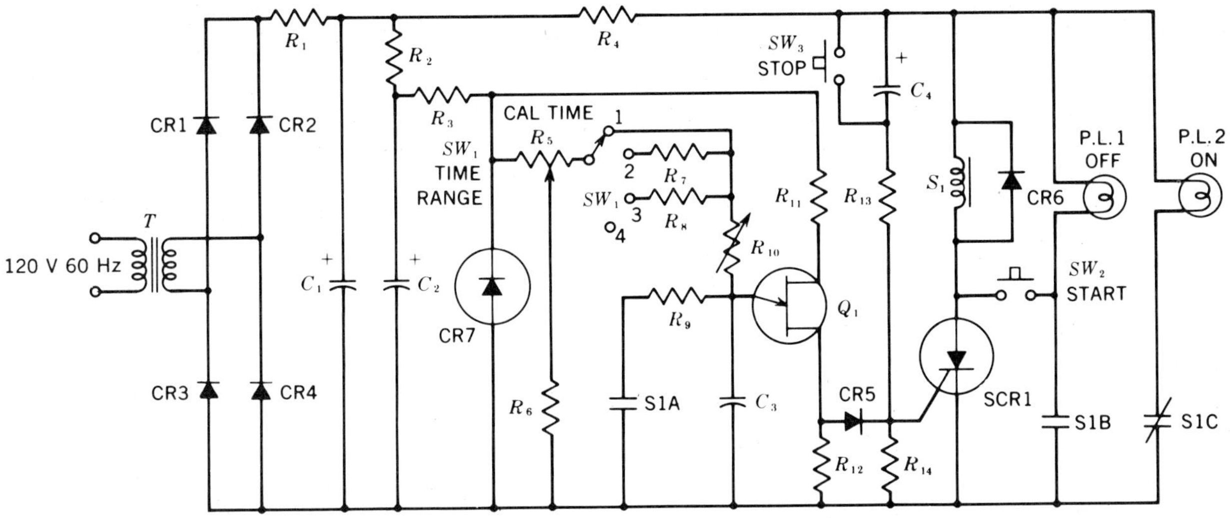

R_1- 2 OHMS, 1 WATT	R_{11}- 150 OHMS, 1/2 WATT	CR1-CR6-GE A13A
R_2- R3-330 OHMS, 1/2 WATT	R_{12}- 18 OHMS, 1/2 WATT	CR7-18 VOLTS, 10% 1 WATT ZENER
R_4- 35 OHMS, 5 WATT	R_{13}- 1,200 OHMS, 2 WATT	Q_1-GE 2N1671B
R_5- 2,500 OHMS, LINEAR POT	R_{14}- 100 OHMS, 1/2 WATT	S1-GE CR2791G122A4
R_6- 25,000 OHMS, 1/2 WATT	C_1-500μFD, 50 VOLTS	4PDT RELAY
R_7- 100,000 OHMS, 1/2%, 1/2 WATT	C_2-100μFD, 50 VOLTS	PL1, PL2-GE 1447, 24 VOLTS LAMP
R_8- 200,000 OHMS, 1/2%, 1/2 WATT	C_3-100μFD, 20 VOLTS TANTALUM	T-TRANSFORMER, PR1-115 VOLTS, SEC-25 VOLTS 1A
R_9- 10 OHMS, 1/2 WATT	C_4-10μFD, 50 VOLTS	
R_{10}- 100,000 OHMS, 10 TURN HELIPOT	SCR1-GE C15F OR C11F	

Fig. 17-3. Variable time-control circuit. *(General Electric)*

2. In Fig. 17-1, when S_1 is first closed, the relay __________ .
3. The timing interval begins when contacts *A* and *B* __________ .
4. The length of time lamp 2 is ON is determined by the time constant C_1 X(__________).
5. The purpose of R_9 (10 Ω) in Fig. 17-3 is to limit the current through contacts S_{1A} when they __________ .
6. In Fig. 17-3, the SCR is turned OFF by shorting its __________ to its __________ .
7. The purpose of C_{R7} is to __________ the timed interval.
8. C_{R6} across the relay coil prevents the counter emf generated when the relay is deenergized from __________ the SCR.
9. In Fig. 17-3, lamp __________ serves as a "pilot light."
10. The operation of this timer is temperature-compensated, partially, by the presence of *R* __________ .

MATERIALS REQUIRED

- Power supply: 120-V ac source; variable regulated low-voltage (high-current) dc source
- Equipment: EVM; VOM; 0- to 250-mA dc milliammeter
- Resistors: ½-W 33-, 150-, 560-, 2200-Ω; 1-W 33-, 51-Ω; 5-W 5-, 250-Ω
- Capacitors: 100-μF 25-V tantalum; two 50-μF 50-V
- Semiconductors: 2N1596 (SCR); 2N2160 (UJT); 18-V 1-W zener diode; five diodes, type 1N5625 or equivalent
- Relay: DC, SPDT, 400-Ω field (approximate), 1-A contacts, 7.0-mA pickup current (approximate)
- Miscellaneous: Two 24-V pilot lights, type 1477 or equivalent and sockets; two SPST switches; power transformer 120-V primary, 25.0-V 1-A secondary; 500,000-Ω 2-W potentiometer

PROCEDURE

DC Time-Delay Relay

1. Connect the circuit shown in Fig. 17-1. Set V_{AA} at + 28 V. S_1 is open. Set R_2 at 100,000 Ω. Note that C_1 is a low-leakage tantalum capacitor.
   Connect an EVM set on 30-V dc range across C_1. Connect a VOM set on 1-V dc range across R_4. A_1 is a 0- to 250-mA milliammeter.

2. Close S_1. Observe and record in Table 17-1, in column "SCR OFF," whether load 1 or load 2 is energized at the instant S_1 is closed. Also observe the voltage across C_1, as this timing capacitor charges toward the peak-point voltage of the UJT. Observe and record the voltage across C_1, just at the peak-point voltage of the UJT (expiration of the time-delay interval). At this point, power to the two loads switches as the relay is deenergized and the SCR turns ON. Indicate in Table 17-1, in column "SCR ON," which load is ON *after* the relay has switched loads. Measure and record the voltage across C_1 at this time.

3. Leave S_1 closed for a minute and note the stability of the circuit. Now, recycle the relay by opening and closing S_1. Measure and record in column "SCR OFF," in Table 17-1, the voltage across R_4 just at the time switch S_1 is closed and the relay is energized. Measure also and record, in column "SCR ON," the voltage across R_4 after the relay has reset and power to the two loads has switched. Record also when load 1 is ON, when load 2 is ON.

4. Recycle the relay by opening and closing S_1. Measure and record in column "SCR OFF," in Table 17-1, the voltage across the zener diode V_{D1}, at the moment S_1 is closed. Also measure and record V_{D1} after the SCR has turned ON. Record also when load 1 is ON, when load 2 is ON.

5. Recycle the relay. Record in column "SCR OFF" the current through R_5, at the moment S_1 is closed; record the current in column "SCR ON" after the SCR has turned ON.

6. Set R_2 at minimum resistance. Recycle the relay. Measure and record in Table 17-2 the time delay introduced by the relay (that is, the time required by the relay, after S_1 is closed, to switch power to the two loads).

7. Set R_2 at maximum resistance. Recycle the relay. Measure and record the time delay introduced by the relay for this value of R_2. Recycle the relay. Measure and record the UJT peak-point voltage across C_1 (i.e., voltage across C_1, just at the moment before SCR turns ON).

8. Power OFF. Remove the dc power source in Fig. 17-1 and construct the ac rectified supply shown in Fig. 17-2. Connect the output of the rectified supply, as in Fig. 17-2, to the time-delay circuit shown in Fig. 17-1. Power is still OFF. Connect the high junction of loads 1 and 2 through R_8 to point X, the junction of R_6 and R_7.

TABLE 17-1. Voltage and Current in a DC Time-Delay Relay

			SCR OFF		*SCR* ON	
Step	*Condition*	*Test Point*	*Load (1 or 2)* ON	*Voltage*	*Load (1 or 2)* ON	*Voltage*
2	S_1 closed and relay cycled	Voltage across C_1		Peak point		
3	Relay recycled	Voltage across R_4				
4	Relay recycled	V_{D1}				
5	Relay recycled	X	*Current through* R_5, mA		*Current through* R_5, mA	

TABLE 17-2. Time Delay in DC Relay

Step	*Condition*	*Time Delay,* s	*Peak-Point Voltage across C_1*
6	R_2 minimum		X
7	R_2 maximum		

9. Set R_2 at maximum resistance. Turn S_2 and S_1 ON. Measure and record in Table 17-3 the dc voltage output of the rectified supply (that is, the voltage across C_3) at the start of the delay interval.

10. After the relay has cycled and has turned ON the SCR, measure and record in Table 17-3 the voltage output of the rectified supply.

TABLE 17-3. Voltage in Rectified Power Supply

Step	*Condition*	*Output,* V
9	S_1 ON S_2 ON Start of delay interval	
10	S_1 ON S_2 ON Relay cycle completed	

11. Open S_1. Leave S_2 closed. Set R_2 at 100,000 Ω. Connect an EVM set on 30-V dc range across C_1. Connect a VOM set on 1-V range across R_4.

12. Repeat step 1 and record your measurements in Table 17-4.

13. Repeat step 3 and record your measurements in Table 17-4.

14. Repeat step 4 and record your measurements in Table 17-4.

15. Repeat step 5 and record your measurements in Table 17-4.

16. Repeat step 6 and record in Table 17-5.

17. Repeat step 7 and record in Table 17-5. Power OFF.

TABLE 17-4. Voltage and Current Measurements in AC-Powered Time-Delay Relay

			SCR ON		*SCR* OFF	
Step	*Condition*	*Test Point*	*Load (1 or 2)* ON	*Voltage*	*Load (1 or 2)* ON	*Voltage*
12	S_1 closed and relay cycled	Voltage across C_1				
13	Relay recycled	Voltage across R_4				
14	Relay recycled	V_{D1}				
15	Relay recycled	X	*Current through* R_5, mA		*Current through* R_5, mA	

QUESTIONS

1. From your measurements, what is the peak-point voltage of Q_1 in Fig. 17-1? in Fig. 17-2? If there is a difference, explain why.

2. Explain why the voltage across D_1 (the zener diode) is not maintained at the zener value after the SCR fires.

3. Explain why load 1 turns ON and load 2 turns OFF after the SCR fires.

4. How many time constants of time delay occur when R_2 is set for minimum resistance? maximum resistance? Show all computations.

5. From what voltage level does C_1 charge when switch S_1 is first closed? Confirm your answer by referring to the measured data in Tables 17-1 and 17-4.

6. What current does A_1 measure when (*a*) switch S_1 is first closed and the relay is ON? (*b*) switch S_1 is closed and the relay is OFF?

7. In Fig. 17-2, why is the junction of L_1 and L_2 connected to point X rather than to V_{AA}?

8. What advantages, if any, does the circuit shown in Fig. 17-2 have over that shown in Fig. 17-1?

Answers to Self-Test

1. deenergized/OFF; deenergized/OFF
2. energizes/pulls in/sets
3. open
4. $R_1 + R_2$
5. close/discharge C_1
6. anode; cathode
7. stabilize (against changes in the line voltage)
8. damaging
9. P_{L1}
10. 11

TABLE 17-5. Range of Time Delay in AC Relay

Step	*Condition*	*Time Delay,* s	*Peak-Point Voltage Across* C_1
16	R_2 minimum		X
17	R_2 maximum		

EXPERIMENT

18

ELECTRONIC VOLTAGE-REGULATED POWER SUPPLY

OBJECTIVES

1. To study the characteristics of a zener diode
2. To study the characteristics of an electronic dc voltage-regulated power supply

BASIC INFORMATION

A voltage-regulated power supply is one whose output voltage remains constant, despite variations in input and/or load. Of course, there are limits within which this regulation is maintained. However, within these limits, the output is held at a predetermined level.

Many electronic devices require regulated power supplies. For example, we find voltage-regulation circuits in motor controls, in computers, in critical-timing circuits, and in other applications. In this experiment, we will study the operation of several types of voltage-regulated supplies. The general principles involved apply to other regulatory devices.

A Simple Voltage Regulator

The output voltage of a rectifier may be held constant by a device as simple as that illustrated in Fig. 18-1. Variable resistor *R*, in series with a load, is connected at the output of a rectifier. The load current flows through *R* and causes an *IR* drop across it. Hence, *R* and the load constitute a dc voltage divider. We can adjust *R* to maintain a desired voltage across the load when variations occur in input voltage to the rectifier or in load current. For example, if the load current should increase, a larger *IR* drop will appear across *R*, thus reducing the voltage across the load. We can then reduce the value of *R*, thus lowering the voltage across *R*, until we have restored the voltage across the load to its required value. Similarly, we can adjust *R* to compensate for changes in input or output which would normally affect the load voltage.

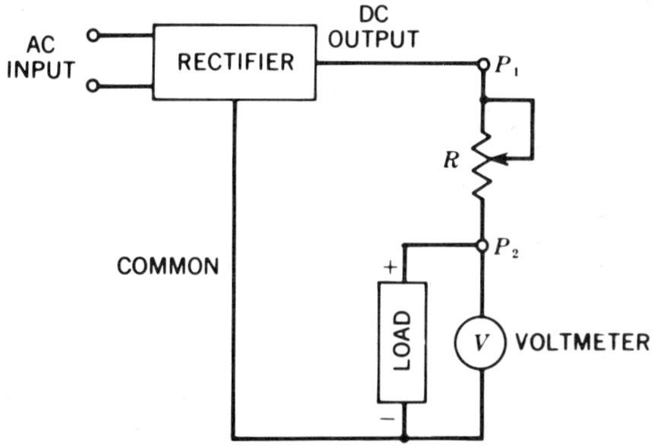

Fig. 18-1. A simple voltage regulator.

To maintain a constant voltage across the load in Fig. 18-1, the following conditions must be met: (1) a reference source, (2) a device for comparing the output voltage with the reference voltage, and (3) feedback of information to correct the output to the desired level. In the circuit shown in Fig. 18-1, the voltmeter acts as the reference level. The technician monitoring the voltmeter fulfills conditions 2 and 3.

Of course, the limitation of this circuit is that it must be manually operated. Our concern must be with a circuit which will provide automatic compensation. However, we will find that the basic principle of control illustrated in Fig. 18-1 is incorporated in automatic voltage-control systems.

Zener Diode Regulator

The constant-voltage characteristic of a reverse-biased zener diode is frequently used in voltage regulation, as we have seen in earlier experiments. Let us briefly review the operating principles involved. Figure 18-2 is the graph of a typical current-voltage characteristic of a zener diode. When forward-biased, the diode acts as a closed switch. Forward current increases with an increase in applied voltage and is limited mainly by the

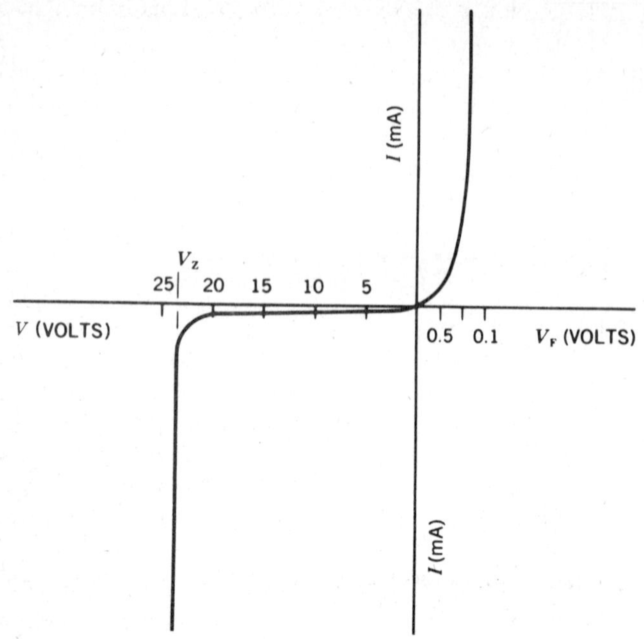

Fig. 18-2. Zener-diode characteristic.

external parameters of the circuit. When the diode is reverse-biased, a small reverse current I_S, called the *saturation current,* flows. I_S remains relatively constant, despite an increase in reverse bias, until the zener breakdown region, in the vicinity of the zener voltage V_Z, is reached. Near V_Z, reverse current starts rising rapidly because of avalanche effect. Finally, zener breakdown (characterized by a sharp increase in current) occurs when the zener voltage V_Z is reached.

In this region, small voltage changes result in large current changes. Obviously, there are dramatic changes in effective resistance at the PN junction in this region.

Zener breakdown need not result in the destruction of the diode. As long as current in the diode is limited by the external circuit to a level within its power-handling capabilities, the diode continues to function normally. Moreover, by reducing reverse bias below the zener voltage, the diode can be brought out of its breakdown level and restored to the saturation-current level. The process of switching the diode between its zener and nonzener current states can be repeated again and again, without damaging the diode.

Zener-diode ratings are given in manufacturer specification sheets. Ratings include zener voltage, tolerance range of zener voltage, zener current limits, maximum power dissipation, maximum operating temperature, maximum zener impedance in ohms, thermal derating factor given in milliwatts per degree Celsius, and reverse leakage current. The material from which it is constructed and the kind of application for which it is intended are frequently given.

The breakdown voltage in a zener diode is dependent on the material from which it is constructed and on the nature of the doping. Zener diodes have been designed to deliver zener voltages from one to several hundred volts. The circuit designer has a wide variety of diodes from which to select the one whose characteristics match circuit requirements.

Figure 18-3 shows the circuit of a zener diode used as a shunt regulator. The diode is in parallel with the load R_L, and this parallel combination is in series with voltage-dropping resistor R. The zener maintains a constant voltage across R_L, within a range of input dc voltage and output load current changes. Within the range of regulation, the internal resistance R_Z of the zener diode varies so that the product $I_Z \times R_Z$ (where I_Z is the current in the diode) remains constant. This product is the required output voltage across the diode.

Consider the operation of the circuit shown in Fig. 18-3 when the voltage source V_{AA} is constant but the load current I_L changes. Assume that a constant output voltage V_{out} is required across the load. The two currents $I_L = V_{out}/R_L$ and $I_Z = V_{out}/R_Z$ combine to form the total current I_T. That is,

$$I_T = I_L + I_Z \qquad (18\text{-}1)$$

The voltage V_R, across the series regulating resistor R, is equal to the product of I_T and R. Thus

$$V_R = I_T \times R \qquad (18\text{-}2)$$

But

$$V_{AA} = V_R + V_{out} \qquad (18\text{-}3)$$

Hence, if V_{AA} is constant and it is required that V_{out} remain constant, V_R must remain constant. Therefore, the total current I_T must remain fixed, despite variations in load current. This can be accomplished only by compensating for changes in I_Z. That is, I_Z must change in the manner shown in Eq. 18-4, assuming that I_T is constant and that I_L can vary.

$$I_Z = I_T - I_L \qquad (18\text{-}4)$$

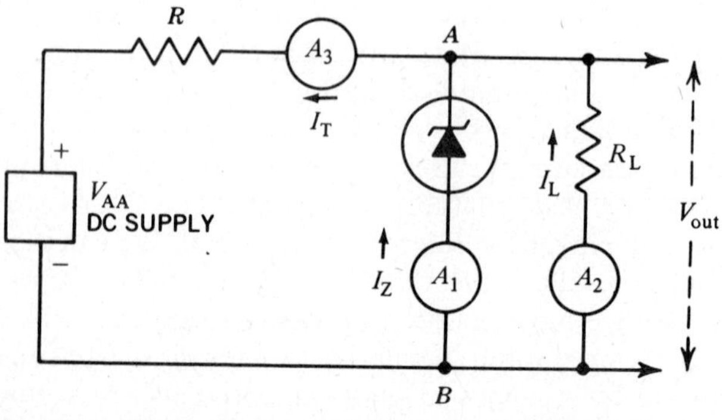

Fig. 18-3. Zener diode used as a voltage regulator.

It is apparent from our discussion that the zener diode selected must be one whose zener voltage $V_Z = V_{out}$. Moreover, the zener diode chosen must have voltage and current characteristics that will fit the requirements of the circuit, and this diode must be operated at the right point in its characteristic.

The circuit shown in Fig. 18-3 can also be used to absorb changes in dc supply voltage when the load resistor R_L remains constant, assuring a constant output voltage V_{out} and hence a constant load current I_L. Assume the circuit is operating properly for a dc supply voltage level V_{AA}. If V_{AA} should now increase, the output voltage V_{out} would tend to rise, thereby increasing zener current I_Z (and load current I_L). It is evident from Eq. 18-1 that I_T would now increase, thus increasing the voltage drop V_R across R. If the regulator circuit has been properly designed, the increased voltage ΔV_R across R should equal (approximately) the increase in supply voltage ΔV_{AA}, and V_{out} would drop back to its original value. The increased zener current would maintain ΔV_R across R, assuring a constant output voltage V_{out}. Similarly, a decrease in V_{AA} would result in a decrease in I_Z and hence in I_T. V_R would be reduced, and V_{out} would return to its predetermined level.

The value of R chosen to achieve proper regulation will depend on the characteristics of the diode and on the conditions of variations of V_{AA} and I_L.

In addition to the regulator diode, whose operation has just been described, there are voltage-reference diodes whose zener voltages are so stable that they can be used as laboratory standards or as reference voltages in a more complex voltage-regulator circuit.

Transistor Regulators

Figure 18-4 is the circuit diagram of a simple transistor regulator. In this circuit, Q_1 has taken the place of the manually variable series resistor R of Fig. 18-1. The advantage of this regulator over that in Fig. 18-1 is that the emitter-collector resistance of Q_1 varies automatically to compensate for changes in input voltage. Now, how does this change in resistance of Q_1 take place, and how is the voltage across the load maintained at a constant level?

In Fig. 18-4, the initial emitter-to-collector resistance of Q_1 is determined by the setting of R_1, the bias control on the base of Q_1. Let us suppose that R_1 has been set to place a desired voltage across the load. Assume that at this desired setting the load voltage (voltage from emitter to ground) is + 10 V and that the voltage from base to ground is + 10.5 V. The bias on this NPN transistor Q_1 is therefore + 0.5 V; that is, the base is 0.5 V more positive than the emitter. Now, if the unregulated dc voltage from the rectifier increases, the voltage across the load rises, say, to + 10.3 V. Since the base voltage tends to remain relatively constant (base current is much smaller than emitter current), the rise in voltage at the emitter (load) makes the base less positive than it originally was, thus reducing current in Q_1, and increasing the emitter-collector resistance of Q_1. Accordingly, there is a greater voltage drop across the transistor, and Q_1 takes up the increase in dc input voltage. This arrangement, therefore, tends to stabilize the load voltage.

In this regulator circuit, the setting of R_1 (which is connected across V_{BB}, a fixed-voltage source) is the reference source, thus providing a certain level of voltage across the load. The feedback, provided by changes in load voltage, affects the bias on Q_1 and therefore affects the internal emitter-to-collector resistance of Q_1. The change in emitter-to-collector resistance is in a proper direction to restore the voltage across the load to its preset level.

In Fig. 18-4, the base voltage is maintained at a desired level by a potentiometer R_1 and a fixed battery V_{BB}. It is possible to eliminate the battery by use of a simple zener diode regulating circuit. In Fig. 18-5, the zener diode maintains the base at a constant dc level, determined by the zener characteristic of the diode. This reference voltage and the circuit constants determine the output voltage across the load.

Series Transistor Regulator

A circuit arrangement which provides better regulation than those previously discussed is shown in Fig.

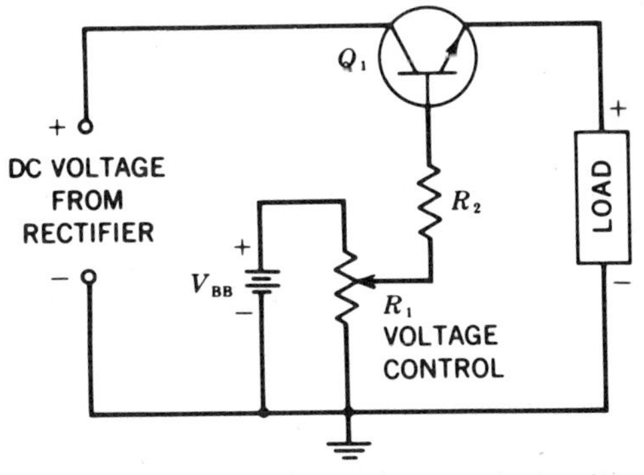

Fig. 18-4. A transistor regulator whose output voltage can be varied.

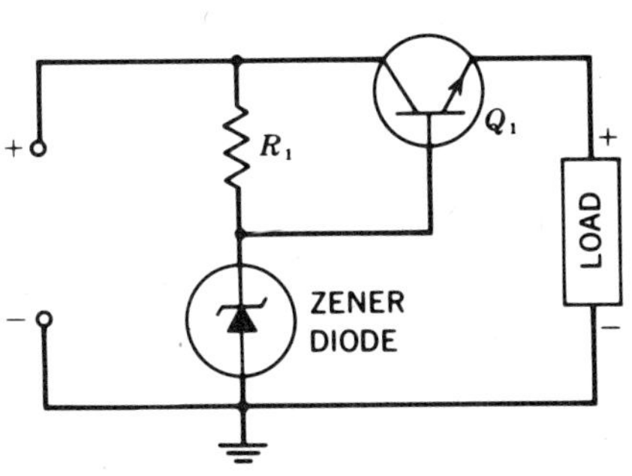

Fig. 18-5. Zener diode used to stabilize the base voltage in a transistor regulator circuit.

18-6. Transistor Q_1 acts as a variable resistance in series with the load connected in its emitter circuit. The emitter voltage across the load is the desired regulated voltage. A high-gain transistor Q_2 is used to amplify small changes in output voltage and feed them back as correcting voltages to Q_1. The emitter-to-collector resistance of Q_1 varies in the proper direction with the feedback voltage to compensate for changes in output.

Let us analyze the operation of this circuit. The emitter of Q_2 is maintained at a fixed potential by the zener diode D_1. This is the reference voltage for amplifier Q_2. The base of Q_2 is connected to the arm of potentiometer R_3. The voltage-control potentiometer in series with R_2 constitutes a dc voltage divider across the regulated output voltage supply. The setting of R_3 determines the bias on Q_2 and will, as we will see, determine the bias on Q_1. Hence, the setting of R_3 determines the emitter-to-collector resistance of Q_1 and therefore the output voltage. Observe that D_1 in series with R_1 constitutes a simple zener regulator which maintains a constant voltage on the emitter of Q_2. Observe also that the collector of Q_2 is connected by R_4 to the positive terminal of the unregulated dc input. Any change in Q_2 collector current will affect Q_2 collector voltage—hence, Q_1 base voltage.

We can now follow the manner in which a constant voltage is maintained across the load despite changes in unregulated input dc voltage. Assume that R_3, the voltage control, is set to deliver + 25 V to the load. Assume further that D_1 is a 12-V zener and that the voltage at the base of Q_2 is + 12.5 V. Now what happens if the regulated output voltage increases? This change will drive the base of Q_2 more positive (while its emitter voltage is constant), thus increasing collector current in Q_2. The increased collector current will increase the *IR* drop across R_4, thus driving the collector of Q_2 and simultaneously the base of Q_1 in a negative direction. The resultant change in bias on the emitter-base junction of Q_1 will reduce current in Q_1, thus reducing the voltage across the load (which is in series with Q_1) and restoring it to its preset level.

Similarly, a decrease in load voltage will decrease bias on Q_2, thereby reducing its collector current and driving its collector in a positive direction. The base of Q_1 is simultaneously driven more positive, thus increasing Q_1 collector current. The increase in Q_1 collector (hence emitter) current will increase the *IR* drop across the load, thus restoring the output voltage to its preset level.

Before completing our discussion of this circuit, we should note the following: (1) Q_2 must be a high-gain transistor, in order to stabilize the output for even small variations; and (2) Q_1 must be a high-current-carrying transistor, since it must handle all the load current, plus the emitter voltage-divider current, and the zener current. It may be necessary to parallel two or more transistors to achieve the high-current-carrying requirements of Q_1.

Series Regulator Variation

Figure 18-7 shows a more sensitive series regulator than the circuit in Fig. 18-6. However, this somewhat more complex arrangement still utilizes the basic principles of the series regulator we have studied. The unregulated dc voltage from a rectifier is applied to power transistor Q_3 in series with the load. The load voltage is determined by the setting of potentiometer R_6. High-gain feedback is provided by transistors Q_5, Q_4, Q_1, and Q_2. Zener diode D_1 provides the reference voltage. Q_3 may be a single transistor or a parallel combination of two or more transistors, depending on the load current requirements.

Suppose that R_6 is set for, say, 24-V regulated output. If the output voltage rises, for any reason, the increase in voltage is distributed across R_8, R_4, and R_6. A proportionate rise in voltage therefore appears at the base of Q_5, and its collector current increases. This increases emitter current and drives the top of emitter

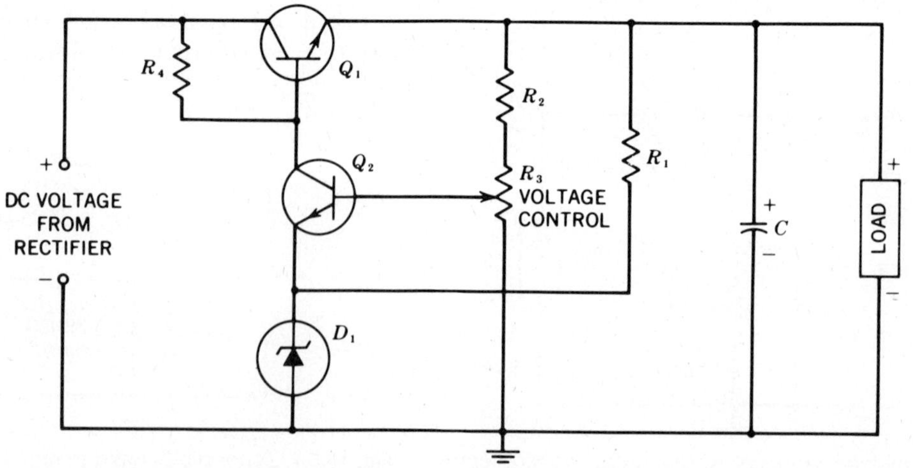

Fig. 18-6. Series-type dc voltage regulator. *(General Electric)*

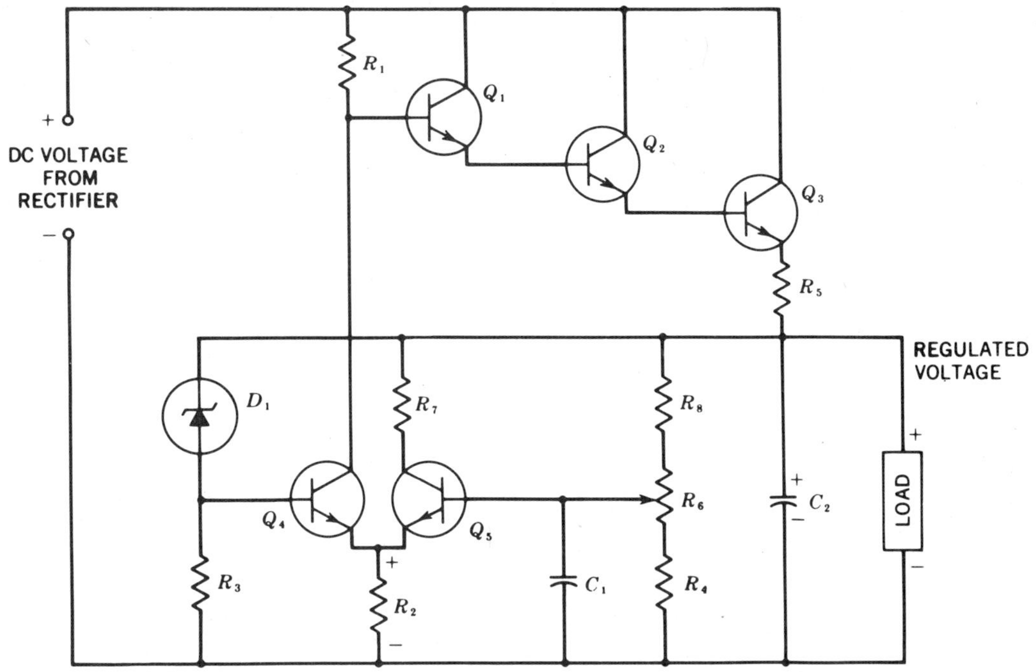

Fig. 18-7. High-gain series voltage regulator.

resistor R_2 slightly more positive. This resistor is common to the emitter of Q_4 and Q_5. As a result, the emitter of Q_4 becomes somewhat more positive. Now, consider the base of Q_4. The *total* increase in regulated voltage appears at the base of Q_4 because of the constant voltage characteristic of zener diode D_1. Since the positive increase at the base is greater than the positive increase at the emitter of Q_4, current flow in Q_4 is increased. The increased collector current, in R_1, drives the base of Q_1 in a negative direction, reducing collector current in Q_1, also reducing Q_1 emitter voltage. As a result, the base of Q_2, which is directly connected to the emitter of Q_1, is driven negative, thus reducing collector current in Q_2 and making the emitter of Q_2 more negative. The base of Q_3 therefore goes more negative, thus reducing current in Q_3. Since the load is in series with Q_3, a reduction in Q_3 current reduces load current also. The voltage drop across the load therefore decreases, and if the circuit was properly designed, the load voltage returns to its preset level. Thus the original increase in regulated voltage output sets up a negative feedback reaction whose total effect is to decrease the regulated voltage to its preset level. Similarly, it can be shown that a decrease in regulated voltage, for whatever reason, will cause feedback to increase current in Q_3, thus restoring voltage across the load to its original level.

Shunt-Type Voltage Regulator

The simple but effective shunt-type regulator in Fig. 18-8 provides a constant dc output of 28 V for load currents up to 0.5 A and dc inputs ranging from 45 to 55 V. Within this range of conditions, the reference zener diode D_1 maintains 27 V across itself, and any changes in output voltage are reflected across R_2. Let us assume, for example, that the output voltage has risen, because of an increase in unregulated input, to 28.5 V. This 0.5-V increase appears as an increased positive voltage on the base of Q_1, thus increasing the collector-to-emitter current flow in Q_1 and driving the emitter of Q_1 and hence the base of Q_2 more positive. A more positive Q_2 base increases current in Q_2. The current increase in Q_1 and Q_2 causes an increased voltage drop across R_1. If the circuit has been properly designed, this increased voltage across R_1 just offsets the 0.5-V rise in regulated output, and the regulated voltage drops back to 28.0 V.

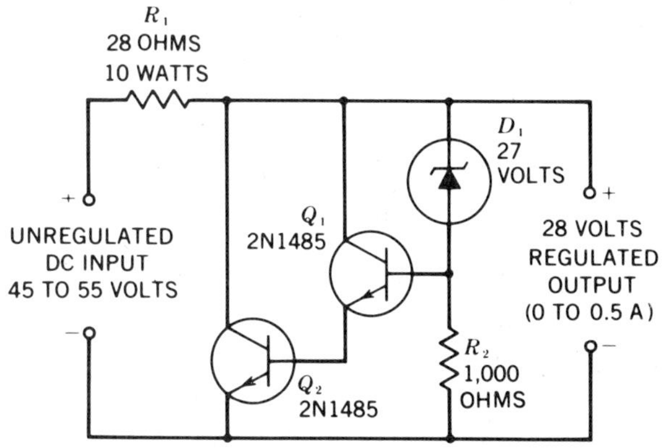

Fig. 18-8. Voltage regulator, shunt type. *(RCA)*

Similarly, it can be shown that a drop in regulated

output voltage, caused by a drop in input voltage, will decrease current flow in Q_1, Q_2, and hence in R_1. The reduced IR drop across R_1 will offset the initial voltage drop across the load, and the load voltage will be restored to 28.0 V.

The circuits discussed so far, which use discrete components, will be encountered for some time in industrial equipment. However, modern integrated circuit (IC) regulators are available as complete units for low-power applications; for high-powered applications, these IC regulators can be used with external power transistors to provide almost any required value of voltage and/or current. Typical devices include the LM309K (5 V/1 A) and 7805 three-terminal IC voltage regulators.

Power Supply Measurements

Any dc power supply can be considered as an ideal dc voltage source connected in series with a resistor representing its source impedance (resistance). A well-regulated supply will have a very low source impedance. The ability of a voltage regulated supply may be expressed either in terms of its internal resistance or of its percentage voltage regulation. Within its ability to regulate, the source impedance may be calculated by measuring the change in output voltage as the load is changed:

$$Z = \frac{dV}{dI}$$

where Z = internal impedance of supply, Ω
dV = V(no load) − V(with load): "delta V", V
dI = change in load current: "delta I", A

For well-regulated supplies, Z may be a very small fraction of an ohm. When the power supply is described by its percentage voltage regulation, the following relationship is used:

$$\%V_R = \left(\frac{V_1 - V_2}{V_2}\right) \times 100\%$$

where $\%V_R$ = percentage voltage regulation
V_1 = voltage with no load
V_2 = voltage under load

A small number, less than 1 percent, indicates good regulation. A common error is to divide by V_1 rather than V_2.

SUMMARY

1. A voltage-regulated power supply is one whose output remains constant despite variations in input and/or load.
2. A reverse-biased zener diode maintains a relatively constant voltage across itself when it is operated in the zener region. Zener current must be kept within safe limits by the external circuit to prevent diode burnout.
3. The zener (or breakdown) voltage of such a diode depends on the material from which it is constructed and on the nature of the doping.
4. A zener diode may be used as a voltage regulator for a load requiring a voltage equal to the zener voltage. Figure 18-3 illustrates such a regulator circuit.
5. Transistors are more sensitive to load changes and input voltage variations than are zener diodes. They therefore provide better regulation than do zener diodes used alone. The more sensitive regulators use transistors and zener diodes in regulator circuits, as does the series regulator in Fig. 18-6.
6. In the series-type voltage regulator (Fig. 18-6) the emitter-to-collector resistance of Q_1 is in series with the load and acts as an automatic variable resistance to offset load changes.
7. More complex regulator circuits are used when it is required to provide a more highly regulated voltage output than that produced by Fig. 18-6. The circuit of Fig. 18-7 is one such high-gain circuit.
8. In shunt-type regulators (Fig. 18-8) the regulator is in parallel with the load and again acts to offset either increases or decreases in load and input voltage. As in series regulators, a zener diode acts as a bias reference for the regulator.
9. Voltage regulators include (*a*) a voltage reference source, (*b*) a device for comparing the output voltage with the reference voltage, and (*c*) feedback to correct the output to the desired level.

SELF-TEST

Check your understanding by answering these questions.

1. A disadvantage of the simple regulator shown in Fig. 18-1 is that it is not ___________ .
2. The ___________ (forward/reverse) bias characteristic of a zener diode is used in electronic voltage regulators.
3. A zener diode cannot be destroyed because it is self-regulating. ___________ (true/false)
4. In Fig. 18-3, assume that V_{AA} remains constant but V_{out} increases. If the circuit has been properly designed, I_Z will ___________ in order to bring V_{out} back to its required voltage.
5. The regulator in Fig. 18-4 is in ___________ (series/parallel) with the load.

6. In the regulator of Fig. 18-5, an increase in load voltage will __________ the negative bias on Q_1, thus __________ the emitter-to-collector resistance of Q_1. The voltage divider action of the emitter-to-collector resistance and the load will bring the load voltage down to its desired level.

7. An advantage of the regulator in Fig. 18-6 over that in Fig. 18-5 is that the former is __________ __________ to changes in load voltage.

MATERIALS REQUIRED

- Power supply: Autotransformer, variable from 0 to 130 V
- Equipment: EVM; VOM; two 0- to 50-mA milliammeters
- Transformer: 120-V primary, 25-V secondary
- Resistors: ½-W, 470-Ω; two 1000-, 1800-, 2200-, 2700-, 3300-, 10,000-, 12,000-, 33,000-, 120,000-Ω; two 20-W, 1250-Ω
- Capacitors: 100-μF 50-V; 25-μF 50-V
- Semiconductors: 18-V 1-W zener diode; four 1N5625 (silicon diode rectifiers) or the equivalent; 2N3397 (transistor) or equivalent; D40E5 (transistor) or equivalent
- Miscellaneous: Two SPST switches; 1000-Ω 2-W potentiometer

PROCEDURE

Effect of Input Voltage Variation on Regulator Output

1. Connect the circuit shown in Fig. 18-9. R_L is an 1800-Ω resistor. S is open. Set V_{AA} at O V as measured by voltmeter V. A milliammeter A_1 measures zener diode current I_z. Milliammeter A_2 measures total current I_T. An EVM measures the voltage V_{AB} across the zener (and across the load).

2. Close S. Increase the output of V_{AA} until I_Z = 5 mA. Measure and record in Table 18-1 V_{AB}, I_T, and V_{AA}.

3. Set I_Z in turn to each of the values shown in Table 18-1. Measure and record $V_{AB,}$ V_{AA}, and I_T.

TABLE 18-1. Zener Response to Variation in Input Voltage

Step	V_{AB}, V	I_Z, mA	I_T, mA	V_{AA}, V
2		5.0		
3		7.5		
		10.0		
		12.5		
		15.0		

Effect of Load Variation on Regulator Output

4. Reset V_{AA} so that I_Z = 5 mA. Measure and record in Table 18-2 I_T, V_{AB}, and V_{AA}. Do not vary V_{AA} in steps 5 and 6.

5. Open S. Replace R_L with a 2200-Ω resistor. Measure and record I_Z, V_{AB}, I_T, and V_{AA}.

6. Repeat step 5 for every value of load resistor shown in Table 18-2.

TABLE 18-2. Effect of Load Change on Zener Regulator Output

Step	R_L, Ω	I_Z, mA	I_T, mA	V_{AB}, V	V_{AA}, V
4	1,800	5.0			
5	2,200				
6	3,300				
	12,000				
	120,000				

7. Connect the circuit of Fig. 18-10. The primary of transformer T is connected to the output of a variable autotransformer set for 120 V, as shown. The secondary winding of T supplies power to the bridge rectifier D_1 through D_4. Capacitor C_1 filters the unregulated dc voltage developed by the rectifier. Capacitor C_2 acts as a bypass for the regulated dc output.

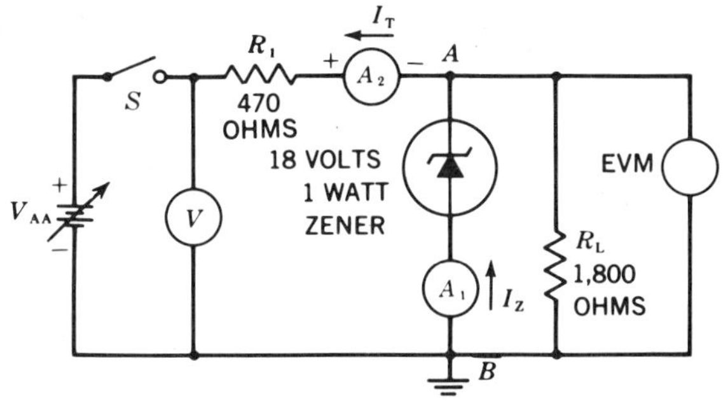

Fig. 18-9. Experimental zener diode shunt regulator circuit.

8. Close S_1. Power ON. S_2 is open. Set R_3 in the middle of its range. Measure and record in Table 18-3 the *no-load* unregulated voltage (point A to ground), the regulated voltage (point P to ground), the bias on Q_2 (point B to D), and the bias on Q_1 (point P to F).

TABLE 18-3. Series-Regulated Voltage Supply Measurements

Step		DC, V *No Load*	DC, V *Load*	*Load Current,* mA
8,9	*A* to ground (unregulated)			
	P to ground (regulated)			
	B to *D* (bias on Q_2)			
	F to *P* (bias on Q_1)			
10	*P* to ground, maximum output			
11	*P* to ground, minimum output			
12	Minimum ac voltage ________ V		Full load regulated dc output ________ V	
13	Maximum ac voltage ________ V		Full load regulated dc output ________ V	

9. Close S_2. The two 1250-Ω loads now draw current from the supply. Repeat the four voltage measurements in step 8. Also measure and record the load current.

10. Set R_3 for *maximum* dc output from the regulated supply. Measure and record in Table 18-2 the *no-load* regulated dc output and the *loaded* regulated dc output. Also measure and record load current.

11. Set R_3 for *minimum* dc from the regulated supply. Repeat the measurements in step 10 and record in Table 18-2.

12. Adjust the output of the variable autotransformer to the *lowest* ac voltage at which the regulated dc output voltage is maintained within 0.5 V of its value in step 11, under load. Measure and record this ac voltage. Open S_2. Observe the effect on load voltage. Close S_2.

13. Adjust the output of the variable autotransformer to the *highest* ac voltage at which the regulated dc output voltage is maintained within 0.5 V of its value in step 11, under load. Measure and record this ac voltage. Open S_2. Observe the effect on load voltage.

14. Power OFF.

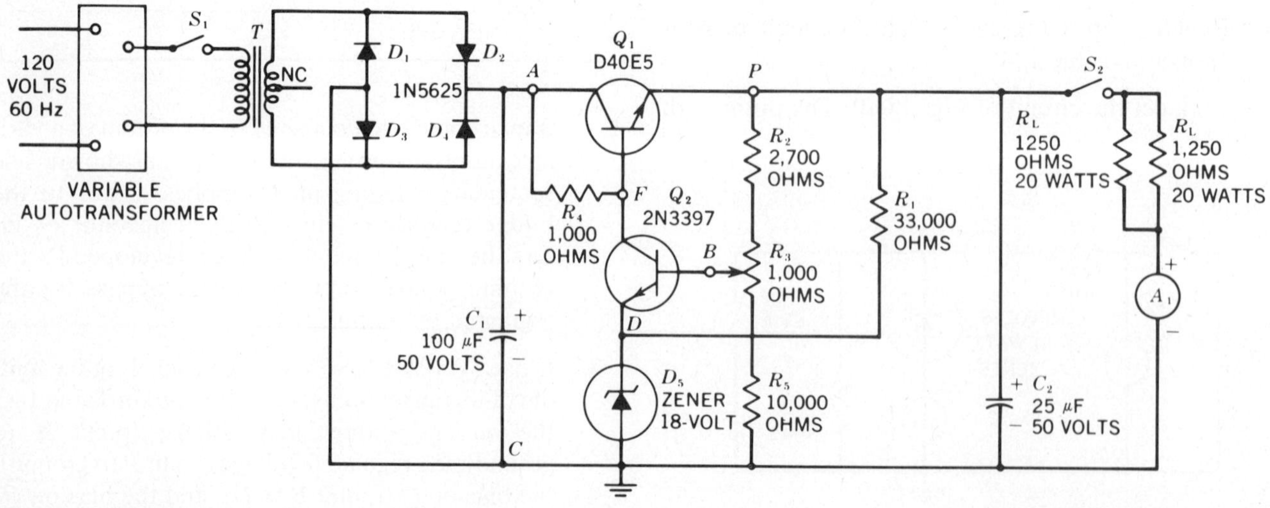

Fig. 18-10. Experimental voltage regulator.

Extra Credit—Optional

15. List the circuit change(s) that would be required to increase, safely, the range of variation of the regulated output voltage of the supply in Fig. 18-10.

16. Modify the experimental supply, following the proposed changes in step 15. Determine, experimentally, the minimum and maximum output voltage both with and without load.

QUESTIONS

NOTE: *Refer to your experimental data, whenever possible, for your answers to these questions.*

1. Explain the current-voltage characteristic of a zener diode.
2. What characteristic of the zener diode makes it particularly useful in voltage regulator circuits?
3. What is the relationship between I_Z, I_L, and I_T in Fig. 18-3?
4. In the regulated supply of Fig. 18-10, how can you determine the limits of load current variation over which the dc output holds within ± 2 percent of its no-load value?
5. In the regulated supply of Fig. 18-10, how can you determine the limits of ac input voltage variation over which the dc output holds within ± 2 percent of its value at 120 V ac?
6. What is the purpose of D_5 in Fig. 18-10?
7. What is the purpose of R_1 in Fig. 18-10?
8. What is the purpose of R_4 in Fig. 18-10?
9. What is the purpose of R_2 and R_3 in Fig. 18-10?
10. A 20-V zener diode is rated at 1 W. What is the maximum permissible zener current it should draw?
11. How can you determine the range of V_{AA} over which the regulator of Fig. 18-3 will hold the output voltage V_{AB} constant within ± 2 percent?

Answers to Self-Test

1. automatic
2. reverse
3. false
4. decrease
5. series
6. increase; increasing
7. more sensitive

EXPERIMENT

19

THREE-PHASE HALF-WAVE RECTIFIER

OBJECTIVES

1. To observe the phase relations between the voltage waveforms in a three-phase supply
2. To observe the load waveforms and their phase in a three-phase half-wave rectifier with resistive load

BASIC INFORMATION

Phase and Amplitude Relations in a Three-Phase System

Rectifiers convert alternating current to direct current. In basic electronics, we studied the operation of single-phase half-wave and full-wave rectifiers. Single-phase rectifiers are used in applications requiring small amounts of dc power, such as radios, audio amplifiers, and television receivers.

The dc power requirements of industry are usually much greater than that which can be economically supplied by a single-phase rectifier system. To meet this demand for large quantities of dc power, industry may employ rotary converters. But these impose severe problems of transmission of the dc power. A more desirable solution is supplied by the use of electronic polyphase rectifier systems.

Polyphase rectifiers produce less ripple voltage per equivalent filter section than do single-phase rectifiers. Moreover, even if the load does not require a filtered dc output, polyphase rectifiers are more efficient in that the dc output is smoother and hence contains less wasted ac power. Since polyphase rectifiers are normally required to supply large amounts of power, heavy-duty, solid-state rectifiers or gas- or vapor-filled tubes are used.

Let us briefly review the phase relationships in a three-phase system. In Fig. 19-1*a*, a four-wire system is used as a source of three-phase power. If an oscilloscope were connected from line 1 to common, the resulting sine wave would appear as *A* in Fig. 19-1*b*. This is our reference waveform. Waveform *B* from line 2 to common is delayed by 120° or one-third of a cycle. That is, there is a 120° phase lag between waveforms *A* and *B*. Similarly, waveform *C* from line 3 to common starts 120° after the start of waveform *B*. Figure 19-1*b* is a composite of the three waveforms, all plotted on a common time axis to show the phase relationships among them. These waveforms further show that the peak amplitude and the frequency of each of the voltages are the same.

It is interesting to note that single-phase power may be taken from any two of the lines in Fig. 19-1*a*. It should be noted, however, that the voltage between any two hot lines, say, 1 and 2, is greater than the voltage between any one line and common. Thus, in a balanced system, the voltage between any two lines, not including common, is equal to the voltage from any one line to common $\times \sqrt{3}$. For example, if voltages *A*, *B*, and *C* in Fig. 19-1*b* are each equal to 120 V, voltage *A* from line 1 to line 2 is $120 \times \sqrt{3} = 208$. Similarly, age *B* from line 2 to line 3 is 208 V, and *C* from line 3 to line 1 is 208 V. Moreover, there is a 120° phase difference between voltages *A*, *B*, and *C*.

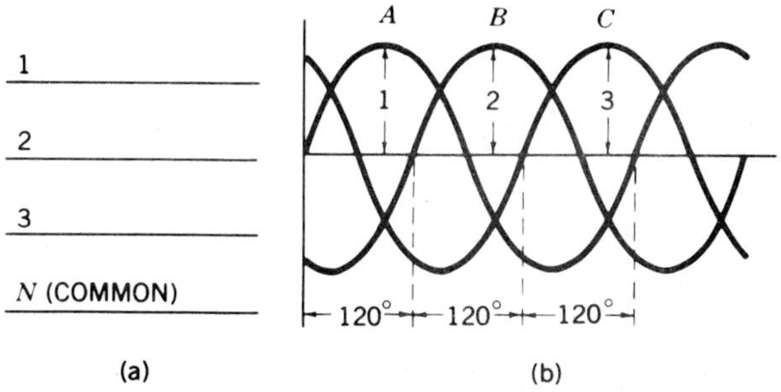

Fig. 19-1. Phase relations in a three-phase system.

Three-Phase Half-Wave Rectifier without a Transformer

Figure 19-2 is the circuit diagram of a transformerless three-phase half-wave rectifier supplying a resistive load R_L. When switches S_1, S_2, and S_3 are closed, lines 1, 2, and 3 supply the anode voltages of rectifiers D_1, D_2, and D_3, respectively. The cathodes of the three rectifiers are connected together. The load R_L is connected between the cathodes and the common return N of the three-phase supply.

The operation of the circuit may be demonstrated by considering the action of each individual rectifier. Assume in Fig. 19-2 that switches S_2 and S_3 are open and that switch S_1 is closed. Diode D_1 then operates as a single-phase half-wave rectifier, and the waveforms from anode to common and across the load are similar to those we observed in a previous experiment involving a single-phase half-wave rectifier. We know, of course, that D_1 conducts only during the half-cycle when its anode voltage is positive with respect to the cathode.

When switches S_1 and S_2 are both closed and S_3 is open, we have a circuit which schematically resembles a single-phase full-wave rectifier but is actually different. The difference arises from the fact that in a single-phase full-wave rectifier the voltages on the anodes of the two rectifiers are 180° out of phase, whereas in the circuit shown in Fig. 19-2 the voltages on the anodes of D_1 and D_2 are 120° out of phase. Figure 19-3*a* shows the voltages A and B, applied to the anodes of D_1 and D_2, respectively, superimposed on a common time base. We see further that there is a period in each cycle (t_1 and t_2) when the positive alternations of voltages A and B overlap, as indicated by the shaded portions of the graph. It might appear as though each of the rectifiers D_1 and D_2 conducts during the overlap interval. To facilitate our explanation of the circuit, we will make an approximation and an assumption: Only one rectifier conducts at any one time, and this is the diode whose positive anode voltage is highest during the interval. Thus the overlap interval of A and B may be divided into two equal periods t_1 and t_2. During the time t_1, voltage A is more positive than voltage B. During the interval t_2, voltage B is more positive than voltage A. Accordingly, D_1 conducts up to the end of interval t_1, when it is cut off. D_2 starts conducting at the end of interval t_1 and is cut off when its positive anode voltage drops to zero (approximately).

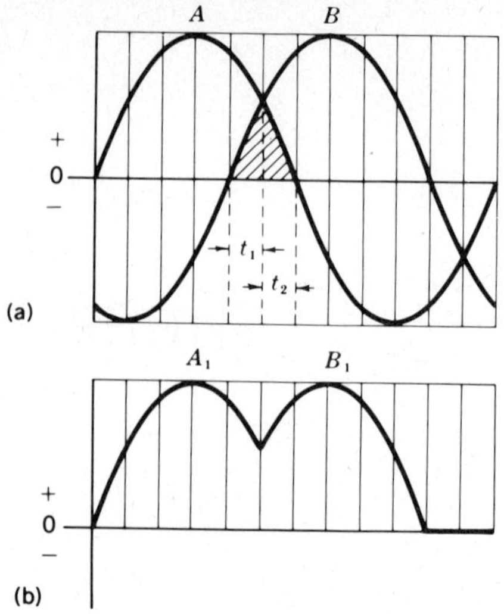

Fig. 19-3. Anode voltage and load-current waveforms in three-phase half-wave rectifier.

The reason for the phenomenon we have just described becomes apparent if we assume that a rectifier acts as a closed switch (short circuit) while it is conducting. From this assumption, it follows that the cathode and anode of a rectifier which is conducting are at the same potential. (Of course, we know that this is only an approximation to help us in an analysis of the circuit.) Now, in the interval t_1, D_1 is conducting. Hence A is the voltage at the anode and also at the cathode of D_1. A must therefore also be the voltage at the cathode of D_2, which is directly connected to the cathode of D_1. Now, the voltage B which is applied to the anode of D_2 is lower than the voltage A during the interval t_1. Hence D_2 cannot conduct during the time t_1 because its anode is negative with respect to its cathode during this interval.

Figure 19-3*b* shows the voltage waveform which would appear across the load resistance R_L. Current through the resistive load is in phase with the voltage, and hence the waveform in Fig. 19-3*b* is also the current waveform. We see that current through the load rises to a peak twice during a complete cycle of input voltage, and that these current peaks are 120° apart. Current through the load ceases for approximately 60° when the anodes of both rectifiers are negative with respect to their cathodes.

Now, if we close switch S_3, thus applying voltage C to the anode of D_3, the circuit of Fig. 19-2 becomes a three-phase half-wave rectifier. We can analyze the operation of this circuit in a manner similar to that of the preceding rectifier. Thus D_1, D_2, and D_3 will conduct

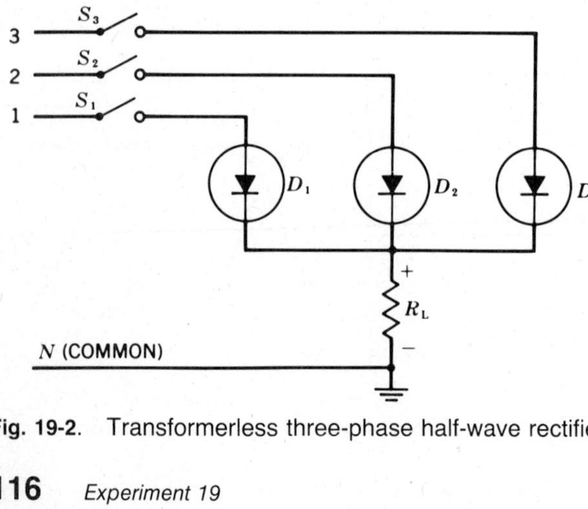

Fig. 19-2. Transformerless three-phase half-wave rectifier.

only when their anode voltages are positive with respect to their cathode voltages. Moreover, each of the three diodes will conduct in turn for one-third (120°) of a complete cycle.

Figure 19-4*a* shows the voltage waveforms that each of the rectifier anodes "sees," superimposed on a common time base. Thus D_1 receives voltage A, D_2 receives voltage B, and D_3 receives voltage C. As in our previous analysis, there are evidently periods during which the positive alternations of voltages A and B, B and C, C and A overlap. These overlap periods are again shaded in Fig. 19-4*a*. As we demonstrated previously, just one diode conducts during the overlap interval, and this is the rectifier whose positive anode voltage is highest during the interval. Thus, in the first overlap interval, voltage A is higher than voltage B during time t_1. Voltage B is higher than voltage A during time t_2. Accordingly, D_1 conducts up to the end of time t_1 and is then cut off. D_2 begins conducting at the start of interval t_2.

The envelope of the waveforms shown as A_1, B_1, and C_1 in Fig. 19-4*a* represents the voltage level during which rectifiers D_1, D_2, and D_3 are conducting, respectively. The current waveform across the resistive load is shown by Fig. 19-4*b*. Observe that current through the load never drops to zero. Hence the ripple excursions are smoother than they would be in the case of a single-phase half-wave or full-wave rectifier.

We can see that a three-phase half-wave rectifier supplies a voltage wave to a resistive load which never falls to zero. The output of a three-phase half-wave rectifier therefore has a smaller ripple factor than that of a single-phase rectifier and a higher average dc voltage. An arithmetical comparison will bring this fact out clearly. Thus the average dc voltage V_{av} across the resistive load of a single-phase full-wave rectifier is

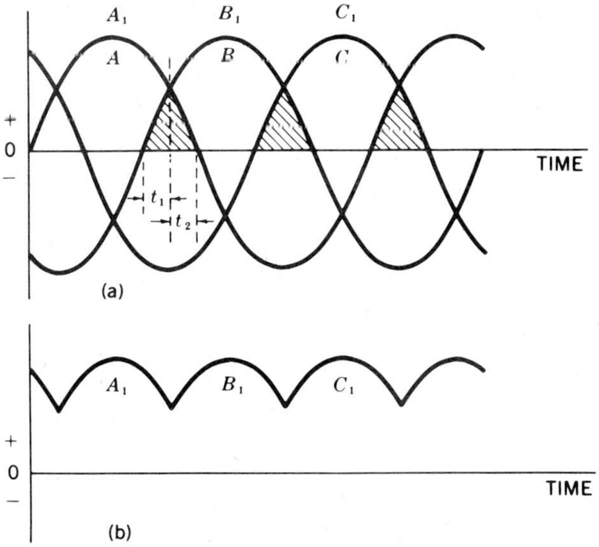

Fig. 19-4. Anode voltage and load-current waveforms in three-phase half-wave rectifier.

$$V_{av} = 0.636\ V_{max} \qquad (19\text{-}1)$$

where V_{max} is the peak of the ac input voltage, assuming zero voltage drop across the rectifier. The average dc voltage of a three-phase half-wave rectifier with resistive load is

$$V_{av} = 0.831\ V_{max} \qquad (19\text{-}2)$$

As the number of phases increases, the average dc output voltage and current increases, and the ripple decreases.

Three-Phase Half-Wave Rectifier with Transformer

A transformerless rectifier has the disadvantage that it is not line-isolated. A rectifier employing a transformer with isolated primary and secondary windings overcomes this disadvantage. Moreover, since the voltage across the secondary can be stepped either up or down, a greater range of dc output voltage is possible. The circuit shown in Fig. 19-5 illustrates a three-phase half-wave rectifier using a transformer. The transformer primary is connected in delta, the secondary in wye (star). Secondary windings Q_1, Q_2, and Q_3 relate respectively to primary windings P_1, P_2, and P_3. Lines 1, 2, and 3 apply a three-phase voltage to the primary. Assume that switches S_1, S_2, and S_3 are closed. Then voltage A appears across Q_1 and is applied to the anode of D_1. Voltages B and C appear across Q_2 and Q_3, respectively, and are applied, respectively, to the anodes of D_2 and D_3. The cathodes of D_1, D_2, and D_3 are connected together. The resistive load R_L, common to the three rectifiers, is connected from cathode to the common return.

Secondary winding voltages A, B, and C are 120° apart, as are the primary voltages. Let us assume that the respective transformer windings are related in a 1:1 ratio. Then, if 120 V appears across each of the primary windings, 120 V will also appear across each of the secondary windings. Operation of the circuit in Fig. 19-4 is then identical with that in Fig. 19-2. The

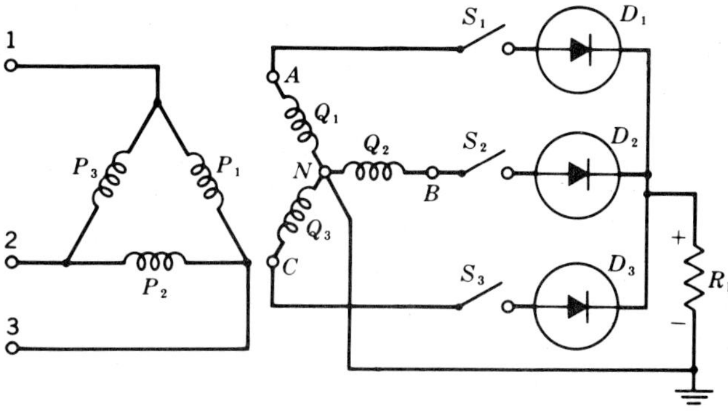

Fig. 19-5. Three-phase half-wave rectifier using transformer.

waveforms in Figs. 19-3 and 19-4 apply also to the circuit of Fig. 19-5. The only advantage we have gained in the latter circuit is that of isolation from the line.

SUMMARY

1. Electronic polyphase rectifiers meet industry's heavy, dc power requirements economically and efficiently.
2. Polyphase rectifiers produce less ripple voltage per equivalent filter section than do single-phase rectifiers.
3. Even though filtered dc power may not be required for a specific application, polyphase rectifiers provide smoother output than do single-phase rectifiers and therefore waste less ac power.
4. In a four-line three-phase system, there is a 120° phase difference between each hot line, relative to the common line.
5. Single-phase power may be taken from any two lines of a three-phase system. But the voltage between any two hot lines is greater than that between any hot line and common.
6. In a balanced three-phase half-wave rectifier, with common resistive load in the cathode, only one rectifier conducts at a time—that one whose anode-cathode positive voltage is highest. Each rectifier conducts for 120° of the cycle of applied voltage.
7. In a balanced three-phase half-wave rectifier with a common resistive load in the cathodes, the output current in the load never drops to zero.
8. The average dc voltage of a three-phase half-wave rectifier with a common resistive load in the cathode is

$$V_{av} = 0.831\ V_{max}$$

SELF-TEST

Check your understanding by answering these questions.

1. The phase difference between the peaks of waveforms A and C in Fig. 19-1 is __________ °.
2. Assume in Fig. 19-2 that switches S_1 and S_2 are closed, but S_3 open and that Fig. 19-3*a* represents the anode voltages on rectifiers D_1 and D_2, respectively. Which rectifier(s) is (are) conducting in the interval t_2? __________
3. Same assumption as in question 2. Rectifiers D_1 and D_2 constitute a single-phase full-wave rectifier and together conduct a total of 360°. __________ (true/false)
4. Switches S_1 through S_3 in Fig. 19-2 are closed, and the output voltage waveform across R_L of this balanced system is as shown in Fig. 19-4. If the peak of A_1 is 100 V, then the average voltage across R_L is __________ V, assuming the resistance of each rectifier when conducting is zero.
5. The primary of the three-phase transformer in Fig. 19-5 is connected in __________ and the secondary in __________ .
6. The secondary winding Q_2 in Fig. 19-5 is associated with primary winding __________ .

MATERIALS REQUIRED

- Power supply: Source of three-phase 6-Hz power; isolation transformer
- Equipment: Oscilloscope; VOM (20,000-Ω/V); 0- to 100-mA milliammeter
- Resistor: 20-W 2500-Ω
- Rectifiers: Three 1N5625 or equivalent
- Miscellaneous: Three SPST switches; three-prong to two-prong adapter plug

PROCEDURE

CAUTION: *Hazardous voltages. Observe safety precautions*

Phase Relations in a Three-Phase Supply

NOTE: *Since low power is required in this experiment, low-current rectifiers are employed. The principles of operation are the same with low- or high-current rectifiers.*

1. Connect the circuit shown in Fig. 19-6. Lines 1, 2, and 3 are the three hot lines of a four-line three-phase 60-Hz power source, each of which measures 120 V with respect to the fourth common line. A is a 0- to 100-mA milliammeter or the equivalent range on a 20,000-Ω/V VOM. Switches S_1, S_2, and S_3 are open. Three-phase power ON.
2. Isolate the oscilloscope from line ground, following the instructions in Experiment 1, Fig. 1-8. Turn scope ON. Center the trace of the oscilloscope. Calibrate the vertical dc amplifier of the oscilloscope for a deflection sensitivity of 150 V/cm. Use the vertical dc amplifier of the oscilloscope in the measurements that follow.

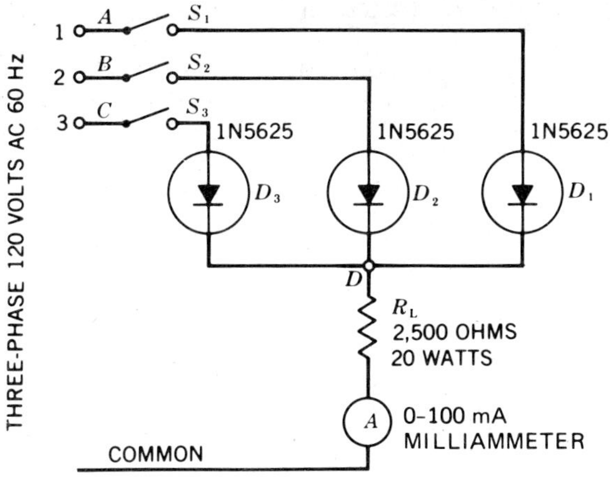

Fig. 19-6. Experimental three-phase half-wave rectifier.

3. Connect the vertical input lead of the oscilloscope to point *A* (line 1). The ground of the oscilloscope is connected to the common of the three-phase source.

4. Adjust the sweep and sync controls so that one sine wave appears on the oscilloscope with the phase and the polarity shown in Table 19-1, waveform 1. The positive alternation is first, followed by the negative alternation. Use line or external sync. External sync may be provided by connecting any line through a 0.001-μF capacitor to the jack marked "external sync."

If you cannot obtain the waveform with the phase shown in Table 19-1, connect the vertical input of the oscilloscope to *B* (line 2) or *C* (line 3). One of these lines will give the required wave. Designate the line which gives the required waveform as "line 1."

Designate as "line 2" the line whose voltage lags the voltage of line 1 by 120°. Designate as "line 3" the line whose voltage lags the voltage of line 2 by 120°. Do *not* vary sweep, horizontal centering, and sync controls for the remainder of this experiment.

Measure and record the p-p amplitude of waveform 1 in Table 19-1. NOTE: Record all waveforms and measurements in Table 19-1.

5. Observe waveforms *B* and *C* in turn, measured from line 2 and line 3 to ground. Draw these waveforms 2 and 3, in proper phase with reference waveform 1. Record p-p amplitude of each.

6. Connect the vertical lead of the oscilloscope to point *D*, and keep the ground lead connected to common. Close switch S_1.

7. Observe and record the waveform 4*A* across R_L and its p-p amplitude. Draw the waveform, in proper phase with relation to reference waveform 1. Measure and record the current in R_L. Measure the average dc voltage and record.

TABLE 19-1. Waveforms and Voltages in Three-Phase Half-Wave Rectifier

Waveform Number		*Waveform*		*Voltage,* V p-p	*Average DC Voltage,* V	*Load Current,* mA
1	+ 0 −					
2	+ 0 −					
3	+ 0 −					
4 *A,B,C*	+ 0 −		*A*			
			B			
			C			
5	+ 0 −					
6	+ 0 −					
	Amplitude of ripple voltage (p-p) =					

8. Open switch S_1 and close S_2. Observe, measure, and record waveform 4B across R_L and its p-p amplitude. Draw this waveform on the same time base as waveform 4A, in proper phase with relation to 4A. Measure and record the average dc voltage across R_L and the load current.

9. Open switch S_2 and close S_3. Repeat step 8 for this waveform 4C.

10. Close switches S_1 and S_2, keeping S_3 open. Now D_1 and D_2 are in the circuit.

11. Repeat step 7. This is waveform 5.

12. Close switch S_3, keeping S_1 and S_2 closed. Now all rectifiers are in the circuit.

13. Repeat step 7. This is waveform 6.

14. Measure the p-p amplitude of the ripple voltage and record.

15. Power OFF.

QUESTIONS

1. What is meant by a three-phase supply?

2. What are the phase relations between each of the hot lines and common in a four-wire three-phase supply?

3. What are the phase relations between the hot lines 1-2, 2-3, 3-1, in that order, using the voltage between lines 1-2 as reference?

4. Explain the operation of the circuit used in steps 6 and 7. Refer to the appropriate waveforms in Table 19-1.

5. Explain the phase of the waveforms in steps 8 and 9.

6. Explain the operation of the circuit used in steps 10 and 11. Refer to the appropriate waveforms in Table 19-1.

7. Explain the operation of the circuit used in step 13. Refer to the appropriate waveforms in Table 19-1.

8. What is the advantage of a three-phase half-wave rectifier over a single-phase half-wave rectifier? Refer to your experimental data.

9. What is the disadvantage of the transformerless rectifier (Fig. 19-6)?

Answers to Self-Test

1. 240
2. D_2
3. false
4. 83.1
5. delta; wye
6. P_2

EXPERIMENT

20

THREE-PHASE FULL-WAVE BRIDGE RECTIFIER

OBJECTIVE

To study the operation of a three-phase full-wave bridge rectifier

BASIC INFORMATION

Six-Phase Half-Wave Rectifier

In Experiment 19 we studied the operation of a three-phase half-wave rectifier frequently used in industry. The advantage of this type of rectifier over a single-phase circuit is that it can supply higher load currents with a lower ripple factor than can the single-phase rectifier.

The principles of operation of the three-phase half-wave rectifier can be extended to explain the six-phase half-wave rectifier in Fig. 20-1*a*. T is a three-phase transformer whose primary windings are P_1, P_2, and P_3. Q_1Q_4 is the center-tapped secondary of P_1; Q_2Q_5 is the center-tapped secondary of P_2; and Q_3Q_6 is the center-tapped secondary of P_3. N is the common to which the center tap of each of the secondary windings is connected.

The secondary windings of transformer T have been drawn to indicate the phase relationships of voltage which exist across each half-winding with respect to common. Thus we know that the sine waves of voltage appearing across Q_1, Q_2, and Q_3 are 120° out of phase with each other. Note also that the voltage across Q_1 is 180° out of phase with that across Q_4, the voltage across Q_2 is 180° out of phase with that across Q_5, and the voltage across Q_3 is 180° out of phase with that across Q_6. The graph in Fig. 20-1*b* is a composite of these voltages, plotted on a common time axis. From this graph, we see that the voltages across Q_1, Q_6, Q_2, Q_4, Q_3, and Q_5 are 60° out of phase with one another, in the order shown. If we consider the positive alternations of each waveform, we observe that there is a specific interval (30° on either side of peak) during which each voltage waveform is more positive than any other voltage wave. Thus, during the interval t_1, voltage Q_1 is more positive than any other voltage wave; during the period t_2, voltage Q_6 is more positive than any other voltage wave, and so on.

Refer again to Fig. 20-1*a* and note that the top and bottom of secondary winding Q_1Q_4 are connected respectively to the anodes of rectifiers D_1 and D_4. Similarly, Q_2Q_5 are connected to D_2 and D_5, whereas Q_3Q_6 are connected to D_3 and D_6. The cathodes of D_1 through D_6 are all tied together to form a common junction. The load resistance R_L is connected between the common cathode junction and the common center tap N of the secondary windings.

Operation of the six-phase half-wave rectifier is similar to that of the three-phase half-wave rectifier we studied in Experiment 19. Only one of the six rectifiers will conduct at any one time to supply current to the load. This is the diode whose positive anode voltage is highest during the interval. For example, D_1 will conduct during the interval t_1, when the voltage across Q_1 is most positive. D_6 will conduct during the interval t_2, and so on. Thus, in the six-phase half-wave rectifier we have just described, each rectifier conducts during 60° of the entire cycle and is cut off for 300°. When it is conducting, each diode supplies the full load current.

Figure 20-1*c* is the waveform which appears across R_L from current through the load resistance. We see that current through the load is smoother; that is, it has a lower ripple factor, than in the case of a three-phase half-wave rectifier. Moreover, the average dc voltage and the current supplied to the load is higher than in the case of a three-phase half-wave rectifier.

Three-Phase Full-Wave Bridge Rectifier

The rectifier shown in Fig. 20-2*a* is popular in industrial applications because it can operate directly from the three-phase power line without the need of a trans-

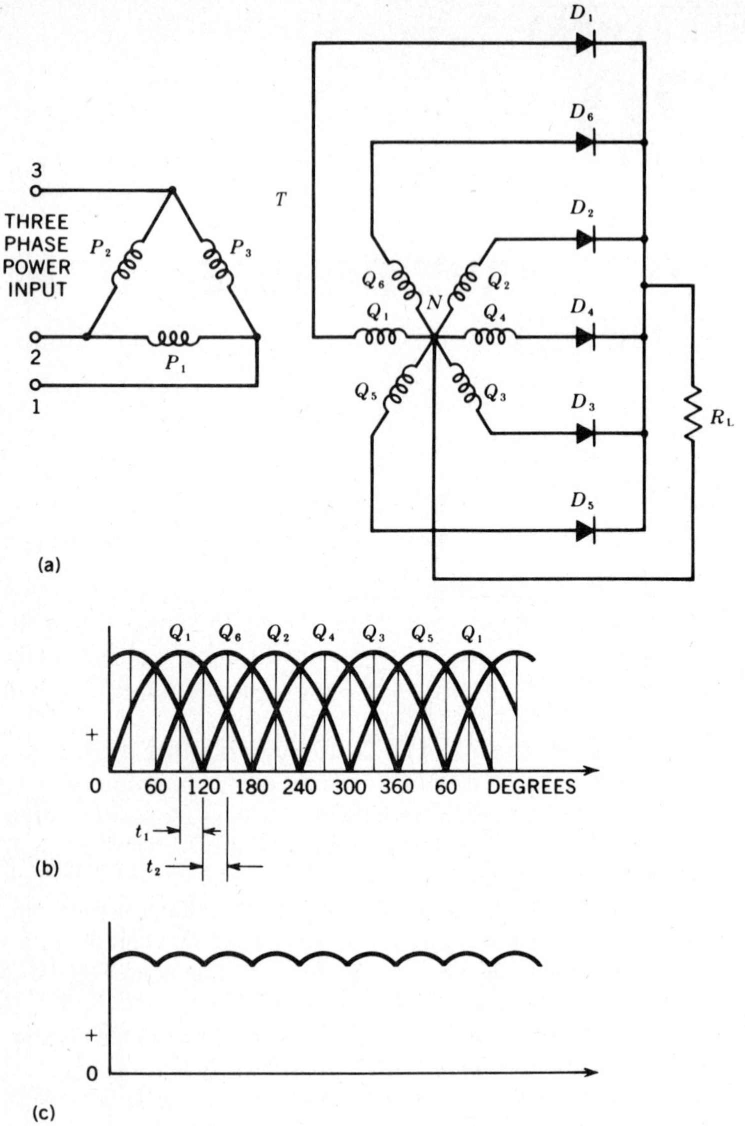

Fig. 20-1. *(a)* Six-phase half-wave rectifier; *(b)* anode-voltage waveforms; *(c)* load-current waveforms.

former. As in the preceding circuit, six rectifiers are employed. However, the circuit in Fig. 20-2*a* is that of a three-phase full-wave bridge rectifier whose operation is quite different from that of the six-diode three-phase half-wave rectifier.

Figure 20-2*a* shows that the cathodes of D_1, D_2, and D_3 are connected at a common point *C*. The anodes of D_4, D_5, and D_6 are tied together at point *A*. The load resistance R_L is connected between *A* and *C*. We note, moreover, that line 1 of the three-phase power source is brought to the junction of the anode of D_1 and the cathode D_4. Similarly, line 2 goes to the junction of the anode of D_2 and the cathode of D_5, and line 3 is connected to the junction of the anode of D_3 and the cathode of D_6. Only the three hot lines of the three-phase source are used. Common is not directly connected to the rectifier circuit.

The waveforms in Fig. 20-2*b* show the phase and amplitude relationships of lines 1, 2, and 3 with respect to common (0). We observe that the sine wave of line 2 lags that of line 1 by 120°, and the sine wave of line 3 lags that of line 2 by 120°. The horizontal axis, marked in 30° units, conveniently shows the phase relationships and also shows the period during which each waveform is the most positive of all three voltages and also the period during which each waveform is most negative. Thus waveform 1 (of line 1) is most positive during the interval $P_1P_2P_3$ and most negative during the interval $N_4N_5N_6$. Waveform 2 is most positive during the interval $P_3P_4P_5$, most negative during the interval $N_6N_7N_8$ (also N_1N_2 of its preceding cycle). Waveform 3 is most positive during the interval $P_5P_6P_7$, most negative during $N_8N_9N_{10}$ (also $N_2N_3N_4$ of its preceding cycle).

We have called the circuit in Fig. 20-2*a* a "bridge-type rectifier" because current flows through the diodes in a manner similar to current through the diodes of a single-phase bridge rectifier. That is, at any one instant of time, load current flows through two rectifiers connected in series. The circuit of Fig. 20-2*a* has

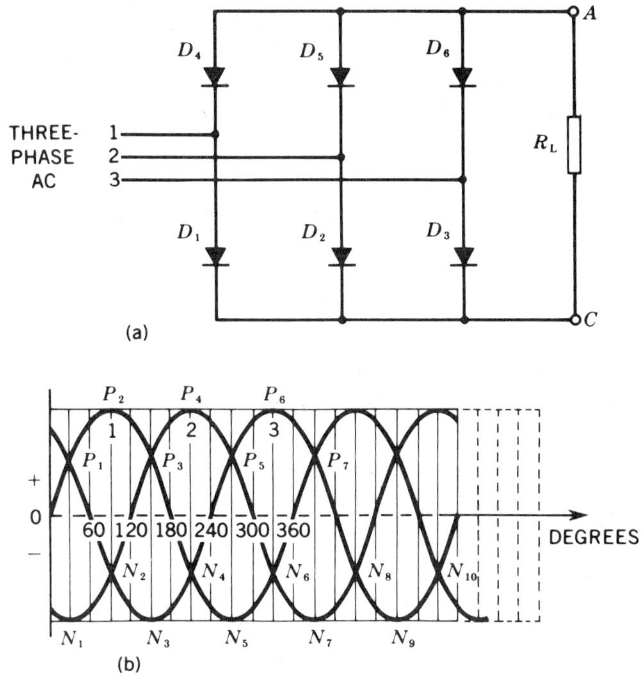

Fig. 20-2. Three-phase full-wave bridge rectifier and applied waveforms shown relative to common.

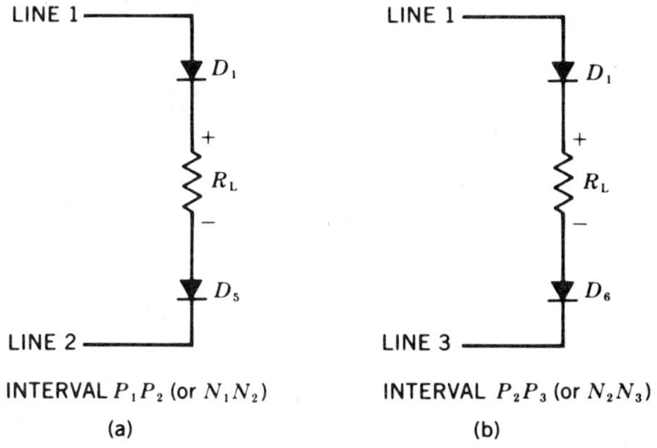

Fig. 20-3. Rectifier action during the intervals P_1P_2 and P_2P_3.

been so designed that at any instant of time the anode voltage of one of the diodes is more positive than that of the others. At the same time, the cathode voltage of another of the six diodes is more negative than any of the others. These two diodes will conduct at that instant of time.

Refer again to Fig. 20-2*a* and 20-2*b*. We note that, during the interval $P_1P_2P_3$, waveform 1 is most positive. Therefore, the anode of D_1 is more positive than the anodes of D_2 and D_3 during this interval. We observe further that during the interval N_1N_2, corresponding in time to P_1P_2, waveform 2 is most negative. Hence, the cathode of D_5 is more negative than the cathodes of D_4 and D_6 during the interval N_1N_2. During the interval N_2N_3, corresponding to P_2P_3, waveform 3 is most negative. Hence, the cathode of D_6 is most negative during the interval N_2N_3. These facts indicate that during the interval P_1P_2 or N_1N_2, diodes D_1 and D_5 in series permit current to flow through the load from lines 1 and 2 of the three-phase source. During the interval P_2P_3 or N_2N_3, D_1 and D_6 in series permit current to flow through the load from lines 1 and 3 of the three-phase source.

A simplified drawing showing circuit action during each of these two intervals is shown in Figs. 20-3*a* and 20-3*b*. From Fig. 20-3 it becomes clear that the voltage applied to the load during this interval is the potential between lines 1 and 2 and lines 1 and 3, in turn, assuming that the rectifiers are effectively short circuits when conducting. If we assume that the voltage from each line to common in the three-phase source is 120 V, then the voltage which the load receives is 208 V.

Similarly, it can be shown that, during the interval P_3P_4 or N_3N_4, D_2 in series with D_6 permits load current to flow from lines 2 and 3 as the source; during the interval P_4P_5, D_2 in series with D_4 permits current to flow through the load from lines 2 and 1 as a source; during the interval P_5P_6 or N_5N_6, D_3 in series with D_4 permits current to flow through the load from lines 3 and 1 as a source; and during the interval P_6P_7 or N_6N_7,D_3 in series with D_5 permits current to flow through the load from lines 3 and 2 as the source. We have now completed one cycle of operation. This sequence is repeated during the next 360°, and so forth.

Figure 20-4 is the waveform of load current in the three-phase full-wave bridge rectifier we have just described. The points P_1P_2, and so on correspond in time to the like-numbered points in Fig. 20-2*b*. At this time, we must note that each rectifier in the three-phase full-wave bridge circuit conducts for one-third of the full cycle, whereas each rectifier in the six-phase half-wave rectifier conducts for one-sixth of the full cycle.

Phase-Shift Control of a Polyphase Rectifier

As in the case of single-phase circuits, it is possible to replace the diode rectifiers used in a polyphase rectifier circuit with SCRs. Where the SCRs are used, we can delay the firing time of each of the rectifiers, thus controlling the output voltage and load current.

Figure 20-5 is the circuit diagram of a three-phase half-wave rectifier in which silicon controlled rectifiers are used. This circuit is similar to that in Fig. 19-5,

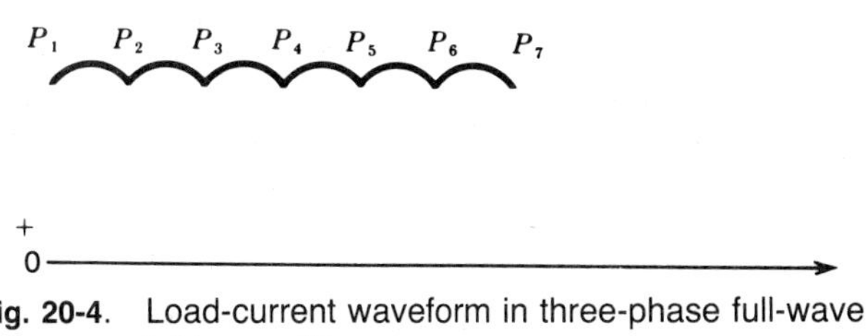

Fig. 20-4. Load-current waveform in three-phase full-wave bridge rectifier.

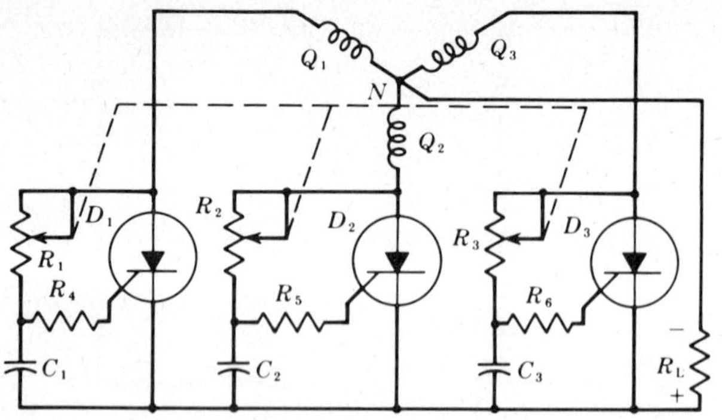

Fig. 20-5. Gate-controlled SCRs replacing diode rectifiers in three-phase half-wave circuit.

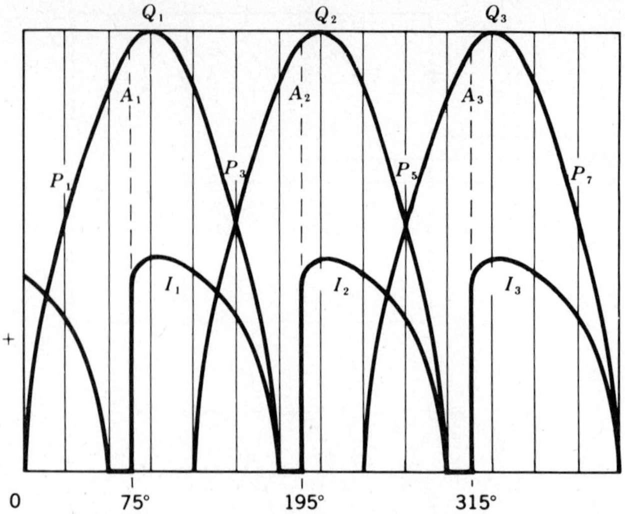

Fig. 20-6. Waveforms in three-phase half-wave silicon-controlled rectifier.

except for the substitution of SCRs for the diodes employed in Experiment 19.

In Fig. 20-5, R_1C_1, R_2C_2, and R_3C_3 constitute the phase-shift networks respectively for D_1, D_2, and D_3. These are identical circuits in which $R_1 = R_2 = R_3$ and $C_1 = C_2 = C_3$. Moreover, R_1, R_2, and R_3 are controls ganged on a single shaft, each of which varies in the same manner as the others as the control shaft is varied. Hence each of the three SCRs receives the same lagging phase-shift voltage, and its firing is delayed by the same time interval. The angle of lag is increased by increasing the resistance of R (i.e., R_1, R_2, and R_3) and is decreased by decreasing the resistance of R.

When $R_3 = 0$, the gate voltage is in phase with the anode voltage, and there is no delay in the firing of D_1, D_2, and D_3. Under this condition, the circuit of Fig. 20-5 is identical to that in Fig. 19-5.

Assume that R is increased to a value high enough to delay the firing of D_1, D_2, and D_3 by 75°. How would current through the SCRs be affected? The conduction characteristics of this circuit are somewhat different from that in Fig. 19-5, as is evident from the current waveforms in Fig. 20-6. The positive alternations of anode voltage on D_1, D_2, and D_3 are shown respectively as Q_1, Q_2, and Q_3. If no firing delay were introduced, D_1 would conduct during the interval P_1P_3, D_2 would conduct during the interval P_3P_5, and D_3 would conduct during the interval P_5P_7. Moreover, conduction would not be affected for any delay angle less than 30°. If a delay angle greater than 30°, say, 75°, were introduced, D_1 would fire at point A_1. Current waveform I_1 shows that D_1 continues to conduct past P_3 for the remainder of the positive alternation of Q_1. The reason D_2 would not take over at P_3 is that the phase-shift voltage applied to its gate by C_2R_2 would not permit it to conduct until its anode voltage reached the level A_2, that is, 75° after the start of its positive alternation. Current waveforms I_2 and I_3 show the conduction of D_2 and D_3 with a resistive load. Note that there is an interval of nonconduction between the firing of each SCR which is determined by the delay angle.

SUMMARY

1. A six-phase half-wave rectifier, such as that in Fig. 20-1, can supply higher load currents with a lower ripple factor (and a higher average dc voltage) than can a three-phase half-wave rectifier operating from the same power source.
2. A six-phase half-wave rectifier, as in Fig. 20-1, requires a three-phase transformer with three center-tapped secondaries connected at the common center tap.
3. In a six-phase half-wave rectifier, only one rectifier conducts at a time for 60° and is OFF during the remaining 300°.
4. A three-phase full-wave bridge rectifier (Fig. 20-2) also uses six rectifiers, as does Fig. 20-1, but now two rectifiers conduct together in series at any one time.
5. In a three-phase full-wave bridge rectifier, the load receives a higher voltage than it does in a six-phase half-wave rectifier, if an equal power source supplies them both. In the full-wave bridge rectifier the load receives a voltage equal to $\sqrt{3}$ times the voltage from any hot line to common.
6. In a three-phase full-wave bridge rectifier each rectifier conducts 120° and is off 240°.
7. The rectifiers used in a polyphase circuit can be either diodes or SCRs.
8. When SCRs are employed as rectifiers in polyphase circuits, the ON time of the rectifiers may be controlled by phase-shift circuits, thus controlling the power delivered to the load.

SELF-TEST

Check your understanding by answering these questions.

1. In the circuit in Fig. 20-1, rectifier __________ is ON during the interval 120° through 180° of the applied sine wave (Fig. 20-1*b*).

2. During the same time interval as in question 1, what other rectifier, if any, is ON?

   __________.

3. In the circuit shown in Fig. 20-1, if the voltage across each entire center-tapped secondary is 240 V, the load R_L receives approximately __________ V.

4. In the circuit shown in Fig. 20-2, if the voltage from each hot line to common is 120 V, the load R_L receives approximately __________ V, assuming the rectifiers have zero resistance when they are ON.

5. In the circuit shown in Fig. 20-2, during the time interval P_2P_3, rectifier D_6 in series with rectifier __________ sends current through the load R_L.

6. In the circuit shown in Fig. 20-2, during the time interval P_3P_4, rectifier D_6 in series with rectifier __________ sends current through the load R_L.

7. In the circuit shown in Fig. 20-2, if a 60-Hz three-phase power source is used, the ripple frequency of the load current is __________ Hz.

8. Equal firing delays for D_1, D_2, and D_3 in Fig. 20-5 are assured by equal time constants __________, __________, and __________.

MATERIALS REQUIRED

- Power supply: Source of three-phase 60-Hz power
- Equipment: Oscilloscope; 0- to 100-mA milliammeter or VOM (20,000-Ω/V); EVM; isolation transformer
- Resistor: 20-W 2500-Ω
- Rectifiers: Six 1N5625 or equivalent
- Miscellaneous: Six SPST switches or equivalent; three-prong to two-prong adapter

PROCEDURE

CAUTION: *Hazardous voltages. Observe safety precautions.*

1. Connect the circuit shown in Fig. 20-7. Lines 1, 2, and 3 are the hot lines of a three-phase supply, each of which measures 120 V with respect to common. D_1 through D_6 are 1N5625 silicon rectifiers or the equivalent. NOTE: *Low-power rectifiers are used here for economy reasons.* In an industrial application, high-current, high-wattage rectifiers would be employed.

   A is 0- to 100-mA milliammeter or the equivalent range on a 20,000-Ω/V VOM. Switches S_1 through S_6 are all open. Three-phase power ON.

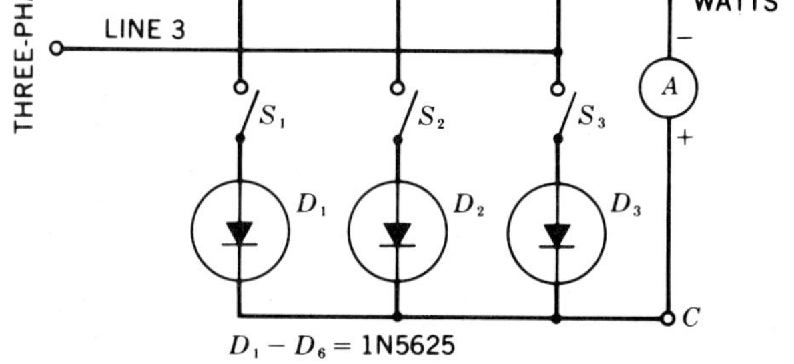

Fig. 20-7. Experimental three-phase full-wave bridge rectifier.

2. "Float" the oscilloscope; that is, isolate it from line ground, following the instructions in Experiment 1, Fig. 1-8. Turn scope ON. Center the trace of the oscilloscope. Calibrate the vertical dc amplifier of the oscilloscope for a deflection sensitivity of 200 V/cm. Use the vertical dc amplifier in the measurements that follow.

3. Connect the vertical input lead of the oscilloscope to line 1. The ground of the oscilloscope is connected to the common of the three-phase source.

4. Adjust the sweep and sync controls so that one sine wave appears on the oscilloscope with the phase and polarity shown in Table 20-1, waveform 1. The positive alternation is first, followed by the negative alternation. Use line or external sync.

TABLE 20-1. Measurement in Three-Phase Full-Wave Bridge Rectifier

Step	Waveform Number		Waveform	Voltage, V p-p	Load Current, mA	DC Voltage
4	1	+ 0 −	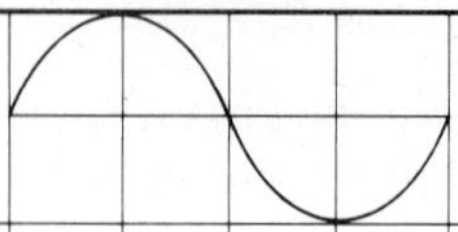		X	X
5	2 and 3	+ 0 −		2 3		
7	4	+ 0 −				
8	5	+ 0 −				
9	6	+ 0 −				
10	7	+ 0 −				
11	8	+ 0 −				
12	9	+ 0 −				
13	10	+ 0 −				

If you cannot obtain the waveform with the phase shown in Table 20-1, connect the vertical input of the oscilloscope to line 2 or line 3. One of these lines will give the required wave. Designate the line which gives the required waveform as "line 1." Designate as "line 2" the line whose voltage lags the voltage of line 1 by 120°. Designate as "line 3" the line whose voltage lags the voltage of line 2 by 120°.

Do not vary sweep centering and sync controls for the remainder of the experiment.

Measure and record in Table 20-1 the p-p amplitude of waveform 1. Record all waveforms and measurements in Table 20-1.

5. Observe waveforms 2 and 3 in turn, measured from line 2 and line 3 to ground. Draw these waveforms 2 and 3, in proper phase with reference waveform 1. Record p-p amplitude of each. Draw all future waveforms in proper phase with reference waveforms 1, 2, and 3.

6. Connect the vertical lead of the oscilloscope to point *C*, the ground lead to point *A*. Close switch S_1.

7. Record oscilloscope waveform if any, and load current, if any, as measured by *A*.

8. Keep S_1 closed and also close switch S_4; all other switches are open. Record oscilloscope waveform, if any, and load current, if any.

9. Open switch S_4 and close switch S_5. Only switches S_1 and S_5 are now closed. All others are open. Observe, measure, and record oscilloscope waveform, its p-p amplitude, and load current. Also measure dc voltage with an EVM and record.

10. Close switch S_6. Only switches S_1, S_5, and S_6 are now closed. All others are open. Observe, measure, and record oscilloscope waveform and its p-p amplitude, load current, and dc voltage.

11. Close switch S_2. Switches S_1, S_2, S_5, and S_6 are now closed. Switches S_3 and S_4 are open. Record

oscilloscope waveform and its p-p amplitude, load current, and dc voltage.

12. Close switch S_4. Switches S_1, S_2, S_4, S_5, and S_6 are now closed. Switch S_3 is open. Record oscilloscope waveform and its p-p amplitude, load current, and dc voltage.

13. Close switch S_3. All switches are now closed. Record oscilloscope waveform and its p-p amplitude, load current, and dc voltage.

QUESTIONS

Refer to the measured data in Table 20-1 to corroborate your answers to the following questions.

1. Which diodes, if any, drew current in step 7? Explain.
2. Which diodes, if any, drew current in step 8? Explain.
3. Explain waveform 6 in step 9. Which rectifiers conducted, and how long (in degrees) did each conduct?
4. Repeat question 3 for waveform 7, step 10.
5. Repeat question 3 for waveform 8, step 11.
6. Repeat question 3 for waveform 9, step 12.
7. Repeat question 3 for waveform 10, step 13.
8. How do load current and load voltage vary with the number of rectifiers in the circuit of Fig. 20-7? Why?

Answers to Self-Test

1. D_6
2. None
3. 120
4. 208
5. D_1
6. D_2
7. 360
8. R_1C_1, R_2C_2, R_3C_3

EXPERIMENT 21

ASTABLE MULTIVIBRATOR

OBJECTIVES

1. To study the operation of a free-running (astable) multivibrator
2. To learn how the frequency of oscillation is determined
3. To show how the multivibrator may be synchronized
4. To demonstrate experimentally the operation of a multivibrator

BASIC INFORMATION

Types of Multivibrators

Many electronic systems involve the use of nonsinusoidal waveforms: square waves, rectangular waves, pulses, saw-teeth, and so forth. One method of generating such waveforms is to use one of the three types of multivibrator (MVB):

1. *Astable MVB* Has no stable state, is "free-running," requires no trigger.
2. *Monostable MVB* Has one stable state, requires a trigger pulse.
3. *Bistable MVB* Has two stable states, trigger pulses cause it to alternate between the two states.

The astable, *RC*-coupled MVB is a form of relaxation oscillator which may free-run at a frequency determined by the values of the *RC* coupling components or maybe synchronized to an external source and therefore operate at a multiple of the external source frequency. In this experiment we will study the astable MVB. The other two types will be considered in later experiments.

An "oscillator" circuit generates a periodic waveform, such as a sine wave, square wave, or saw-tooth. If the "output" of an amplifier is fed back to its "input" with proper amplitude and phase, the amplifier will oscillate. A typical example of this is the well-known howling or ringing of a public-address system when the signal from the loudspeaker comes back to the microphone, is amplified further, and so on, until the familiar high-pitched squealing occurs.

Consider the simple two-stage common emitter amplifier circuit of Fig. 21-1*a*. In this circuit, arrows at the input and output show the direction of the signal at each point. Note that each stage provides 180° of phase inversion as well as amplification. Two stages provide 360° of phase shift, and, therefore, the assumed input and output have the same phase (direction). Thus, if some or all of the output is fed back to the input, this positive feedback may cause the circuit to generate sustained oscillations. To avoid such unwanted oscillations in a high-gain amplifier, the output is isolated from the input.

On the other hand, to produce oscillations in the astable MVB circuit, the output is deliberately connected to the input as shown by the dotted lines in Fig. 21-1*a*. The standard representation of this MVB is shown in Fig. 21-1*b*.

Operation of the Multivibrator

Let us assume that the circuit is "symmetrical," that is, Q_1 and Q_2 are the same type of NPN transistor, the corresponding base bias and collector load resistors have the same values, and C_1 and C_2 are also equal.

When S_1 is closed, it is most likely that one transistor (say, Q_1) will begin to conduct before the other, thus producing a large negative-going signal at its collector. This negative-going signal is coupled to the base of Q_2 by the capacitor C_1, tending to turn Q_2 OFF by driving its base negative. Since Q_2 is being turned OFF, a positive-going signal is generated at its collector. This

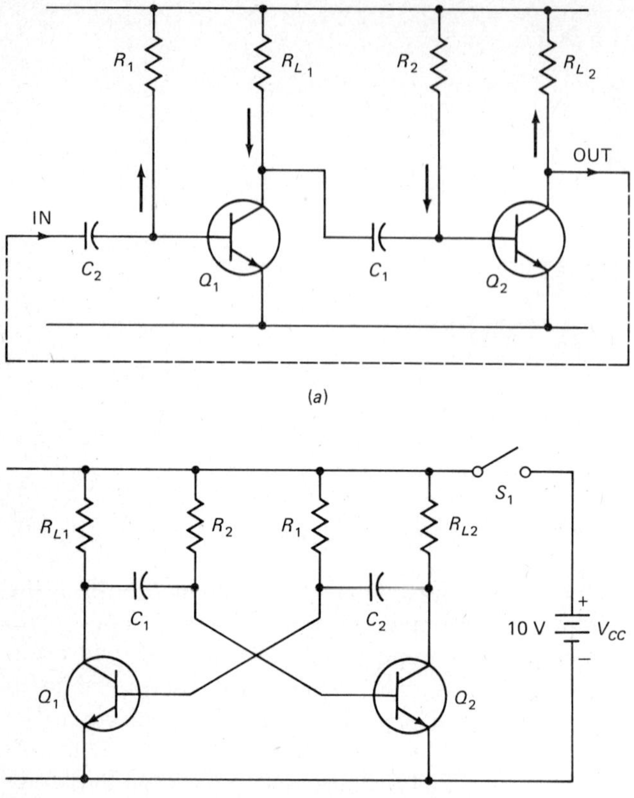

Fig. 21-1. *(a)* Two-stage common emitter amplifier; *(b)* collector-coupled astable MVB.

signal is coupled to the base of Q_1 and helps to turn it on hard (drive it into saturation). These actions take place very rapidly, and at this point, Q_1 is fully ON and Q_2 is fully OFF.

The time constant R_2C_1 (approximately) determines how long Q_1 remains ON and Q_2 remains OFF. Assume (1) that $V_{cc} = +10$ V and (2) that Q_1 saturates immediately after power is applied. V_{ce} of Q_1 is practically zero and its collector-to-emitter resistance is very small. Since the voltage across a capacitor cannot change instantaneously, the 10-V negative-going change at the collector of Q_1 drives the base of Q_2 negative with respect to ground, cutting it off. The simplified diagram of Fig. 21-2 shows these relationships.

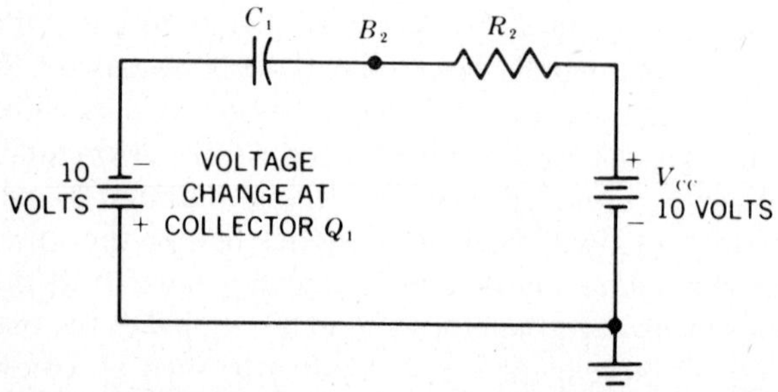

Fig. 21-2. Voltages that *C*1 "sees" at the instant that *Q*1 saturates.

Now capacitor C_1 starts to charge from -10 toward $+10$ V through R_2 as shown in Figs. 21-3 and 21-4. When the voltage at the base of Q_2 becomes slightly positive, Q_2 begins to conduct and quickly turns Q_1 OFF, thus beginning the second half-cycle of operation with R_1C_2 determining the time that Q_1 is OFF and Q_2 is ON.

When Q_1 is ON, as in Fig. 21-5, capacitor C_2 is being charged toward $+10$ V through the resistor R_{L2} of Q_2 and the base-emitter resistance of Q_1. The time constant $R_{L2}C_2$ is usually much shorter than the time constants R_1C_2 and R_2C_1 and accounts for the rounding of the leading edge of V_{c2} in Fig. 21-4.

Figure 21-4 shows that the waveforms at the base and collector of both transistors are the same but displaced in time by one half period. Such waveforms are sometimes said to be "complementary."

Frequency of Operation

The period t of a complete cycle of the square wave at the collector of Q_1 or Q_2 is the interval from A to C in Fig. 21-4; that is,

$$t = t_1 = t_2 \qquad (21\text{-}1)$$

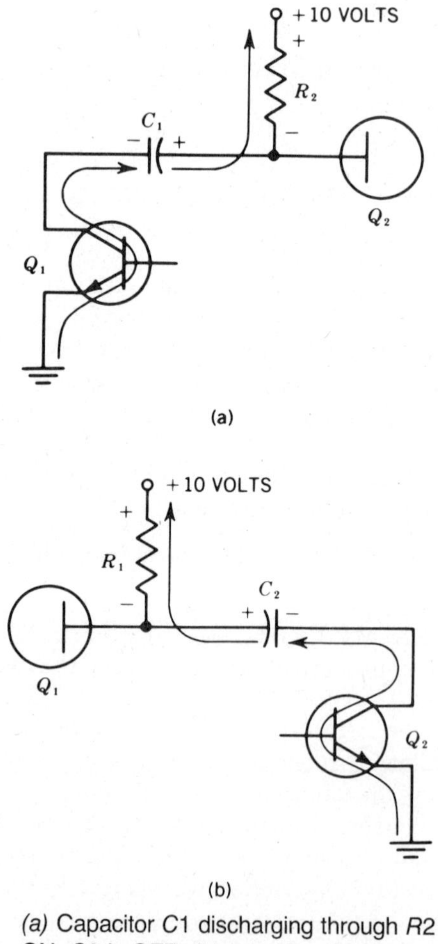

Fig. 21-3. *(a)* Capacitor *C*1 discharging through *R*2 × *Q*2 is ON; *Q*2 is OFF; *(b)* capacitor *C*2 discharging through *R*1 × *Q*2 is ON; *Q*1 is OFF.

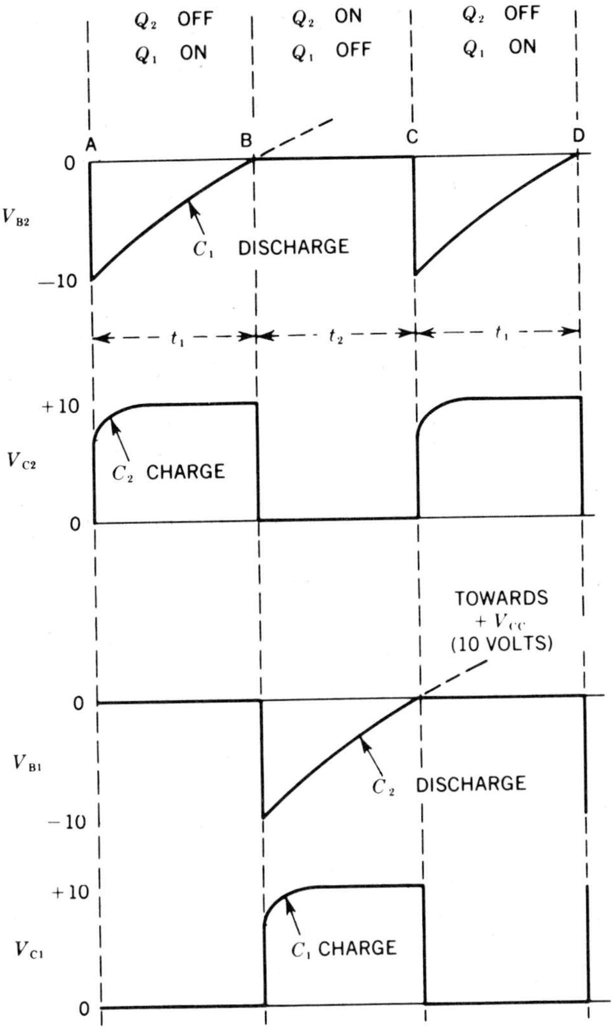

Fig. 21-4. Voltage waveforms in NPN collector-coupled MVB.

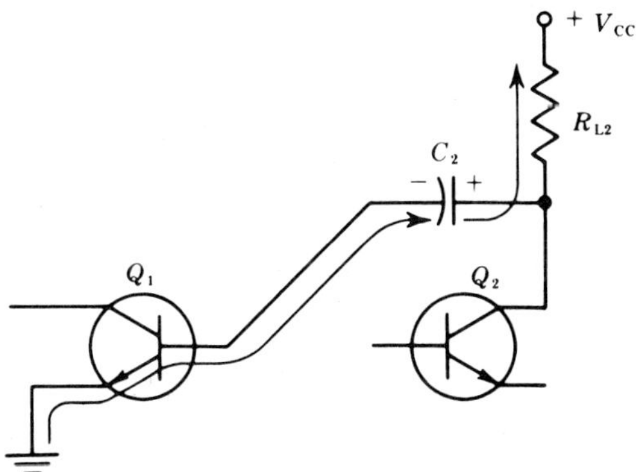

Fig. 21-5. Charge path for $C2 \times Q2$ is cut OFF, $Q1$ is ON.

The free-running frequency of the MVB is therefore:

$$f = \frac{1}{t} = \frac{1}{t_1 + t_2} \qquad (21\text{-}2)$$

In Fig. 21-4, the time required for C_1 to reach a slightly positive voltage is about $0.7R_2C_1$ and for C_2 it is $0.7R_1C_2$. If the MVB is symmetrical, its frequency of operation is approximately

$$f = \frac{1}{0.7\,(2 \times R_1C_2)}$$

or

$$f = \frac{1}{1.4R_bC_c} \qquad (21\text{-}3)$$

Example
Let a symmetrical MVB have the following values:

$$R_1 = R_2 = 500\ \text{k}\Omega$$

$$C_1 = C_2 = 0.1\ \mu\text{F}$$

Then,

$$t_1 = t_2 = 0.7(500\ \text{k}\Omega \times 0.1\ \mu F) = 0.035\ \text{s}$$

and

$$t = 0.070\ s$$

Therefore,

$$f = \frac{1}{t} = \frac{1}{0.07} = 14.29\ Hz$$

A nonsymmetrical MVB has different R_1C_2 and R_2C_1 time constants and therefore generates a "rectangular" waveform rather than a square wave (the ON and OFF periods at each collector are not the same). Equation 21-2 can be used as a guide to determine the approximate frequency of operation of a nonsymmetrical MVB.

Multivibrator Synchronization

If a negative-going pulse is applied to the base of Q_1 through a small capacitor at the correct time, it will be amplified by Q_1 and cause the positive-going waveform at the base of Q_2 to become positive sooner than it would otherwise (see Fig. 21-6). Therefore, the period when Q_2 is cut off becomes shorter and the frequency of operation becomes higher. In the absence of trigger pulses, Q_2 would have started to conduct at point Y. Therefore, if the natural free-running frequency of the MVB is slightly lower than the frequency of the synchronizing pulses the MVB oscillates at (exactly) the frequency of the synch pulses. This technique is used in television receivers to cause the vertical scanning of the picture in the receiver to correspond exactly with the vertical scanning of the picture at the transmitter. The TV vertical "hold" control permits adjustment of the vertical scanning frequency so that synchronization is achieved.

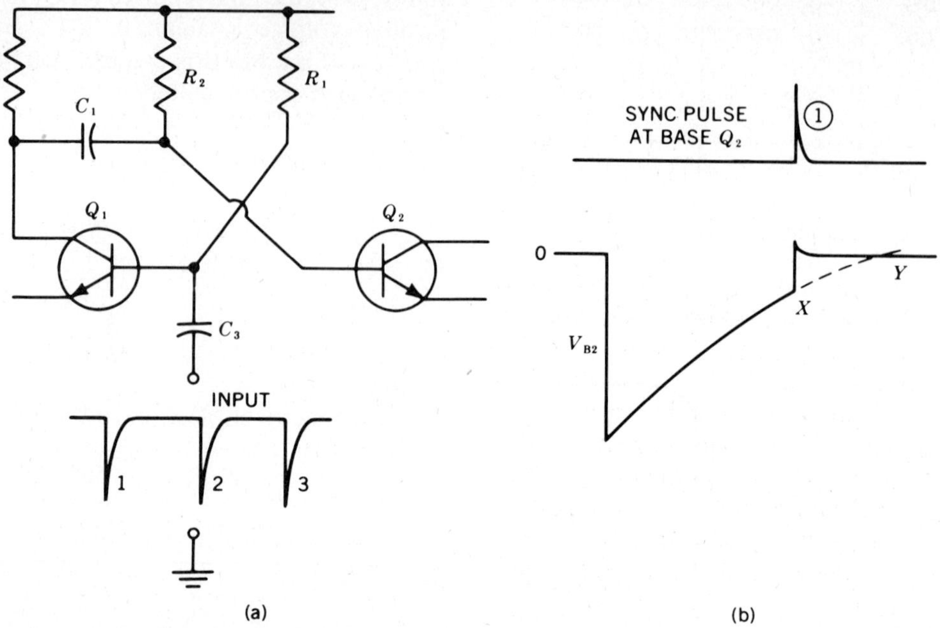

Fig. 21-6. *(a)* Synchronizing a MVB; *(b)* negative synchronizing pulse is amplified by *Q*1 and appears as a positive pulse at base of *Q*2, turning *Q*2 ON.

SUMMARY

1. MVBs, as a class, generate nonsinusoidal waveforms.
2. MVBs are either free-running or driven. A free-running MVB is self-excited and requires no external trigger pulse to operate (except for synchronization.) A driven MVB requires trigger pulses to operate.
3. A free-running, or self-excited, MVB, is called an astable MVB.
4. Positive feedback is an important part of the operation of an MVB.
5. A symmetrical MVB has equal values of corresponding parts and generates a square wave.
6. The timing components of an asymmetrical MVB are unequal and such a MVB generates a rectangular waveform.
7. The frequency of oscillation of a symmetrical MVB is approximately

$$f = \frac{1}{1.4R_bC_c}$$

where R_b is the base-bias resistance and C_c is the value of coupling capacitance.

8. The frequency of a MVB may be synchronized with an external source of pulses by applying them to the base (or collector) of one of the transistors.
9. For proper synchronization, the free-running frequency of the MVB should be lower than the frequency of the synch pulses.

SELF-TEST

Check your understanding by answering these questions.

1. MVBs generate sine waves. __________ (true/false)
2. An astable MVB requires no trigger pulse. __________ (true/false)
3. If corresponding parts of a MVB are the same, the MVB is a __________ MVB.
4. If a symmetrical MVB has the values R_b = 100 kΩ and C_c = 0.02 μF its frequency of oscillation is __________ Hz.
5. If R_b is increased, the frequency of oscillation __________.
6. If an MVB operates at a frequency determined by external pulses, it is said to be __________.

MATERIALS REQUIRED

- Power supply: Variable regulated low-voltage dc source
- Equipment: Oscilloscope (with dc vertical amplifiers, if possible); square-wave generator (AF range)
- Resistors: ½-W, two 6800-, 10,000-, 22,000-, 33,000-Ω, and two 100,000-Ω
- Capacitors: Two 0.01-μF, 47-pF, 250-pF
- Semiconductors: Two 2N6004 transitors; 1N4154 diode or their equivalents; two 2N6005 transistors as required for extra-credit experiment
- Miscellaneous: SPST switch; 2-W 500,000-Ω potentiometer.

PROCEDURE

Symmetrical Multivibrator

1. Connect the circuit shown in Fig. 21-7. Power ON.
2. *Externally* synchronize your oscilloscope with the signal from the collector of Q_1. Use the dc vertical amplifiers of your oscilloscope, if available, and set the zero level of the trace.
3. Observe the waveforms at each of the test points shown in Table 21-1. Draw these waveforms in proper time phase in the table. Measure also the p-p amplitude of each waveform. Record the data in Table 21-1. If you are using a dc (vertical amplifier) oscilloscope, show also the polarity of voltage with reference to zero. If you are using a dual-trace oscilloscope, observe and compare the waveforms at the collector and base of Q_1, at the collector of Q_1 and collector of Q_2, and at the base of Q_1 and collector of Q_2.
4. Compute and record the frequency of your MVB in column labeled "Multivibrator Frequency, Computed." Show your computations.
5. If the trace of your oscilloscope is time-calibrated, measure and record times t_1 and t_2 of a complete cycle. Using the value of $t = t_1 + t_2$, determine the frequency; record this frequency in column labeled "Multivibrator Frequency, Measured."
6. If the trace of your oscilloscope is not time-calibrated, measure and record the frequency of the multivibrator by using (for comparison) as a frequency standard a signal from a calibrated, variable-frequency, audio sine- or square-wave generator.

TABLE 21-1. Symmetrical Multivibrator Measurements

Test Point	*Waveform*				*Volts, p-p*
Collector of Q_1					
Base of Q_1					
Collector of Q_2					
Base of Q_2					
	Time Duration, s				
	t_1	t_2	t_1	t_2	X
	Multivibrator Frequency, Hz				
Computed			*Measured*		

Asymmetrical Multivibrator

7. Power OFF. Replace R_2, the Q_2 base-bias resistor, with the series combination of R_3, a 10,000-Ω resistor, and R_4, a 500,000-Ω potentiometer connected as a rheostat (see Fig. 21-8). Set R_4 for *minimum* resistance.
8. Power ON. Observe the waveforms at the test points shown in Table 21-2. Draw these waveforms in proper time phase in the table. Also measure and record the p-p amplitude of the waveforms. Measure and record the frequency of the MVB.
9. Set R_4 for maximum resistance. Repeat step 8. Use Table 21-3.

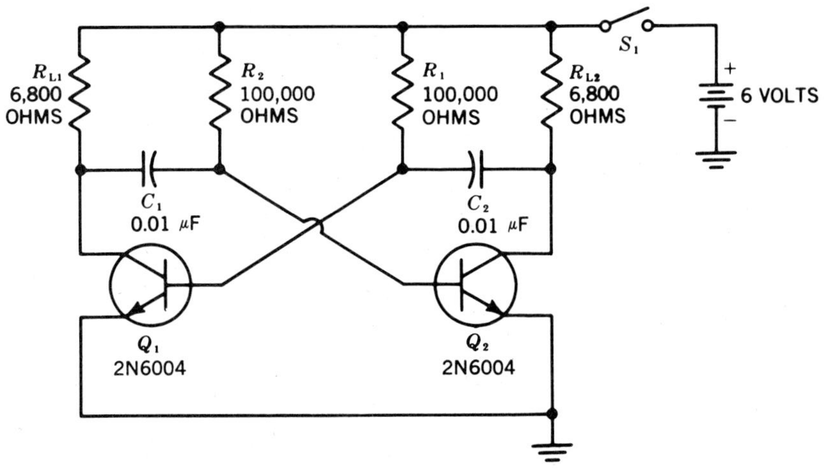

Fig. 21-7. Experimental MVB.

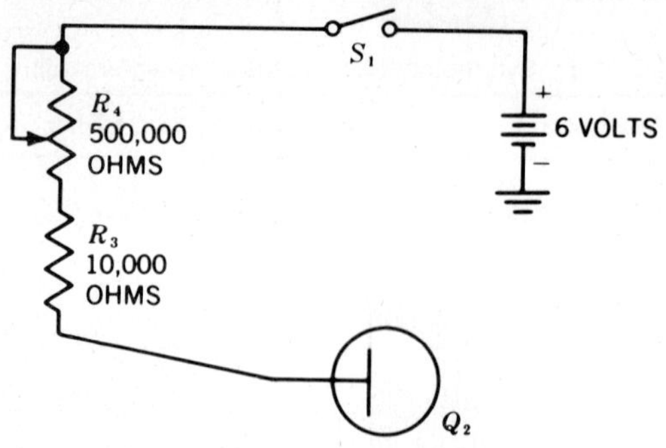

Fig. 21-8. *Q2 base-bias circuit for asymmetrical MVB.*

TABLE 21-2. R_4 Minimum Resistance

Test Point	*Waveform*	*Volts, p-p*
Collector of Q_1		
Base of Q_1		
Collector of Q_2		
Base of Q_2		
Frequency, Hz (measured):		

TABLE 21-3. R_4 Maximum Resistance

Test Point	*Waveform*	*Volts, p-p*
Collector of Q_1		
Base of Q_1		
Collector of Q_2		
Base of Q_2		
Frequency, Hz (measured):		

10. Reset R_4 for a symmetrical square wave. Measure and record its frequency: f = ____________ Hz. Leave R_4 at this setting.

Synchronizing the Multivibrator

11. Connect the output of a square-wave generator to the input of the circuit in Fig. 21-9. The output of

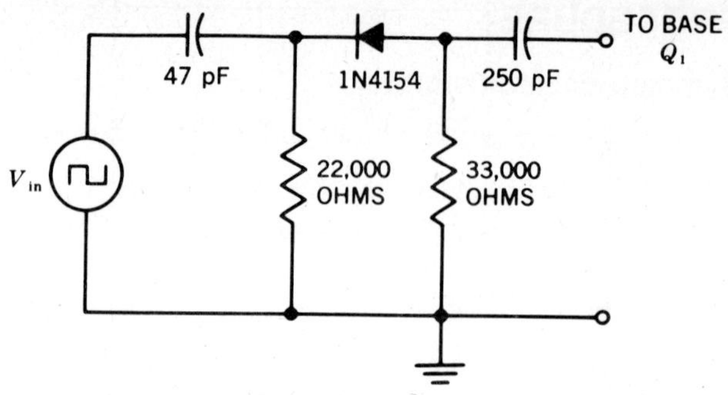

Fig. 21-9. MVB synchronizing circuit.

the synchronizing circuit is connected to the base of Q_1 as shown. Set the square-wave-generator frequency 200 Hz higher than the measured frequency f of the MVB (see step 10). Set the output of the square-wave generator at 6 V p-p.

12. Externally synchronize the oscilloscope with the output of the square-wave generator. Observe the waveform at the base of Q_1, at the collector of Q_1, and at the base of Q_2. Draw these waveforms in Table 21-4.

13. Measure and record the frequency of the MVB: ____________ Hz.

14. Slowly reduce the frequency of the square-wave generator to f. Observe and record effect on shape and frequency of waveform while monitoring the output at the collector of Q_1.

Extra Credit (Optional)

15. Experimentally determine and explain the operation and purpose of the components in the synchronizing circuit of Fig. 21-9. State your procedure.

16. (*a*) Experimentally determine what effect, if any, the level of the synchronizing pulse has on synchronization. Explain. (b) What is the minimum-level synchronizing pulse which will permit stable synchronization of the *symmetrical* MVB?

TABLE 21-4. Multivibrator Synchronization

Test Point	*Waveform (Synchronized at f + 200)*
Base of Q_1	
Collector of Q_1	
Base of Q_2	

17. Draw the circuit diagram of a symmetrical collector-coupled MVB using 2N6005 (PNP) transistors. Show polarity of battery voltage.

18. Connect the circuit. Observe the waveforms at the collector and the base of each transistor. In a specially constructed table, draw the waveforms in proper time phase.

QUESTIONS

1. Comment on the waveforms in Table 21-1. Do they agree with the theoretical waveforms in Fig. 21-4?
2. In the circuit in Fig. 21-7, what determines how long (*a*) Q_1 is cut off? How long is it off? (*b*) Q_1 is conducting? How long is it conducting?
3. How does the computed frequency compare with the measured frequency in Table 21-1? Explain any discrepancies.
4. What is the effect on frequency of the MVB as R_4 (Fig. 21-8) is (*a*) increased in value, (*b*) decreased? Why?
5. Is your answer to question 4 confirmed by the data in your experiment? Refer specifically to the data on which your answer is based.
6. Would there be any appreciable change in frequency if the voltage source of the MVB in Fig. 21-7 were reduced from 6 to 5 V? Why?
7. What is meant by synchronization of a MVB?
8. How is the frequency of a collector-coupled varied in this experiment?
9. What other method may be used for synchronizing the MVB in Fig. 21-8?
10. (*a*) Can PNP transistors be used instead of NPNs in designing a MVB? (*b*) What changes, if any, are required in the circuit?

Extra Credit

11. Will the level of the synchronizing pulse affect synchronization? How?
12. Is it possible to synchronize an MVB with a negative pulse whose frequency is *twice* the frequency of the MVB? Explain.

Answers to Self-Test

1. false
2. true
3. symmetrical
4. 357
5. decreases
6. synchronized

EXPERIMENT

22

UJT-CONTROLLED HYBRID BISTABLE MULTIVIBRATOR

OBJECTIVES

1. To study the operation of a UJT-triggered transistor flip-flop used to generate square waves
2. To observe how the frequency of this multivibrator may be controlled

BASIC INFORMATION

The hybrid bistable MVB in Fig. 22-1 delivers precise square waves which may be used in industrial applications for generating timing or control pulses. In this circuit, two NPN transistors are connected as a flip-flop which is triggered by positive pulses generated across R_6 by the UJT. The pulses delivered by the UJT determine the frequency of the MVB. MVB frequency is actually one-half the UJT pulse frequency. It should be noted that PNP transistors may be used in a similar configuration, but the polarity of the voltage source must then be reversed.

If the connection from the base of Q_3 to R_6 is broken, as in Fig. 22-2, that is, if the UJT triggering pulses are removed, Q_1 and Q_2 will *not* act as a free-running MVB. Instead, one of the transistors, say, Q_1, will be conducting while Q_2 will be cut off. The transistors will maintain this state until some *external* driving pulse, say, a positive pulse applied at R_6, reverses their conductivity, thus turning Q_2 ON and Q_1 OFF. The transistors maintain this state until another driving pulse turns Q_1 ON again and Q_2 OFF.

A bistable MVB, then, is characterized by the fact that it is not free-running but is stable in either of two

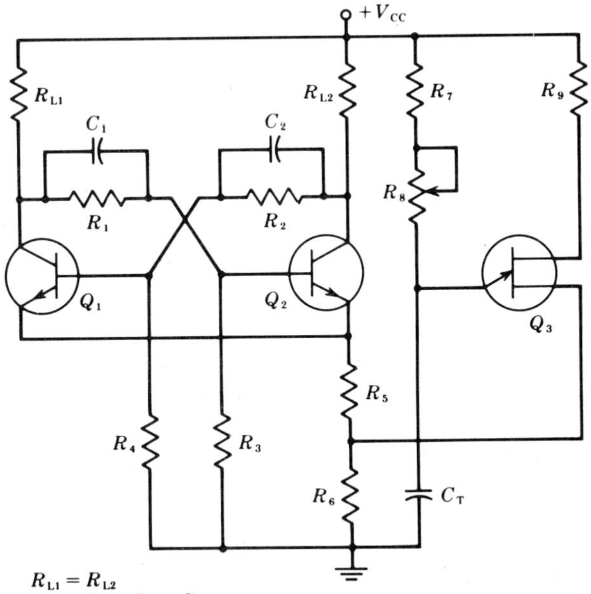

$R_{L1} = R_{L2}$
$R_1 = R_2 = R_3 = R_4$

Fig. 22-1. UJT-controlled bistable MVB.

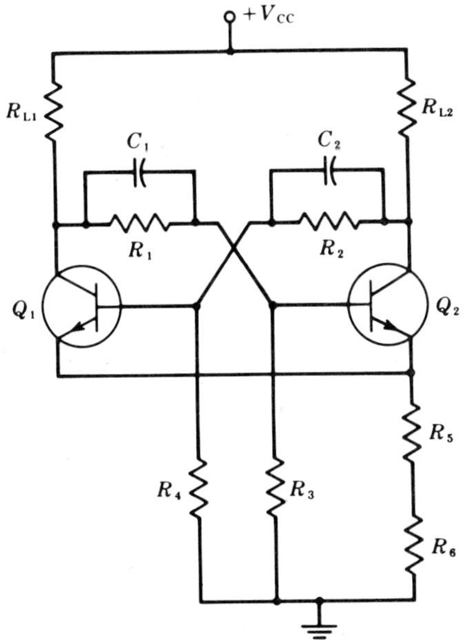

Fig. 22-2. Bistable MVB requires external driving pulses to reverse conduction of Q1 and Q2.

states, requiring an external driving pulse to reverse the states of conduction of the two transistors.

The circuit shown in Fig. 22-2 may be considered two direct cross-coupled dc amplifiers, having common emitter resistors R_5 and R_6. The circuit is so designed that either Q_1 or Q_2 is ON at any one time, but not both. When Q_1 is ON, Q_2 is OFF. When Q_2 is ON, Q_1 is OFF. Circuit design is such that the ON transistor is saturated when it is conducting. This is effected by heavily forward-biasing its base-emitter junction.

R_{L1}, R_1, and R_3 constitute a voltage divider which determines the voltage on the base of Q_2. R_{L2}, R_2, and R_4 form a second voltage divider, setting the voltage on the base of Q_1. The voltage on the emitters of Q_1 and Q_2 is developed by current flow in R_5 and R_6 and is the same for both transistors. Base-emitter bias of Q_1 is determined by the difference between the voltages on the base and the emitter of Q_1. Similarly, base-emitter bias of Q_2 is the difference between the voltages on the base and the emitter of Q_2. For conduction of either Q_1 or Q_2, its base must be positive (NPN transistor) relative to its emitter.

Assume in Fig. 22-2 that Q_1 is conducting heavily. Its collector voltage V_{C1} is therefore lower than V_{CC} because of the *IR* drop across R_{L1}. The voltage at the base of Q_2 is therefore $V_{C1}/2$ (approximately) because of the divider action of equal resistors R_1 and R_3. Circuit design is such that V_E (the voltage at the emitters of Q_1 and Q_2) is more positive than $V_{C1}/2$. Hence Q_2 is cut off. V_{C2}, the collector voltage of Q_2, would be at the supply voltage, except for the divider action of R_{L2}, R_2, and R_4. But the resistance of R_{L2} is here so low, compared with the resistance of R_2 and R_4, that $V_{C2} = V_{CC}$, approximately. Therefore, the voltage at the base of Q_1 is $V_{CC}/2$ (approximately). Again, circuit design is such that $V_{CC}/2 > V_E$, and Q_1 is biased to saturation. It is evident therefore that when Q_1 is turned ON, its collector voltage drops sufficiently to turn OFF Q_2, and the collector voltage at Q_2 rises sufficiently to *maintain* Q_1 at saturation.

Q_1 remains ON and Q_2 OFF until an external pulse reverses their conductivity. In the case of Fig. 22-1, this is a positive pulse applied across R_6. The pulse must be sufficiently large to drive the emitter of Q_2 more positive than its base. (Q_1 is unaffected because it is already saturated.) Q_2 starts conducting. Its collector voltage drops, driving the base of Q_1 *down*, decreasing collector current in Q_1. The collector voltage of Q_1 rises, and this positive change is dc-coupled to the base of Q_2, thus increasing current in Q_2 and driving the collector of Q_2 *down* further. This negative change at the collector of Q_2 again is dc-coupled as a negative change to the base of Q_1, thus decreasing current in Q_1. In short order, by direct collector-base feedback action just described, Q_2 turns ON, turning Q_1 OFF. The two transistors remain in this state until another positive trigger pulse across R_6 reverses their conductivity.

The conductivity of the two transistors may also be reversed by applying a negative (with respect to the emitter) pulse to the base of the transistor which is conducting.

The flipping speed of the transistors in Fig. 22-2 is limited by the action of the voltage dividers. A certain amount of shunt capacitance C_S exists between base and the ground, as in Fig. 22-3. The upper resistors in the voltage dividers and the shunt capacitances produce integrating circuits that resist rapid voltage changes at the bases. The effect of the shunt capacitances can be overcome by adding capacitors of the proper value across the upper resistors in the voltage dividers, as in Fig. 22-4.

The circuit of Fig. 22-1 is perfectly symmetrical in that Q_1 and Q_2 are similar transistors. Moreover, $R_{L1} = R_{L2}$, and $R_1 = R_2 = R_3 = R_4$. The circuit therefore produces a symmetrical square wave. External driving pulses are delivered by Q_3, when it fires. The student will recognize that unijunction transistor Q_3 and its associated circuit act as a free-running oscillator, whose frequency is determined by the time constant $C_7(R_7 + R_8)$. R_8 is made variable to change Q_3 frequency. R_8 therefore controls the frequency of the flip-flop.

Output square waves are taken either from the collector of Q_1 or Q_2 or from both. These waveforms are equal in amplitude, but 180° out of phase.

SUMMARY

1. A bistable MVB consists of two direct cross-coupled dc amplifiers.
2. A bistable MVB, also called a *flip-flop*, is *not* free-running, but requires an external triggering pulse to turn it on and off.
3. In the absence of any external triggering pulse, in the flip-flop in Fig. 22-2, either transistor may be

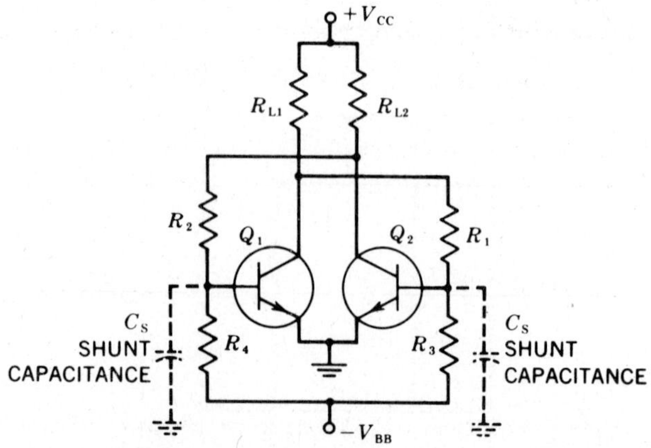

Fig. 22-3. Shunt capacitance limits transistor switching speed.

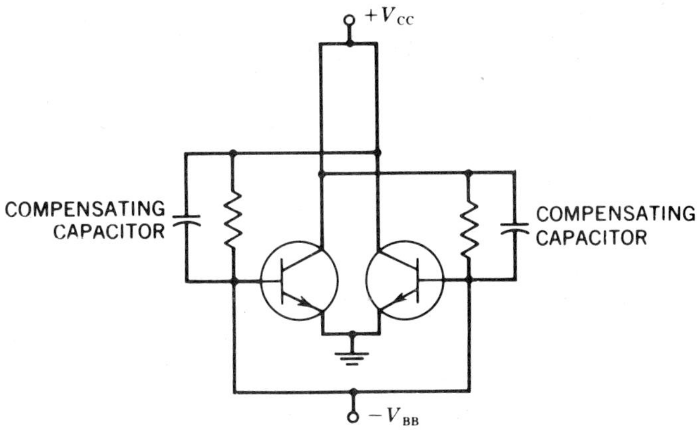

Fig. 22-4. Compensating capacitors increase switching speed.

ON while the other is OFF. This state continues until a driving pulse reverses the conduction of the two transistors.

4. In a flip-flop both transistors cannot be ON at the same time.
5. In the UJT-driven flip-flop in Fig. 22-2, both Q_1 and Q_2 share a common emitter resistor, actually two series-connected resistors R_5 and R_6.
6. Circuit design of the UJT-driven flip-flop is such that the voltage developed across the common emitter resistor by the transistor which is ON is more positive than the voltage on the base of the OFF transistor and hence continues to keep it cut off.
7. To reverse conduction of the two transistors, either a negative pulse of sufficient amplitude must be applied to the base of the ON transistor, or a positive pulse must be applied to the common emitter, or to the junction of the two series-connected emitter resistors (Fig. 22-2).
8. In this experiment, positive pulses from the UJT MVB (Q_3 in Fig. 22-1) will be used to flip the conduction of Q_1 and Q_2.
9. The bistable MVB in Fig. 22-1 will generate square waves whose frequency will be half the frequency of the triggering UJT relaxation oscillator.
10. The output square waves in Fig. 22-1 are taken either from the collector of Q_2 or Q_1.

SELF-TEST

Check your understanding by answering these questions.

1. Flip-flop is another name for a __________ .
2. In the circuit shown in Fig. 22-1, when Q_1 is ON, Q_2 must be __________ .
3. The voltage divider which determines the bias on Q_1 in Fig. 22-1 consists of __________ , __________ , and __________ .
4. The frequency at which Q_1 and Q_2 alternate conduction is equal to the frequency of Q_3 in Fig. 22-1. __________ (true/false)
5. To *continue* reversing the conduction of Q_1 and Q_2 in Fig. 22-2, a negative pulse may be applied to the base of Q_1. __________ (true/false)
6. The purpose of C_1 and C_2 in Fig. 22-2 is to offset the effects of the __________ __________ on the bases of Q_1 and Q_2.
7. In Fig. 22-2, when Q_1 is ON, the dc voltage at the collector of Q_2 is approximately __________ .
8. Flip-flops are __________ , not free-running MVBs.

MATERIALS REQUIRED

- Power supply: Variable regulated low-voltage dc source
- Equipment: Oscilloscope; EVM; VOM (20,000 Ω/V) square-wave generator (AF range)
- Resistors: ½-W 68-, 330-Ω, three 560-Ω, four 15,000-Ω 5 percent, 22,000-Ω
- Capacitors: Two 0.02-μF, 0.1-μF
- Semiconductors: Transistors: two D40E5 or equivalent; UJT transistor 2N2160
- Miscellaneous: Two SPST switches; 2-W 100,000-Ω potentiometer

PROCEDURE

1. Connect the circuit shown in Fig. 22-5. S_1 and S_2 are open. Set the supply voltage at + 20 V. Adjust R_8 to the middle of its range. Connect an EVM to the collector of Q_1 and a VOM to the collector of Q_2, both set to measure dc voltage.
2. Close S_1, power ON. S_2 is still OFF. With an oscilloscope, observe the output waveform, if any, at the collector of Q_1. Is the MVB operating?
3. With the voltmeters measure and record in Table 22-1 the voltages with respect to ground at the col-

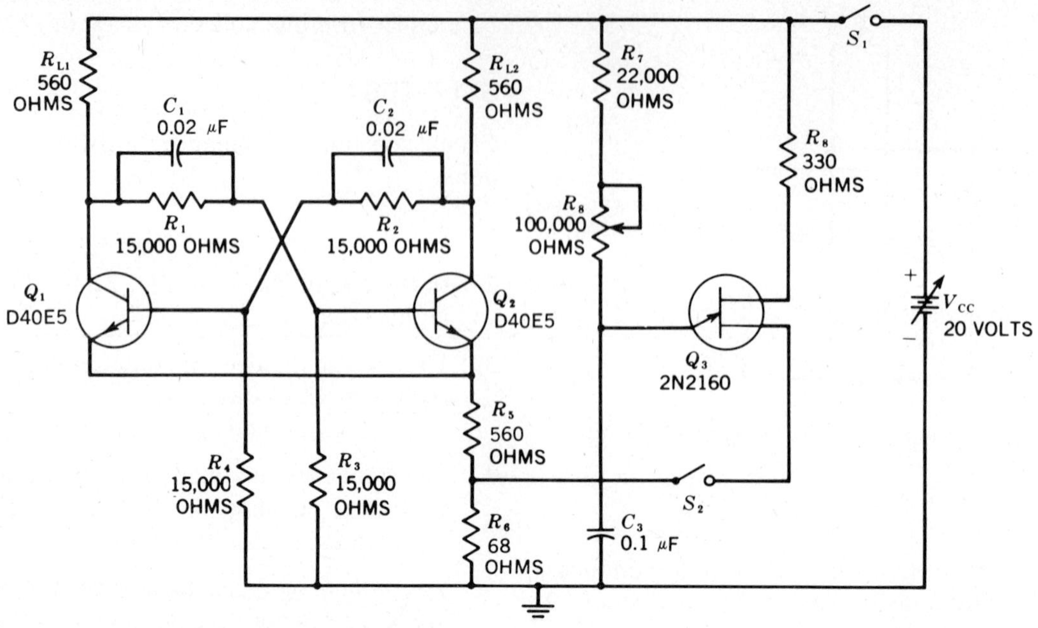

Fig. 22-5. Experimental hybrid square-wave generator.

TABLE 22-1. Voltages in Bistable Multivibrator

Step	*Transistor*	*Collector Voltage,* V	*Status of Q_1 and Q_2,* ON/OFF
3	Q_1		
	Q_2		
5	Q_1		
	Q_2		

lector V_C, of Q_1 and Q_2. Indicate in Table 22-1 whether Q_1 and Q_2 are ON or OFF.

4. How can you reverse the conductivity of Q_1 and Q_2 without closing S_2?

5. Using the method outlined in step 4, flip the conductivity of Q_1 and Q_2. Measure and record in Table 22-1 the voltages with respect to ground at the collector of Q_1 and Q_2. Indicate in Table 22-1 whether Q_1 and Q_2 are ON or OFF.

6. Restore the circuit to its original state and close S_2. Both switches are now closed. With an oscilloscope triggered or synchronized externally by the pulse from the junction of R_5 and R_6, observe and record in Table 22-2 the waveform measured from the emitter of Q_3 to ground. Measure and record also the p-p amplitude of the waveform. The oscilloscope sweep controls should be set so that there are at least four cycles on the screen.

7. Now, observe and record in proper time phase the waveforms and their amplitude at the junction of R_5 and R_6 and at the collector and the base respectively of Q_1 and Q_2.

8. Set R_8 for minimum resistance. Measure and record in Table 22-3 (p. 129) the frequency of the pulse derived from Q_3, and the frequency of the driven multivibrator. NOTE: If your oscilloscope has a calibrated time base, you may determine frequency in hertz by measuring the period t of each cycle in seconds and substituting the value of t in the formula $f = 1/t$. If the base of your oscilloscope is not time calibrated, you may measure frequency by the waveform comparison method, using the output of a square-wave generator as your frequency standard.

9. Set R_8 for the maximum resistance and repeat step 8.

QUESTIONS

1. How does a bistable MVB differ from an astable one?

2. In a nontriggered bistable MVB, how can you determine which transistor is ON? which is OFF? Refer specifically to your measurements in Table 22-1.

3. In a driven bistable MVB, how can you determine which transistor is ON and which is OFF at any instant of time? Refer specifically to your waveforms in Table 22-2.

4. What is the relationship between the frequency of the saw-tooth waveform developed across R_6.

TABLE 22-2. Waveforms in UJT-Controlled Multivibrator

Test Point	*Waveform Number*	*Waveform*	*Volts, p-p*
Emitter of Q_3	1		
Junction of R_5 and R_6	2		
Collector of Q_1	3		
Base of Q_1	4		
Collector of Q_2	5		
Base of Q_2	6		

TABLE 22-3. Pulse and Multivibrator Frequencies

	Frequency, Hz	
Step	*Triggering Pulse from* Q_3	*Multivibrator*
8		
9		

5. What is the relationship between the frequency of the pulse across R_6 and the frequency of the MVB? Why? Refer to the data in Table 22-3.

6. What procedure did you use for measuring the frequencies of the waveforms? Explain in detail.

Answers to Self-Test

1. bistable multivibrator
2. OFF
3. R_{L2}; R_2; R_4
4. true
5. false
6. shunt capacitance
7. $+V_{CC}$
8. driven (triggered)

EXPERIMENT

23

THE ONE-SHOT (MONOSTABLE) MULTIVIBRATOR AND THE SCHMITT TRIGGER

OBJECTIVES

1. To study the monostable MVB and experimentally demonstrate its characteristics
2. To study the Schmitt trigger circuit and experimentally demonstrate its characteristics

BASIC INFORMATION

One-Shot Multivibrator

Like the bistable MVB, the monostable MVB (one-shot MVB) is driven by external pulses. It has only one stable state and will remain in that state until triggered. The trigger pulse will cause it to change states, and it will remain in the new state for a period of time determined by an *RC* time constant. At the end of the timed period, it will return to the first state and remain in that state until another pulse is applied.

The one-shot MVB is used to generate timed intervals following an event, such as a narrow trigger pulse. It is used in many industrial applications as well as in computers, radar, TV, and other pulse and digital systems.

Figure 23-1 shows two transistors cross-connected as in other MVBs, but note that the coupling from Q_2 to Q_1 is direct (dc), while the coupling from Q_1 to Q_2 is through a capacitor (ac).

To analyze the operation of this circuit, assume that C_1 is removed and that there is no coupling from Q_1 to Q_2. Q_2 will therefore be ON because it is forward-biased through R_1 and R_{b2}. The collector-to-emitter voltage of Q_2 will be small and, although directly coupled to the base of Q_1, will be too small to turn Q_1 ON. Thus Q_1 is OFF. This is the stable state of the one-shot MVB.

Now let the capacitor be replaced and the MVB be in the stable state. If a positive trigger pulse is applied to the base of Q_1, Q_1 will conduct and its collector-to-emitter voltage will fall. This negative-going pulse drives the base of Q_2 negative and turns Q_2 OFF. The collector voltage of

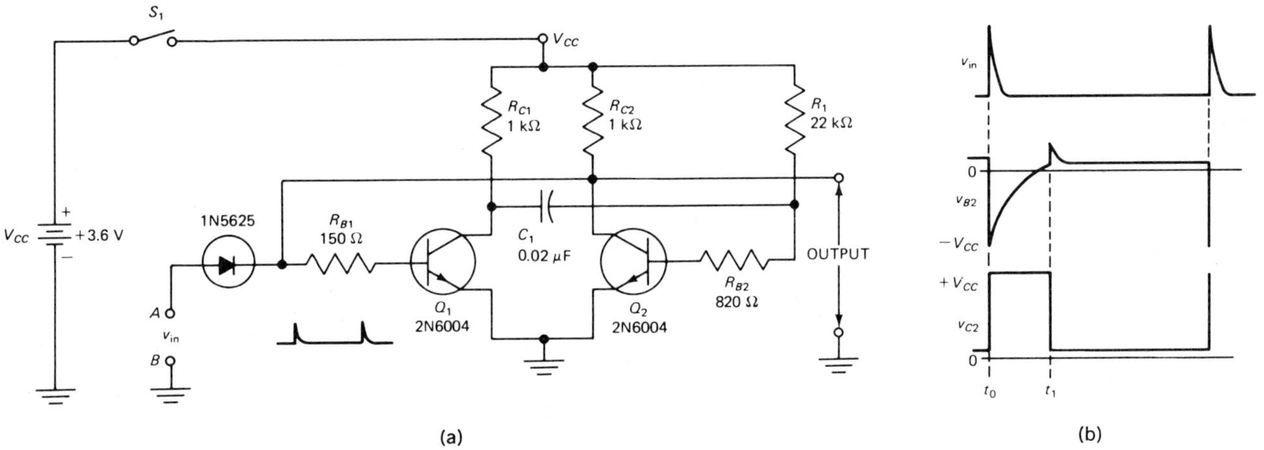

Fig. 23-1. *(a)* Monostable (one-shot) MVB; *(b)* waveforms.

Q_2 goes "high" and provides base current to Q_1, turning it ON.

In the stable state, capacitor C_1 charged through R_{c1}, R_{b2} and the base-emitter resistance of Q_2 to nearly V_{cc} with polarity as shown in Fig. 23-2*b*. When Q_1 is ON, however, C_1 charges toward V_{cc} through R_1 (from $-V_{cc}$) (see Fig. 23-2*a*.) This is the unstable condition of the one-shot.

As the capacitor C_1 charges toward V_{cc} through R_1, the base-emitter voltage of Q_2 (V_{b2}) reaches a slightly positive voltage and causes Q_2 to turn ON. When Q_2 is ON, Q_1 is OFF and the MVB is in its stable state again. The time period during which the MVB is in the unstable state is determined by the time constant R_1C_1.

In Fig. 23-1*a* note that the positive pulse at the collector of Q_2 corresponds with the time during which Q_2 is cut off due to the charging of C_1. Because of the positive feedback in this circuit, the rise and fall times of this pulse are very rapid.

The positive pulses required to trigger the one-shot may be obtained by differentiating a square wave by passing it through an *RC* coupling circuit having a short time constant (Fig. 23-3). The diode in Fig. 23-1*a* permits only the positive pulses produced by this circuit to affect Q_1.

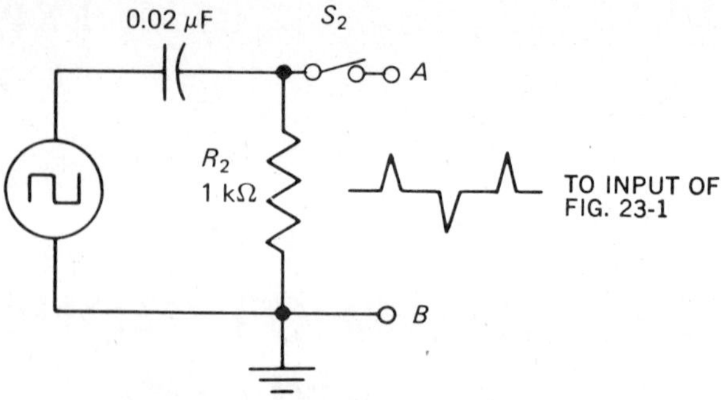

Fig. 23-3. The capacitor and resistor differentiate the square wave. Only the positive pulse triggers the one-shot MVB.

Schmitt Trigger

Figure 23-4*a* shows an emitter-coupled, positive feedback circuit called the *Schmitt trigger*. It has the property that its output is "high" if its input is above a certain voltage level, and its output is "low" if its input is below a certain level. It has many uses, but a common one is to take a slowly varying waveform, such as a sine wave, and generate from it a rectangular waveform with steep edges and flat tops. See Fig. 23-4*b*.

Like the flip-flop, the Schmitt trigger has two stable states, but these states are determined by input voltage levels, not by trigger pulses.

To analyze the operation of the circuit in Fig. 23-4*a*, first assume there is no ac input signal to the base of Q_1 and assume Q_1 is OFF. If Q_1 is OFF, its collector is "high" and this collector voltage is coupled to the base of Q_2 through the voltage divider R_2R_3. Q_2 is then turned ON, and a positive voltage is developed across the emitter resistor R_4 due to current flow through Q_2. Since the base of Q_1 is returned to ground, its base-emitter junction is reverse-biased by the voltage drop across R_4 and Q_1 is OFF, as was assumed.

Now let an input signal to the base of Q_1 drive it more positive than the drop across R_4 (point P_1 in Fig. 23-4*b*). Q_1 now begins to conduct because its base-emitter junction is forward-biased, its collector voltage drops to a value too small to keep Q_2 forward-biased through the $R_2 - R_3$ voltage divider, and Q_2 is OFF. As long as the input voltage to the base of Q_1 is sufficiently positive, Q_1 is ON and Q_2 is OFF.

When the input voltage falls below P_2 in Fig. 23-4*b*, Q_1 is turned OFF and Q_2 is turned ON. The slight difference between P_1 and P_2 is a form of hysteresis. The turn-on point on the positive-going part of the waveform (P_1) is slightly greater than the turn-off point (F_2) on the negative-going part of the waveform. Positive feedback through the emitter resistor causes the ON-OFF transitions to be very rapid.

Note that the width of the positive pulse in Fig. 23-4*b* is determined by the time that the input waveform exceeds P_1. If the input is too small, there is no output. With a large input voltage, the output is nearly a square wave and the circuit is called a *squaring circuit*.

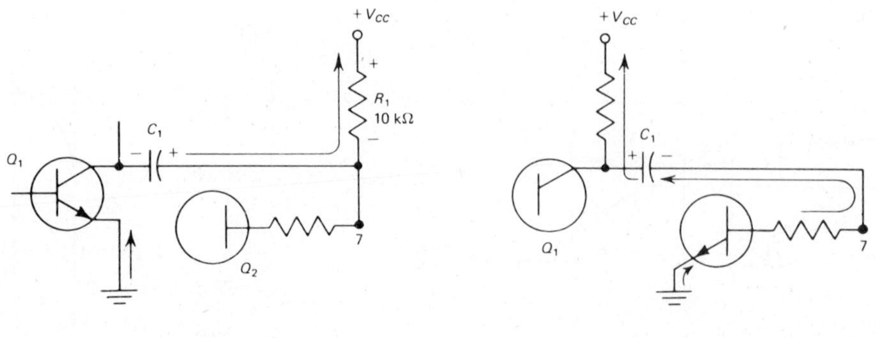

Fig. 23-2. *(a)* Discharge; *(b)* charge path for C1.

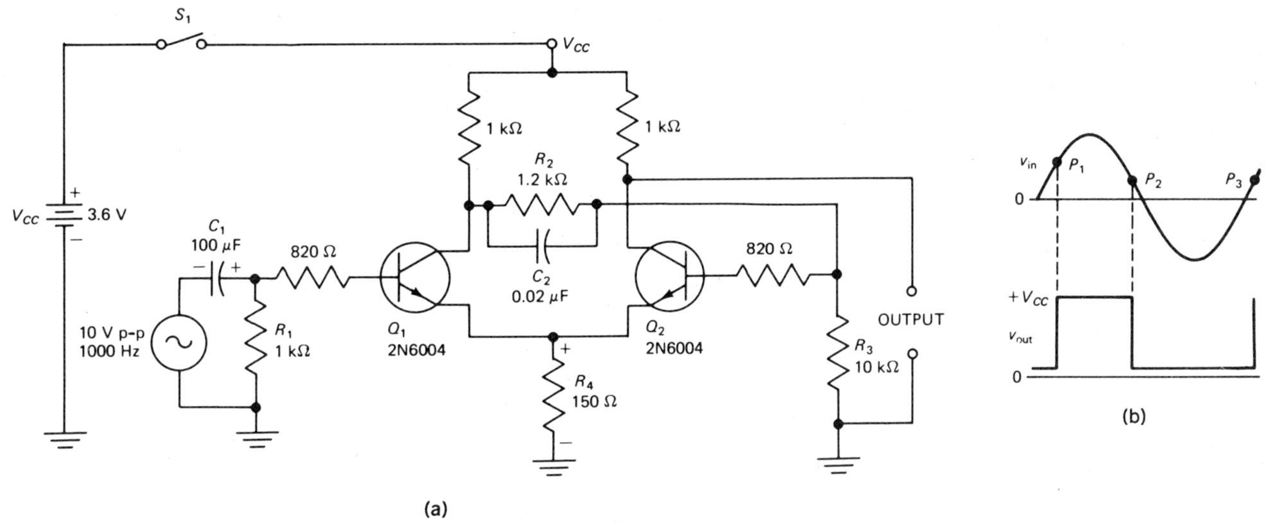

Fig. 23-4. *(a)* Schmitt trigger produces a square wave from a sine-wave input; *(b)* waveforms.

SUMMARY

1. A one-shot, or monostable MVB, has only one stable state and remains in this state until triggered.

2. After being triggered, the one-shot MVB goes into its unstable state and remains there for a time determined by an *RC* time constant.

3. At the end of the unstable state, the one-shot MVB returns to its stable state and will remain there until triggered again.

4. The one-shot produces an output pulse whose duration is determined by an *RC* time constant: the period may be from microseconds to seconds.

5. Only a very brief pulse is required to trigger the one-shot.

6. If the trailing edge of the output pulse from a one-shot is used to trigger another circuit, the one-shot may be called a "delay MVB" because the second circuit is "delayed" from the trigger pulse by the width of the one-shot's pulse.

7. The Schmitt trigger circuit is sensitive to the voltage level of the input: the output is "high" when the input exceeds a given level, and the output is "low" when the input is less than a given level.

8. The Schmitt trigger is useful as a squaring circuit. It will produce a rectangular waveform having steep rising and falling edges from a slowly varying waveform, such as a sine wave. Rectangular waveforms are ideal for triggering pulse and digital circuits. Sine waves, generally, are not.

SELF-TEST

Check your understanding by answering these questions.

1. Another name for the monostable MVB is the ____________ .

2. In a monostable MVB, one transistor is normally ____________ and the other is normally ____________ , in the absence of a trigger pulse.

3. In Fig. 23-1*a*, because Q_1 is normally OFF, it requires a ____________ trigger to cause a change of state.

4. In Fig. 23-1*a*, the components which determine the width of the output pulse are ____________ and ____________ .

5. The voltage level at its input determines the state of the output in a ____________ ____________ circuit.

6. The difference in the trigger points P_1 and P_2 in Fig. 23-4*b* is a form of ____________ .

MATERIALS REQUIRED

- Power supply: Variable regulated low-voltage dc
- Equipment: Oscilloscope; EVM; AF sine/square-wave generator (with external sync output jack, if possible)
- Resistors: ½-W 150-, two 820-, three 1000-, 1200-, 10,000-, 22,000-Ω
- Capacitors: Two 0.02-μF, 100-μF 50-V
- Semiconductors: Two 2N6004 transistors; 1N5625 diode; or equivalent
- Miscellaneous: Two SPST switches

PROCEDURE

One-Shot Multivibrator

1. Connect the circuits in Figs. 23-1*a* and 23-3, connecting *A* to *A* and *B* to *B*. S_1 and S_2 are open. Set V_{CC} at + 3.6 V. Set square-wave generator at 500 Hz, output at 5 V p-p.

2. Externally trigger/sync your oscilloscope with signal at the sync output jack on the generator, or with its square-wave output.

3. With your oscilloscope, observe the square wave at the output of the generator and set the oscilloscope controls for two waveforms, properly centered, as in Table 23-1.

4. Switch S_2 is still open. Observe the waveform across R_2, the 1000-Ω input resistor. In Table 23-1 draw this wave in time phase with the input square wave. Measure and record the p-p amplitude of the waveform.

5. *Close* S_1, applying power to the circuit. S_2 is still open. Observe the waveform, if any, at the collector of Q_2 and record in Table 23-1.

6. *Close* S_2. Both switches are now closed. Observe, measure, and record the waveform at the base of Q_2, in proper phase with the input.

7. Observe, measure, and record the output waveform at the collector of Q_2, in proper phase with the input. Measure and record the width of the positive pulse. NOTE: If there is a spike riding at the leading edge of the output pulse, gradually reduce the output of the generator until the spike disappears.

8. Power OFF. Substitute a 10,000-Ω resistor for R_1.

9. Power ON. Observe and measure the amplitude and pulse width of the output waveform at the collector of Q_2. Draw the waveform in proper phase with the input. Record its amplitude and pulse width. NOTE: If you do not have an oscilloscope with a calibrated time base, measure the pulse width in centimeters or inches.

10. Reduce the frequency of the square-wave generator to 350 Hz. Measure and record the width of the positive output pulse. Has the pulse width changed from that in step 9? What has changed?

Schmitt Trigger

11. Connect the circuit shown in Fig. 23-4*a*. V_{CC} is set at + 3.6 V. Close S_1, applying power to the circuit. Set the *sine-wave* generator frequency at 1000 Hz, output at 10 V p-p.

12. Externally trigger/sync the oscilloscope as in step 2.

TABLE 23-1. Monostable Multivibrator Characteristics

Step	*Test Point*	*Waveform*	*Volts, p-p*	*Pulse Width*
3	Generator output	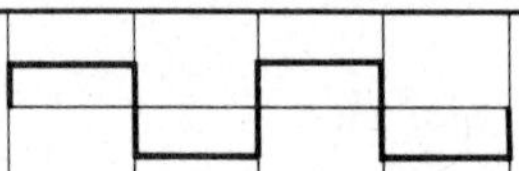	5	X
4	Across R_2			X
5	Collector of Q_2			X
6	Base of Q_2			X
7	Collector of Q_2			
9	Collector of Q_2			
10	Collector of Q_2			

13. With your oscilloscope observe the sine wave at the output of the generator and adjust the oscilloscope controls for two waveforms, properly centered as in Table 23-2.

14. Observe the output waveform (collector of Q_2). Measure and record in Table 23-2, in proper phase with the input, its p-p amplitude and the width of the positive pulse.

15. Reduce the sine wave generator output to 4 V p-p and repeat step 14.

TABLE 23-2. Characteristics of a Schmitt Trigger

Step	*Test Point*	*Waveform*	*Volts, p-p*	*Pulse Width*
13	Generator Output	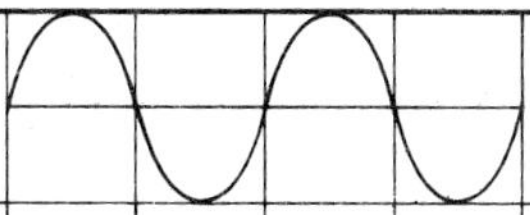	10	X
14	Collector, Q_2			
15	Collector, Q_2			

QUESTIONS

1. Why is the waveform across R_2, step 4, Table 23-1, no longer a square wave?

2. In which step of the procedure did you determine that the circuit of Fig. 23-1 is not a free-running MVB? Refer specifically to the data in Table 23-1 which confirm your answer.

3. In the one-shot MVB, which time constant yielded a wider pulse: 10,000 Ω × 0.02 μF or 22,000 Ω × 0.02 μF? Why?

4. What is the relationship between the waveform on the base of Q_2 and the width of the pulse at the collector of Q_2 in Fig. 23-1*a*? Why?

5. (*a*) In Fig. 23-1*a*, what turns Q_2 OFF? (*b*) How come it does not remain OFF all the time?

6. In the Schmitt trigger, what effect does the amplitude of the input sine wave have on the width of the positive alternation at the collector of Q_2 in Fig. 23-4*a*? Why?

7. What effect does the frequency of the input sine wave have on the width of the positive alternation at the collector of Q_2 (Fig. 23-4*a*)? Why?

Answers to Self-Test

1. one-shot MVB
2. ON-OFF, OFF-ON
3. positive
4. R_1, C_1
5. Schmitt trigger
6. hysteresis

EXPERIMENT

24

DIGITAL INTEGRATED CIRCUITS: THE AND AND OR GATES

OBJECTIVES

1. To learn how integrated circuits are manufactured
2. To introduce the RTL, DTL, TTL, and CMOS logic families
3. To demonstrate the operation of digital AND and OR circuits

BASIC INFORMATION

Integrated circuits (ICs), the microelectronic components which make possible the scientific calculator, microcomputer, and many other modern electronic devices, are also used extensively in industrial control systems: in numerically controlled machines, robotics, and automation of all types, to name a few applications.

ICs have replaced discrete components (resistors, diodes, transistors, etc.) in many situations. A good example is the Motorola MC1533 operational amplifier (op amp). Rather than design an op amp using discrete (individual) components, today the engineer would probably select an IC op amp. The "chip" shown in Fig. 24-1 includes all the resistors, capacitors, diodes,

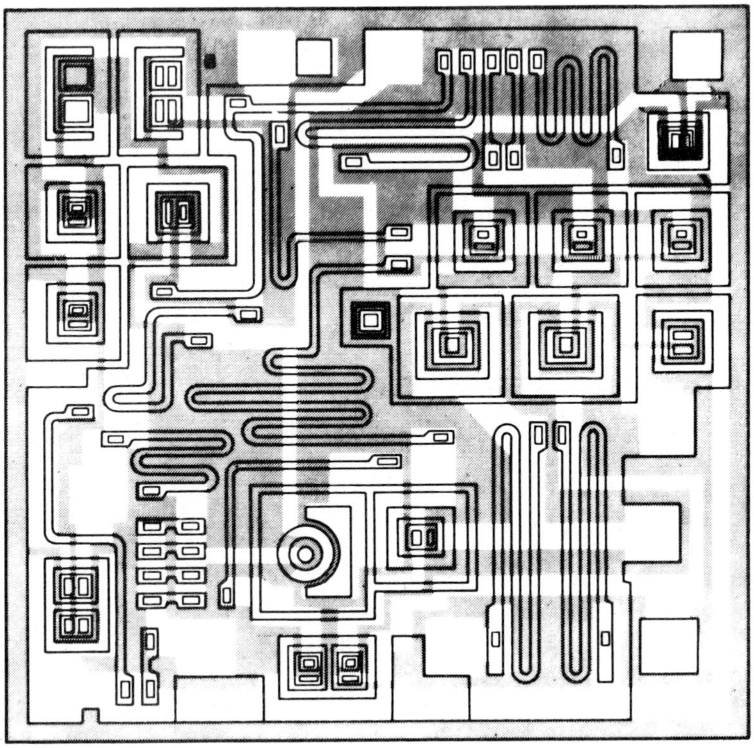

Fig. 24-1. Magnified view of an IC chip of an MC1533.

and transistors required to produce a high-performance op amp.

The chip is not only small (about 0.1-in square and 10 mils thick) but also inexpensive, since many op amps are produced at the same time. The important process by which they are manufactured, *photolithography*, is discussed below.

The op amp is a "linear" device since it is used in applications where its output should be directly proportional to its input: if 10 mV produces 1.0-V output, then a 20-mV input should produce an output of 2.0 V and the voltage gain (or amplification) of the device is 100. A digital device, in contrast, can have only one of two possible outputs in response to its input(s). Its output is either "high" or "low," depending on its input(s). Both linear and digital ICs are produced by the photolithographic process.

IC Manufacture: Photolithography

A thin wafer of silicon, perhaps 3 in or more in diameter, has its surface oxidized in a high-temperature oven as one of the first steps in the manufacture of ICs. Then it is coated with a lacquer or *photosensitive resist*. The surface of the disk is now like a photographic emulsion, and images may be transferred to it by exposing it to light through a precision *mask*. This mask was drawn on a very large scale and then photographically reduced in size. Where the mask has holes in it, light falls on the resist and changes its chemical structure so that it will resist the effect of acids.

The lacquer which was not struck by ultraviolet light is removed by a solvent which develops the image. The hardened resist protects the underlying oxide coating from acid which is used to remove the oxide in specific areas.

After exposure and development of the image on the surface of the silicon wafer, "windows" are etched in the oxide coating on the surface of the disk. Through these windows are diffused doping agents such as boron or arsenic to form P and N regions in the wafer.

The steps briefly described above are repeated as many times as necessary to produce the regions in the wafer that correspond to the required number of resistors, capacitors, diodes, and transistors. The final step is to interconnect the individual components by a thin metal layer on the top surface. Figure 24-2 shows a section of a finished chip.

The masks used in this process have hundreds of repeated patterns corresponding to the device to be produced so that hundreds of ICs are made at one time on one wafer. The final step is to slice the wafer into individual pieces which are then mounted in small cases such as those shown in Fig. 24-3. Thus many ICs have been produced at the same time with much of the process being automated. For these reasons, ICs are relatively inexpensive when made in large quantities.

For a more complete treatment of the planar diffusion, photolithographic process, the student is referred to standard transistor texts which describe it in more detail.

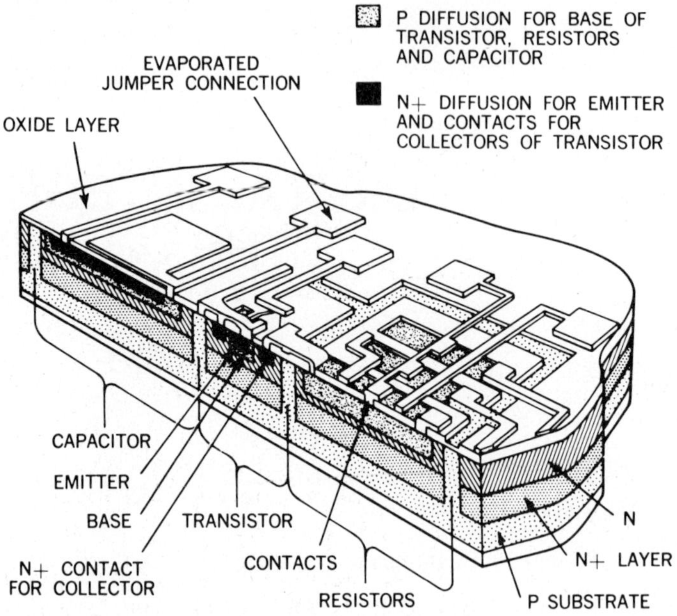

Fig. 24-2. IC chip cross-section. *(RCA)*

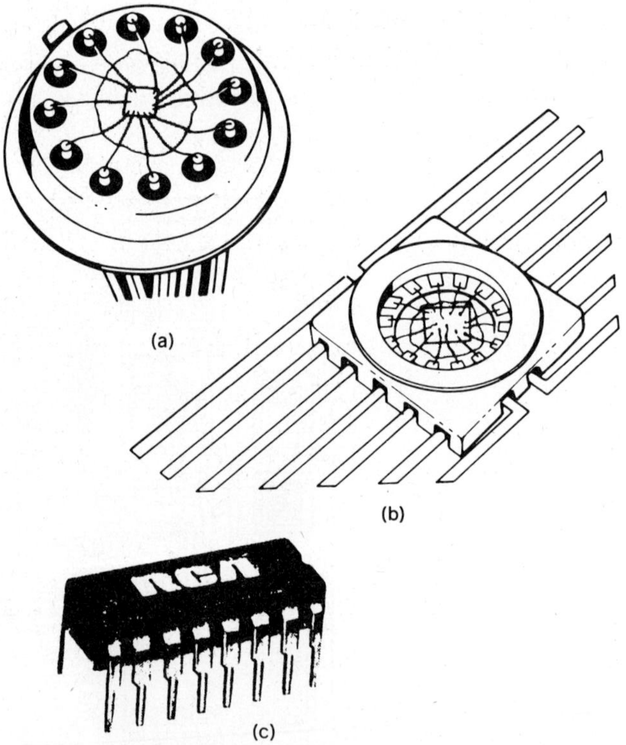

Fig. 24-3. ICs housed in *(a)* transistor-type TO-5 case; *(b)* flat pack; *(c)* dual-line package. *(RCA)*

IC Circuit Configurations

ICs typically include more transistors and diodes than resistors and capacitors because the active components (diodes and transistors) are less costly to produce. Resistors and capacitors occupy more space on the chip. For related reasons, the internal coupling between stages of an IC device is usually direct current rather than alternating current. Figure 24-4 shows the internal components of an RCA CA3018 amplifier and the external components and connections required for its operation.

Figure 24-5 shows the equivalent internal circuit of the MC1533 monolithic (''one stone'') op amp. Manufacturers provide these diagrams as an aid to understanding the characteristics and limitations of their products. The experienced design engineer thinks of the IC op amp as a building block in much the same way as designers dealt with individual vacuum tubes or transistors in the past.

Digital Logic Circuits

Digital logic circuits are the building blocks of devices as familiar as the scientific calculator and as complex as the mightiest digital computer. The basic logical AND and OR functions which are found in almost every digital system can be accomplished in a variety of ways; an elementary method uses the unidirectional characteristics of diodes. Recall that an ideal diode conducts only when it is forward-biased. When reverse-biased, it acts as an open circuit. These properties will be used in the diode AND circuit, or ''gate.''

Digital signals are of the yes or no, up or down, true or false variety: they have two unique possibilities. Logic circuits have either a ''high'' or a ''low'' output. Intermediate values are undefined and undesirable. If the desired state of a light bulb is that it is ON, then the ON state is designated by a logical 1, and the OFF condition by the symbol 0. If the light bulb is operated from a +5-V source, +5 V indicates the 1 condition and 0 V represents the 0 condition.

When the more positive of two voltage conditions represents a logical 1 and the less positive voltage indicates a 0, positive logic is in use. If the more positive voltage equals 0 and the less positive voltage equals 1, then negative logic is in use. The previous example illustrated positive logic.

A simple circuit will illustrate the logical AND function. In Fig. 24-6*a* two SPDT switches are connected in series with a battery and a light bulb. When will the light bulb be ON? Obviously, when both switches are

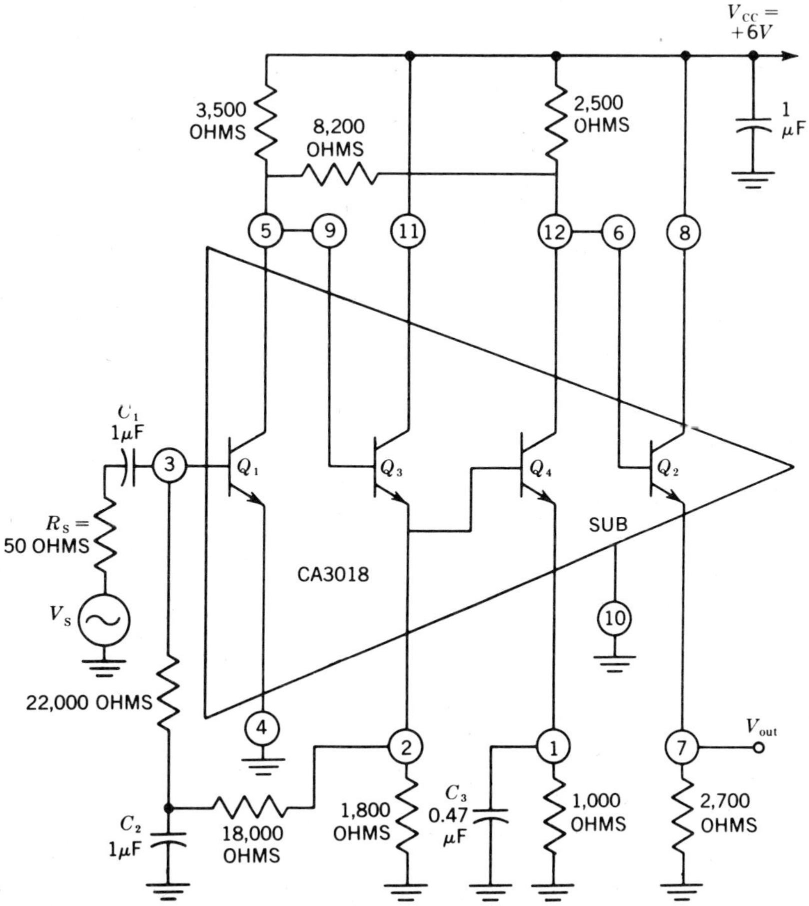

Fig. 24-4. Circuit for a CA3018 broad-band amplifier. *(RCA)*

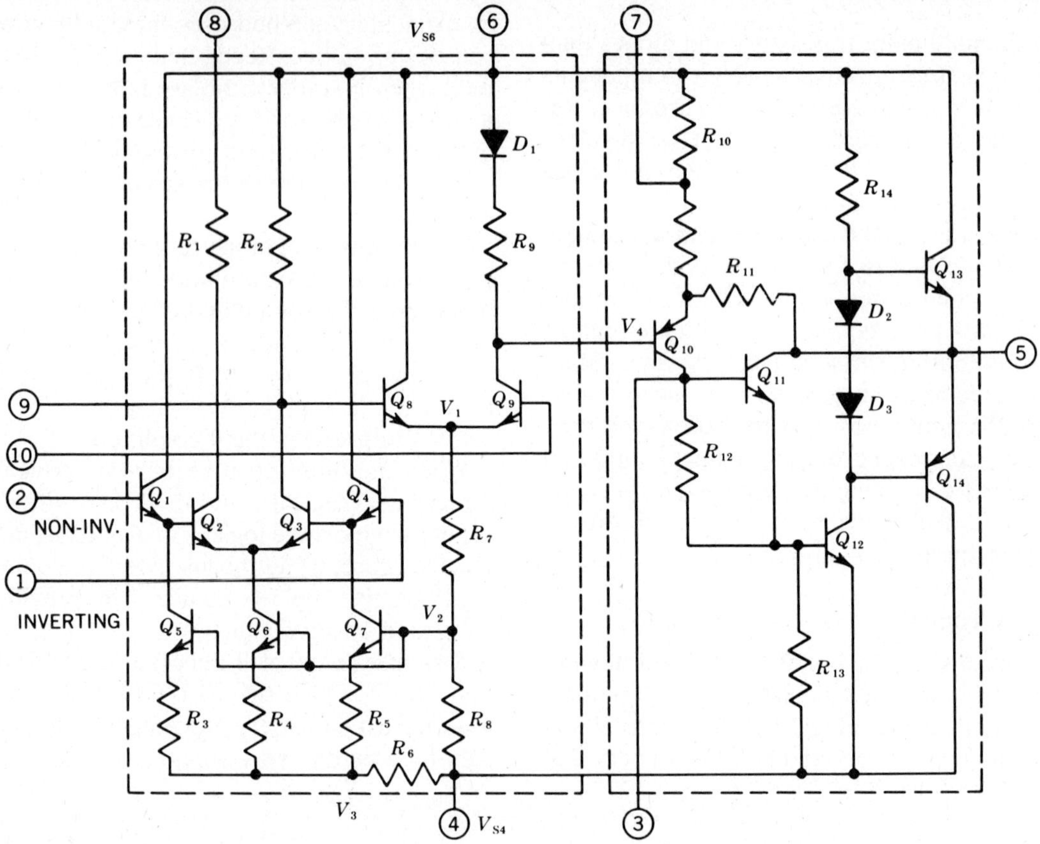

Fig. 24-5. Schematic of MC1533, a high-voltage, monolithic operational amplifier. *(Motorola)*

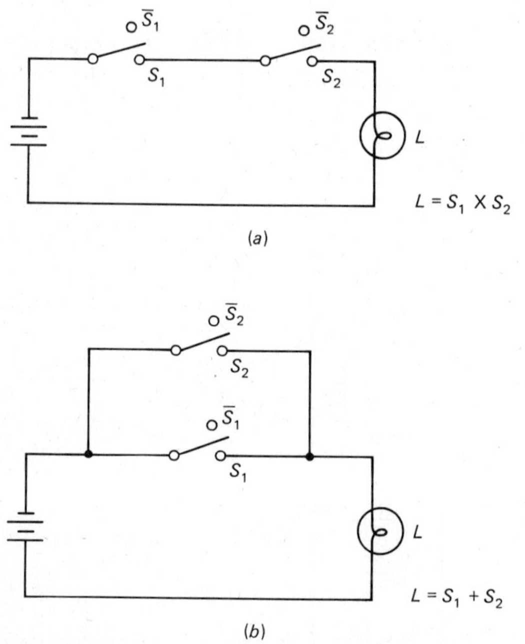

Fig. 24-6. *(a)* An AND circuit; *(b)* an OR circuit.

closed. If either or both of the switches is open, the bulb is OFF. Let S_1 mean that S_1 is closed and $\overline{S}_1$ mean that the switch is open. Similarly, let L mean that the light bulb is ON and $\overline{L}$ mean that it is off. Then the operation of this simple circuit may be described by the boolean algebra expression:

$$\begin{aligned} L &= S_1S_2 \ or \qquad \text{(four ways to show logical AND)} \\ &= (S_1)\,(S_2) \ or \\ &= S_1 \times S_2 \ or \qquad\qquad (24\text{-}1) \\ &= S_1 \cdot S_2 \end{aligned}$$

These relationships may be shown compactly in a *truth table,* which is a standardized representation of a logical function. The truth table for this circuit is given in Table 24-1. Notice that the boolean expression (Eq. 24-1) states what is "true." Every other possible condition is "false." (If both switches are ON, the bulb is ON. Other combinations of the switches turn the bulb OFF.)

TABLE 24–1 AND Truth Table

Inputs		
S_1	S_2	*Output*
0	0	0 (OFF)
0	1	0
1	0	0
1	1	1 (ON)

With appropriate column headings, Table 24-1 is the standard truth table for any two-input logical AND function. Note that the S_1S_2 inputs are arranged like the binary numbers which represent the base 10 numbers 0, 1, 2, 3:

Binary (base 2)	Decimal (base 10)
00	0
01	1
10	2
11	3

For the output of an AND circuit to be true, all of its inputs must be true ("true," "1," or "high" have the same meaning).

The Diode AND Gate

The schematic diagram for a two-input diode AND gate and a standard block diagram symbol for this function are shown in Fig. 24-7. In this circuit the switches provide a convenient method of applying inputs to the AND gate; they are *not* a part of the logic. By choice, the logic levels in this circuit are +5 V = 1 and −5 V = 0.

To analyze this circuit, first consider that both switches are in the −5 V position. What is the output voltage? If perfect diodes are assumed, the output voltage is −5 V because both diodes are forward-biased and are conducting (replace a conducting diode with a short circuit and replace a nonconducting diode with an open circuit).

Let S_2 continue to be connected to −5 V and let S_1 be connected to +5 V. What is the output voltage? It cannot be +5 V because if it were, diode D_2 would be forward-biased by 10 V, an impossibility. But if the output voltage is −5 V, diode D_2 can conduct and diode D_1 is reverse-biased by 10 V; therefore, it is nonconducting and could be removed without affecting the circuit. The output voltage is −5 V.

If the conditions of S_1 and S_2 above were reversed, the result would be the same and the output voltage would again be −5 V.

Let both switches be connected to +5 V. Now both diodes are conducting, and each carries half the current flowing through R, which has 20 V (25 − 5) drop across it. Since the diodes are conducting and connected to +5 V, the output voltage is +5 V.

It should be obvious that each of the four situations above corresponds with the inputs and output obtained from the operation of a two-input AND gate where +5 V = logical 1 and −5 V = logical 0.

The second characteristic of an AND gate is that its output is zero if any input is zero.

The Diode OR Gate

Consider Fig. 24-6*b*. Here two SPDT switches are connected in parallel with each other and in series with a light bulb they control. There are four possible combinations of switch positions: which will turn the bulb ON? It is apparent that if either, or both, switches are ON, the bulb is ON. The light bulb is OFF only if both switches are OFF. We may write the boolean algebra expression as follows:

$$L = S_1 + S_2 \quad (+\text{ means ``or''}) \qquad (24\text{-}2)$$

Spelled out, Eq. 24-2 states, "The light L is ON if S_1 is closed, if S_2 is closed, or if they are both closed."

For an OR gate:

1. If any input is true, the output is true.
2. If all inputs are false, the output is false.

The diode OR gate is shown in Fig. 24-8 together with a standard block diagram symbol for the function. Again the switches provide a convenient way to enter 1s and 0s but are *not* part of the logic circuit. Let us analyze this circuit in the following way.

Let both of the switches provide an input of −5 V. The diodes are effectively connected in parallel and may be considered as one diode. The anode of this diode is returned to −5 V and the cathode is returned to −25 V through the resistor R.

Since the anode is at a more positive potential than the cathode, the diode is forward-biased and conducts. If we replace the conducting diode with a short circuit,

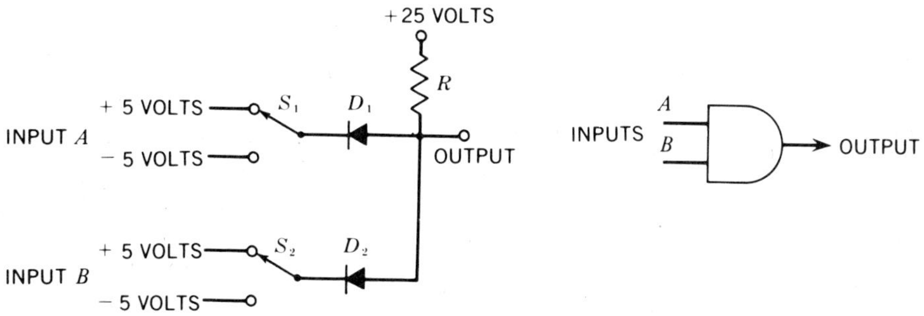

Fig. 24-7. AND circuit.

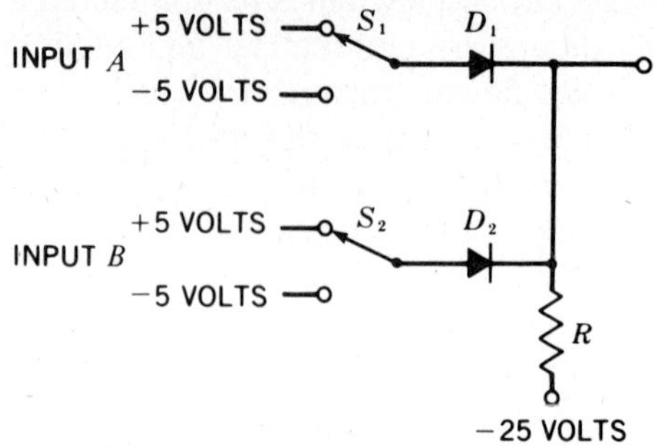

Fig. 24-8. OR circuit.

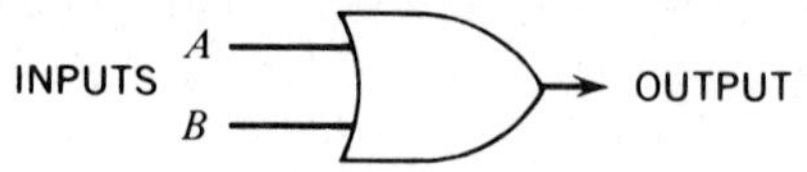

it is apparent that the output voltage is −5 V, or logical 0. The inputs were also logical 0.

Let input A (S_1) be changed to +5 V and input B remains −5 V. D_1 conducts, bringing the output to +5 V. Diode D_2 is therefore negatively biased and is turned off. D_2 can therefore be removed and the output is a logical 1.

The conditions of inputs A and B above could be interchanged, and the output would also be +5 V, or logical 1.

In the final case, let inputs A and B be +5 V. The diodes are effectively connected in parallel and may be treated as one diode. The anode of this diode is returned to +5 V and the cathode is returned to −25 V through R. The anode is more positive than the cathode and the diode conducts. Therefore, replacing the diode with a short circuit, the output voltage is +5 V, or logical 1.

In summary, if any input to the diode OR gate was +5 V, the output was also +5 V, or logical 1. The output was logical 0 only when all inputs were −5 V (logical 0). These relationships are shown in the truth table for the two-input OR gate. Table 24-2.

Combined AND-OR Circuits

The AND and OR circuits just described can be combined to yield more complex functions such as the diagram shown in Fig. 24-9. By combining these two functions, we obtain the new function $AB + C$. The output is true if A and B are true simultaneously; if C alone is true; or if A, B, and C are all true. It is false otherwise.

TABLE 24–2 OR Gate Truth Table

Inputs		
S_1	S_2	*Output*
0	0	0
0	1	1
1	0	1
1	1	1

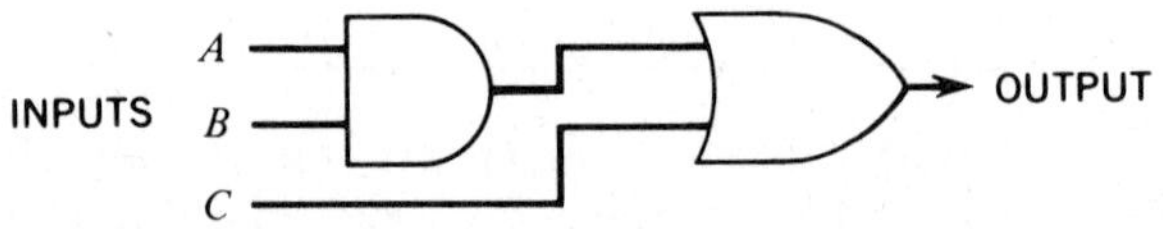

Fig. 24-9. Combined AND-OR logic block diagram.

The diode circuits just described are useful but have some limitations: (1) real diodes have a voltage drop of about 0.7 V, so the ideal logical levels (+5 and −5 V) are not maintained; and (2) since there is no amplification within these circuits, the number of other similar circuits they may drive (the "fan-out") is limited. Therefore, simple diode logic is less common than other types.

Resistors and transistors may be combined to perform logic and to provide amplification as shown in Figs. 24-10 and 24-11. This is an early form of IC logic known as *resistor-transistor logic* (RTL). If diodes are used to perform logic and transistors provide the needed amplification, we have a later logic family, *diode-transistor* logic (DTL). A more modern outgrowth of DTL is *transistor-transistor logic* (TTL), the 7400 family of TTL logic is widely used today (1988). Another popular logic family is CMOS *complementary metal-oxide semiconductor* (CMOS) logic. CMOS is attractive because it uses *field-effect* transistor technology and requires very little operating power. For high-frequency operation, TTL is preferred.

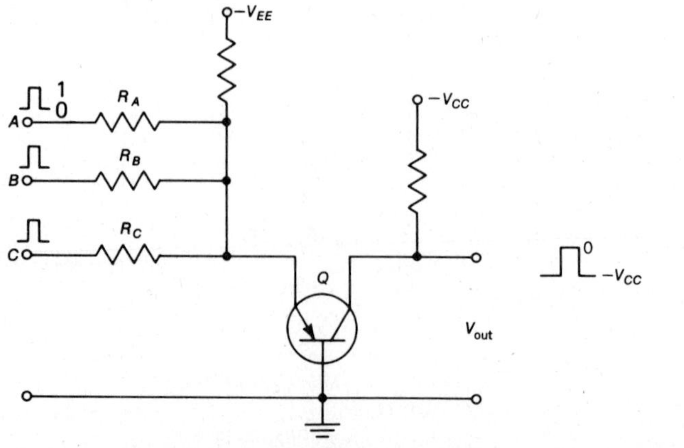

Fig. 24-10. Three-input, ground-base OR gate.

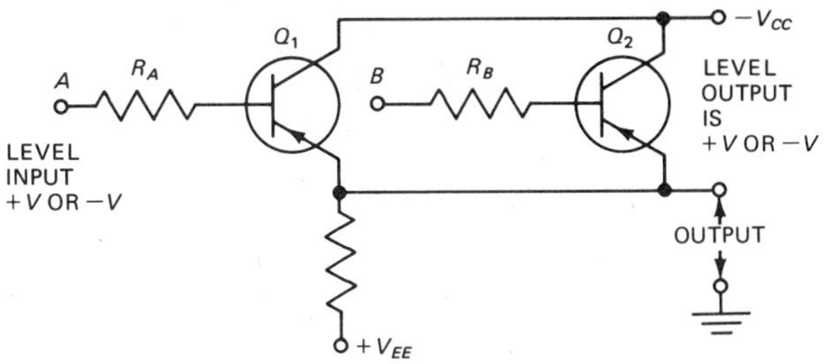

Fig. 24-11. Two-input, grounded-collector AND gate.

In each of these logic families, there are numerous compatible devices: OR, AND, NOR, NAND gates; flip-flops; counters; etc. In general, the various logic families are *not* compatible with each other because their voltage levels and drive requirements/capabilities are different. Most of the desired digital functions are available in each of the logic families, however, and the same standard block diagrams are used to describe the function performed.

Figures 24-12 and 24-13 show the block diagram symbols for two common digital ICs. The TTL 7408 is called a *quad* two-input AND gate because there are four identical AND gates in the package. Similarly, the TTL 7432 is a quad, two-input OR gate. Note that a *top* view of the IC terminals is given. The four gates are independent of each other except for the common power supply connections. These modules are designed to operate between +5 V and ground. The logic levels, 1 and 0, are +5 V and 0 V.

SUMMARY

1. Linear ICs are used where signals may have any value. An op amp is a linear IC.
2. Digital ICs, such as OR or AND gates, operate with signals which may have only one of two possible values. Their inputs and outputs are either *high* or *low*.
3. Both types of IC are made by the planar diffusion/photolithographic process.

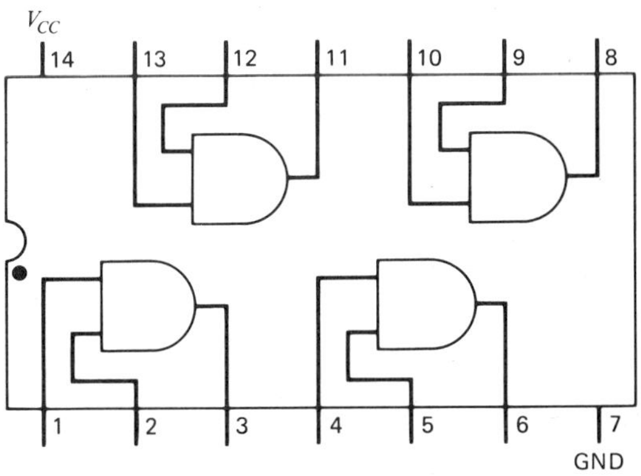

Fig. 24-12. Block diagram of the 7408, quad, two-input AND gates; top view of the terminal connections.

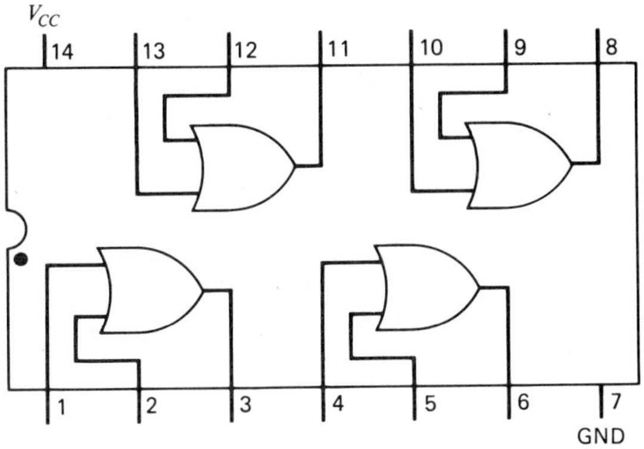

Fig. 24-13. Block diagram of 7432, quad, two-input OR gates.

4. A gate is a circuit which has two or more inputs and one output. The level of the output depends on the levels of the inputs and the type of logic performed.
5. The output of an AND gate is true only if all of its inputs are true. It is false otherwise.
6. The terms *true, high,* and *logical 1* all mean the same thing.
7. The output of an OR circuit is high if any input is high and is low only when all inputs are low.
8. A *truth table* describes the outputs of a logical system in terms of the possible combinations of its inputs.
9. The *fan out* of a digital logic module is the number of members of the same logic family it can drive.
10. TTL logic is fast and widely used, but CMOS logic is preferred when low power consumption is important.

SELF-TEST

Check your understanding by answering these questions.

1. A process which is important to the manufacture of all ICs is ___________ .
2. The two types of IC are ___________ and ___________ .
3. How many inputs of a three-input OR gate must be "high" for the output to be "high"? ___________
4. The plus sign (+) means logical ___________ .
5. The boolean expression $f = ABC$ describes a ___________ -input ___________ gate.

6. Sketch the block diagram symbol for a two-input OR gate.
7. In positive logic, the more __________ voltage represents a 1.
8. TTL gates are designed to operate with a __________ -V power source.
9. In negative logic, +5 V could indicate logical 0 if −5 V represented logical 1. __________ (true/false)
10. The __________ (top/bottom) view of the terminal connections of a digital IC is normally given.

MATERIALS REQUIRED

- Power supply: Variable regulated low-voltage dc source
- Equipment: EVM
- Integrated circuits: 7408 (AND gate); 7432 (OR gate); or equivalent
- Miscellaneous: Three SPDT switches; one SPST switch

PROCEDURE

CAUTION: Do not exceed the +5.0 V required for the ICs. Observe the polarity of the $+V_{cc}$ source.

AND Gates

1. Connect the circuit of Fig. 24-14. S_3 is *open*. Set the power supply for 5.0 V. *Close* S_3. S_1 and S_2 are SPDT switches which will deliver either 0 V (binary 0) or +5.0 (binary 1) to inputs 1 and 2 of the AND gate.

TABLE 24-3. AND Logic

Inputs		Output
A	*B*	*C*
0	0	
1	0	
0	1	
1	1	

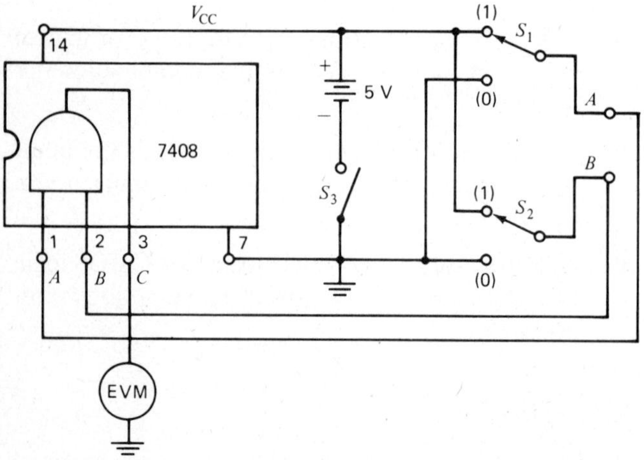

Fig. 24-14. Experimental AND circuit. Terminal numbers on the AND gate refer to terminals on the 7408.

2. Perform the logic and complete Table 24-3. Use an EVM to determine the input and output voltage levels.

OR Gate

3. Connect the circuit shown in Fig. 24-15. The power supply is still set for +5.0 V.
4. Perform the logic and complete Table 24-4. Use an EVM to determine input and output voltage levels.

TABLE 24-4. OR Logic

Input		Output
A	*B*	*C*
0	0	
1	0	
0	1	
1	1	

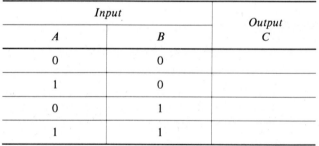

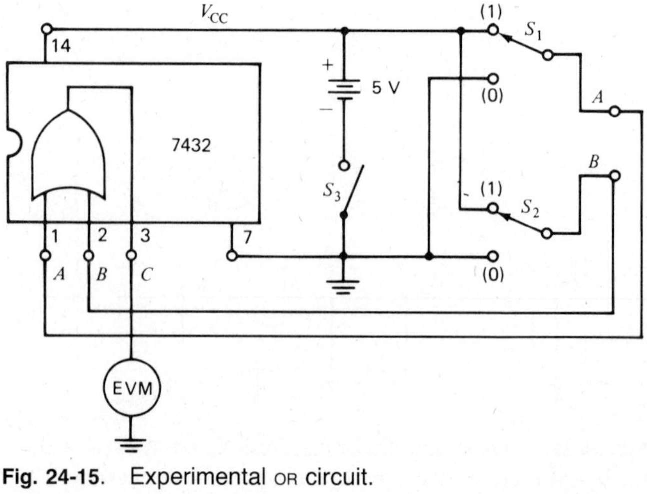

Fig. 24-15. Experimental OR circuit.

Combined AND-OR Gate

5. Connect the circuit of Fig. 24-16: Pin 14 of *both* ICs is connected to +5 V, pin 7 to ground.
6. Perform the logic and complete Table 24-5.

Extra Credit

7. Using any combination of gates, design a three-input circuit such that the output is UP only when all the inputs are UP. Draw the circuit.

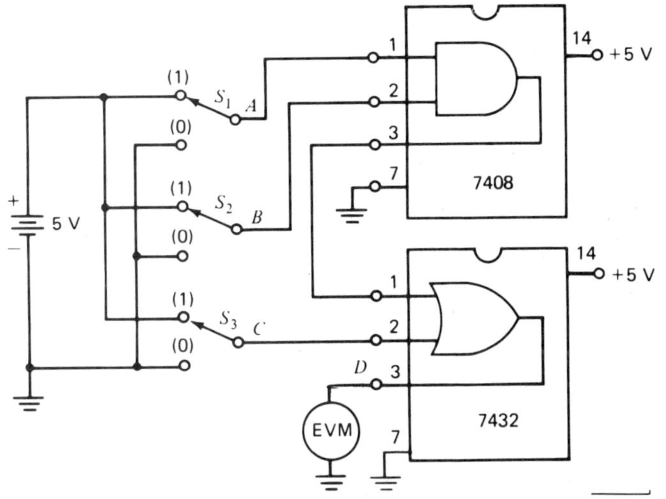

Fig. 24-16. Experimental AND-OR circuit. Terminal numbers on the AND gate and on the OR gate refer, respectively, to terminals on the 7048 and 7432.

TABLE 24-5. Combined AND-OR Logic

Inputs			Output
A	*B*	*C*	*D*
0	0	0	
1	0	0	
0	1	0	
0	0	1	
1	1	0	
1	0	1	
0	1	1	
1	1	1	

8. Verify the circuit experimentally. Record your results in a truth table.

Extra, Extra Credit

9. Design a four-input OR circuit using any combination of gates. Draw the circuit.
10. Verify the circuit experimentally. Record your results in a truth table.

QUESTIONS

1. How does a logical circuit differ from a linear circuit?
2. What does an output of logical 1 mean?
3. Sketch a circuit using three SPDT switches which would satisfy the boolean expression $L=(A+B)(C)$. *L* means a light is ON, and *A*, *B*, *C* mean those switches are closed.
4. Make a truth table for the function of question 3.
5. Referring to the diagram of Fig. 24-17, fill in the blanks in the truth table. Table 24-6.

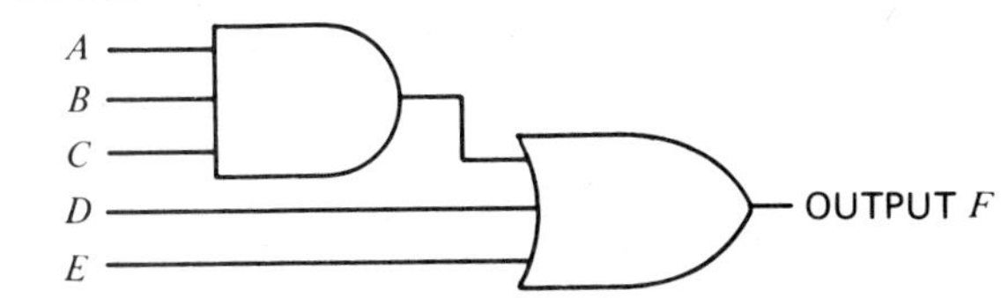

Fig. 24-17. AND-OR diagram for question 5.

Answers to Self-Test

1. photolithography
2. linear; digital
3. at least one
4. OR
5. three; AND
6. (see text)
7. positive
8. +5
9. true
10. top

TABLE 24-6 AND-OR Truth Table

Input					*Output*
A	*B*	*C*	*D*	*E*	*F*
1	0	0	0	0	
1	1		0	0	1
1	1	1		0	1
0	0	0	0		1
	1	1	0	0	0
0	0	0	1	1	

EXPERIMENT

25

DIGITAL ICs: INVERTER, NOR AND NAND GATES

OBJECTIVES

1. To study the inverter and how it may be combined with AND and OR circuits to produce the NAND and NOR functions
2. To demonstrate experimentally how the NOR circuit can be used in a variety of ways

BASIC INFORMATION

In Experiment 24 it was learned that simple circuits can be constructed which will provide "logical" outputs in response to their inputs: they have the ability to make logical decisions. If two bits of information must be "true" for a task to be performed, it will not be performed if only one condition is true. As a safety precaution, for example, a pilot should not be able to retract the landing gear unless he or she (1) is flying, and (2) operates the landing gear switch. Therefore, if the gear retraction is accomplished by an electrical motor, two switches (as in Experiment 24) could be connected in series with the motor to allow it to retract the gear only when the aircraft is flying *and* the pilot operates the landing gear switch.

In a hazardous industrial environment, for example, using a punch press, to avoid injury the operator must depress two switches, one with each hand, for the punch to operate. The switches are in locations which keep the operator's hands clear of the punch.

The Inverter

Perhaps the simplest logic circuit is the *inverter*. In Fig. 25-1*a* a common emitter amplifier is used to reverse, or invert, the polarity of the input pulse at point A. A similar circuit is used in linear applications and is called a *phase inverter*. If the transistor is initially OFF, a positive pulse at its base turns it ON and collector current flows. Current flow through R_L causes the collector

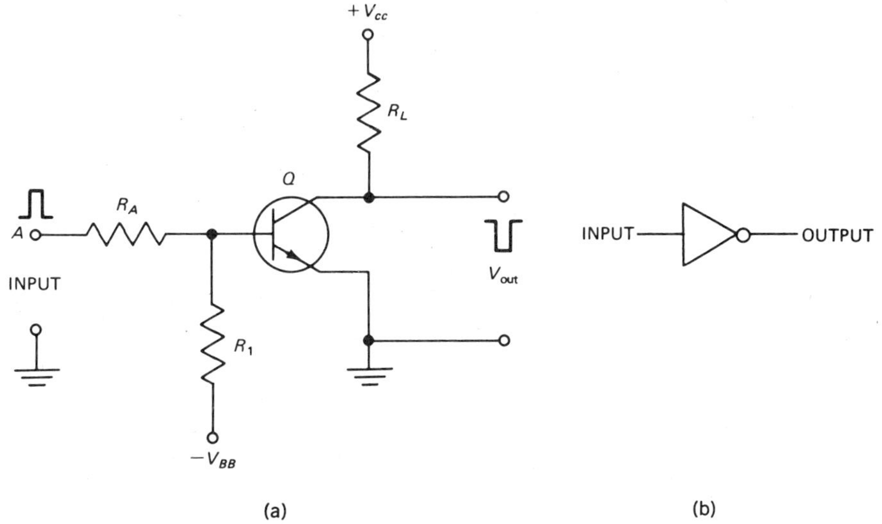

Fig. 25-1. *(a)* NOT or inverter circuit; *(b)* logic symbol.

voltage to drop to a very small value. If there is little collector current, the output is high. Thus a 1 at the input produces a 0 at the output, and a 0 at the input produces a 1 at the output.

In digital applications, if the input to an inverter is a logical 1, the output is a logical 0. If the input is logical 0, the output is logical 1. The logic symbol for the inverter (Fig. 25-1*b*) is the general amplifier symbol, a triangle, with a "ball" at the output. This ball signifies phase inversion, or "active low." Without the ball the symbol would mean a noninverting buffer.

The inverter is sometimes referred to as the NOT circuit. If its input is *A*, its output is "not *A*," which is indicated by placing a bar over *A* ($\overline{A}$). The output of the inverter is the complement of its input. Any function may be complemented by passing it through an inverter.

In Fig. 25-1*a*,

$$\text{If } V_{in} = A \quad \text{then } V_{out} = \overline{A} \qquad (25\text{-}1)$$

The NOR Gate

The NOR circuit may be thought of as an OR circuit followed by an inverter. A NAND gate may be thought of as an AND gate followed by an inverter.

Recall that the output of an OR gate is high if any input is high. An inverter following the OR gate would have a low output if any input to the OR gate is high.

To illustrate the NOR function, consider Fig. 25-2*a*. This is the RTL two-input NOR gate. If R_a and R_b are both grounded, there is no base current to either transistor: they are nonconducting and there is no voltage drop in R_L. The output is high. If a positive pulse is applied to input *A*, for example, Q_1 is turned ON and the current flow through R_L causes the output to go low. Similarly, if a pulse is applied to input *B*, Q_2 will be turned ON and the output will go low. If positive inputs are applied to *A* and *B*, the output will also be low. The truth table for this function is given in Table 25-1 and the logic symbol is shown in Fig. 25-2*b*.

If we label the output *C*, the boolean expression for the NOR gate is

$$C = \overline{A + B} \qquad (25\text{-}2)$$

TABLE 25-1. Two-input NOR Gate

Input *A*	Input *B*	Output *C*	
0	0	1	$\overline{A+B} = C$ or $\bar{A} \cdot \bar{B} = C$
0	1	0	
1	0	0	
1	1	0	

By using DeMorgan's theorem, Eq. 25-2 can also be written:

$$C = \overline{A} \times \overline{B} \qquad (25\text{-}3)$$

Equation 25-2*b* is an AND function in terms of the complements of the variables and suggests another way to produce the NOR function.

The NAND Gate

An RTL NAND gate using PNP transistors is shown in Fig. 25-3*a* and the general NAND gate logic symbol is given in Fig. 25-3*b*. The diagram resembles the NOR gate diagram, but the operation is totally different because of the opposite polarities required by PNP transistors.

Here both transistors are ON if inputs *A* and *B* are both low because they are forward-biased by R_1 and R_2, which are connected to the voltage $-V_{cc}$. (A PNP transistor is turned ON by making its base negative with respect to its emitter.) Therefore, if input *B* is low and input *A* receives a positive pulse, Q_1 will be turned OFF. Q_2 will continue to conduct, and the output will be logical 1. (For the output of this circuit, logical 1 is 0 V and logical 0 is $-V_{cc}$. It is still positive logic.)

If Q_2 receives a positive pulse while input *A* is low, the same result is obtained.

Now let both inputs *A* and *B* receive positive inputs. Both transistors are turned OFF, there is no current through R_L, and the output goes to $-V_{cc}$ (low). The output then is logical 0.

Recall that for the AND circuit to have a true output,

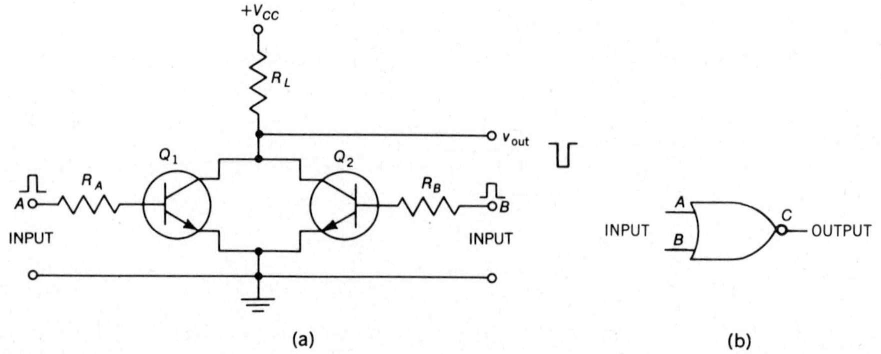

Fig. 25-2. Two-input NOR gate; *(b)* logic symbol.

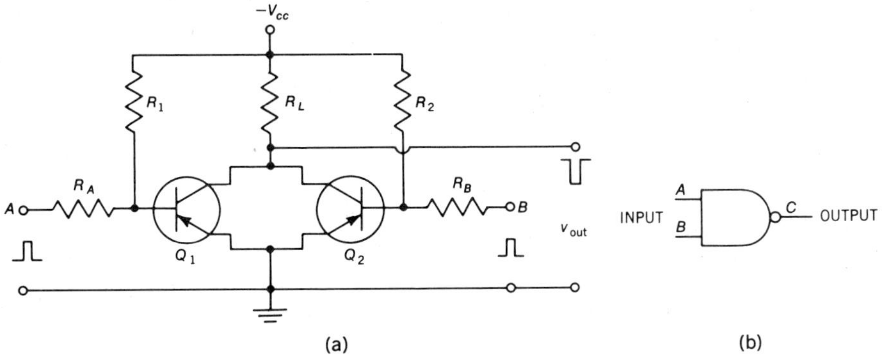

Fig. 25-3. *(a)* Two-input NAND gate; *(b)* logic symbol.

all of its inputs must be true. The NAND gate's output is false if all inputs are true and is true if any input is false. Table 25-2 is the truth table for the two-input NAND gate.

TABLE 25-2. Two-input NAND Gate

Input		*Output*	
A	*B*	*C*	
0	0	1	$\overline{A \cdot B} = C$ or $\bar{A} + \bar{B} = C$
0	1	1	
1	0	1	
1	1	0	

The boolean expression for the two-input NAND gate is

$$C = \overline{A \times B} \qquad (25\text{-}4)$$

and by De Morgan's theorem:

$$C = \overline{A} + \overline{B} \qquad (25\text{-}5)$$

Equation 25-3*b* says that the NAND function may be considered as an OR function of the complements of the input variables and suggests another method for producing the NAND function.

TTL Logic

The equivalent circuit diagram of a TTL 7427 triple three-input NOR gate is given in Fig. 25-4, and a block diagram and pin connection diagram are given in Fig. 25-5. Each of the three NOR gates in this package are independent except for the power supply inputs to pins 7 (ground) and 14 (V_{cc}). The IC requires V_{cc} = +5 V and this value should not be exceeded.

With reference to Fig. 25-4 (and all TTL ICs), all inputs do not have to be used, but an unused input effectively has a 1 applied to it. It is best to do the following: either connect unused inputs to inputs which

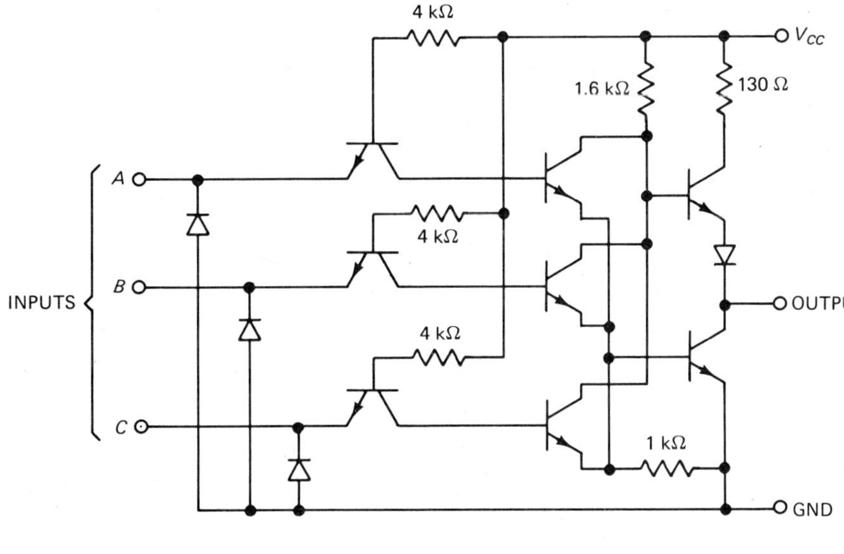

Fig. 25-4. Circuit diagram of one of the triple three-input NOR gates in IC 7427.

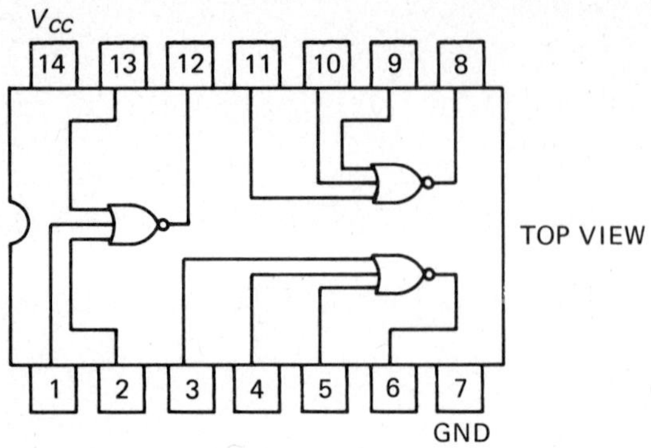

POSITIVE LOGIC: $Y = \overline{A + B + C}$

Fig. 25-5. Top view and block diagram of IC 7427.

are actually used or return the unused inputs to ground or V_{cc}, depending on the type.

1. For AND and NAND gates, return unused inputs to V_{cc}.
2. For OR and NOR gates, return unused inputs to ground.

For example, a three-input NOR gate becomes a two-input NOR gate if one of its inputs is either grounded or connected to a used input. If all the inputs are connected in parallel, the NOR gate can be used as an inverter.

These modifications are used in the procedures which follow.

SUMMARY

1. The most basic digital IC building blocks are the inverter and the AND, OR, NAND and NOR gates.
2. The NOR and NAND gates are like OR and AND gates followed by an inverter.
3. Gates may have any number of inputs, but unused inputs must be connected to an appropriate point.
4. A bar placed over one or more letters representing a logical function means the function has been inverted, or complemented.
5. DeMorgan's theorem provides an alternative expression for a complemented function.
6. The boolean expression for a two-input NOR gate is

$$C = \overline{A + B} = \overline{A} \times \overline{B}$$

7. The boolean expression for a two-input NAND gate is

$$C = \overline{A \times B} = \overline{A} + \overline{B}$$

SELF-TEST

Check your understanding by answering these questions.

1. If a logical 1 is provided to each of its inputs, the output of a three-input NAND gate would be logical __________ .
2. If an unused input of a three-input OR gate were permanently connected to logical 1, the output of the gate would always be logical __________ .
3. The output of the __________ is always the complement of its input.
4. The relationships between the inputs and the outputs of a logical device are shown in a __________ .
5. To say $\overline{A \times B} = \overline{A} + \overline{B}$ is an example of the use of __________ theorem.

MATERIALS REQUIRED

- Power supply: Regulated low-voltage dc source
- Equipment: EVM or VOM
- Semiconductors: IC 7427 (three-input NOR gate) or equivalent; 2N6004 or the equivalent
- Resistors: ½-W 820-Ω, two 1000-, 1500-, 1800-, and 4700-Ω
- Miscellaneous: Three SPDT switches

PROCEDURE

NOTE: Use the IC 7427 shown in Fig. 25-5.

NOR Gate Connected as Inverter

1. Connect one of the NOR gates in the IC 7427, as in Fig. 25-6. The voltage divider provides the proper dc input levels for the gate. In this IC, a binary 1 is represented by +2.4 to +5.0 V (approximately) and binary 0 is represented by 0 to +0.4 V (approximately). S_1 delivers either 0 or +3.2 V to the input of the gate. In this and subsequent steps pin 14 of the IC is connected to +5 V, and pin 7 is connected to ground.
2. Complete the truth table of Fig. 25-6.

NOR Gate Logic

3. Connect the circuit shown in Fig. 25-7 and complete the truth table.
4. Connect the circuit shown in Fig. 25-8 and complete the truth table.

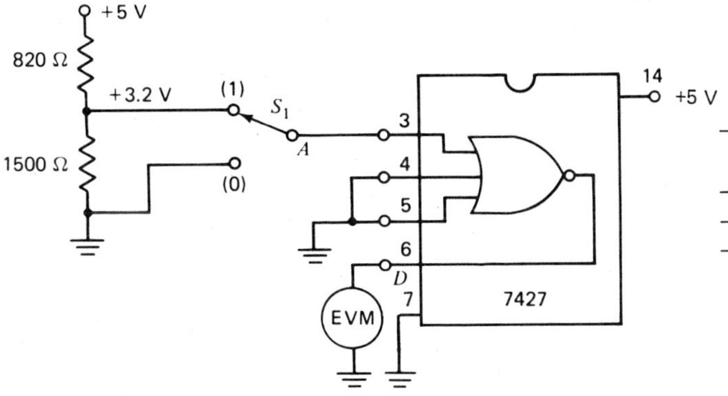

Input A	Output D
0	
1	

Fig. 25-6. Experimental circuit 1 and truth table.

Input			Output
A	B	C	D
0	0	0	
1	0	0	
0	1	0	
0	0	1	
1	1	0	
1	0	1	
0	1	1	
1	1	1	

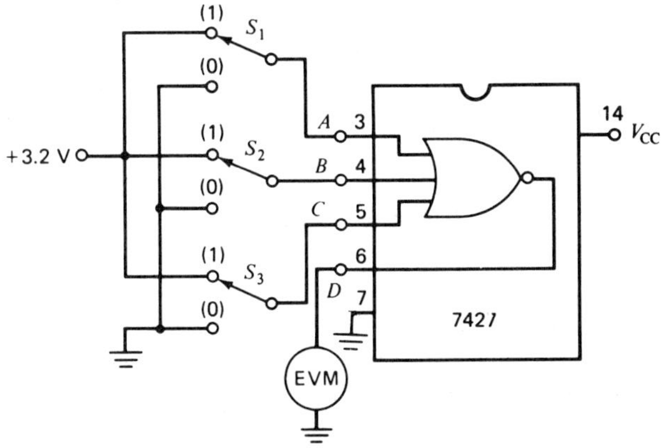

Fig. 25-7. Experimental circuit 2 and truth table.

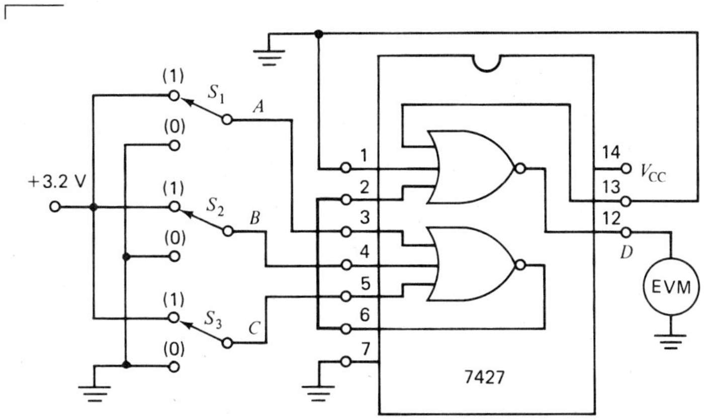

Input			Output
A	B	C	D
0	0	0	
1	0	0	
0	1	0	
0	0	1	
1	1	0	
1	0	1	
0	1	1	
1	1	1	

Fig. 25-8. Experimental circuit 3 and truth table.

5. Connect the circuit shown in Fig. 25-9 and complete the truth table.
6. Connect the circuit in Fig. 25-10 and complete the truth table.

Extra Credit

7. Design a three-input AND gate utilizing the IC 7427 and the 2N6004. Draw the circuit showing all connections, numbering all terminals.
8. Connect the circuit and complete a truth table.

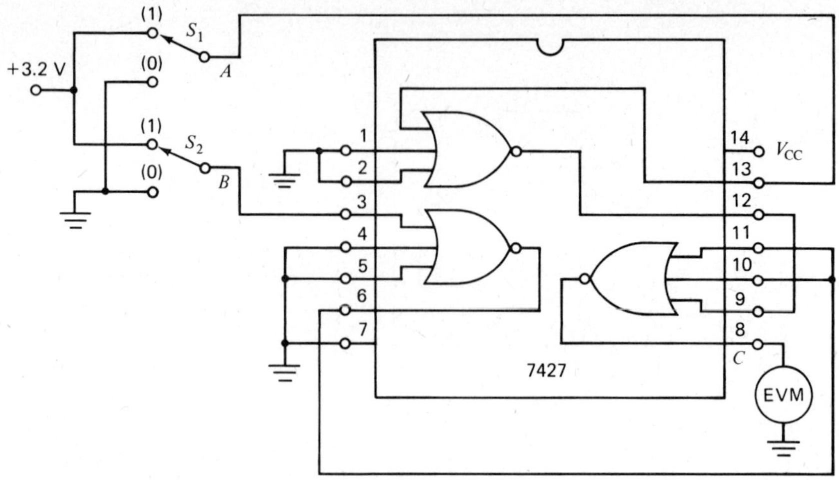

Input A	Input B	Output C
0	0	
1	0	
0	1	
1	1	

Fig. 25-9. Experimental circuit 4 and truth table.

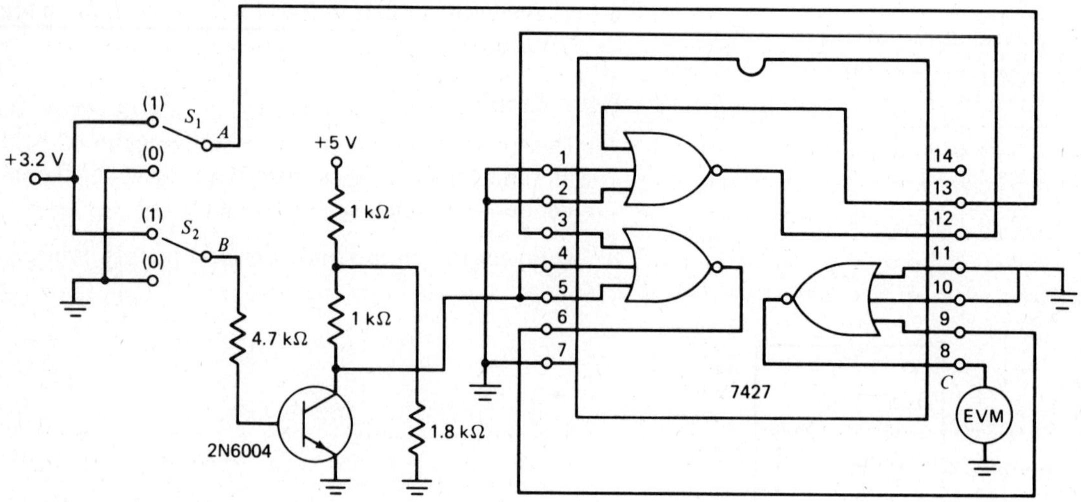

Input A	Input B	Output C
0	0	
1	0	
0	1	
1	1	

Fig. 25-10. Experimental circuit 5 and truth table.

QUESTIONS

1. What are the characteristics of an inverter?
2. What are the characteristics of a NOR gate?
3. What are the characteristics of a NAND gate?
4. Identify each of the experimental circuits by its logic name (inverter, three-input NOR gate, etc.).
   (a) Circuit 1 (Fig. 25-6) __________ ;
   (b) Circuit 2 (Fig. 25-7) __________ ;
   (c) Circuit 3 (Fig. 25-8) __________ ;
   (d) Circuit 4 (Fig. 25-9) __________ ;
   (e) Circuit 5 (Fig. 25-10) __________ .
5. (a) What is the truth table of the function $A \cdot B$?
   (b) How does this compare with the truth table of a two-input NOR gate?
6. (a) What is the truth table of the function $A + B$?
   (b) How does this compare with the truth table of a two-input NAND gate?
7. Why is the NOR gate a universal building block?
8. Sketch a simplified logic diagram for each of the experimental circuits 1 through 5. For example, circuit 1 is simply an inverter.

Answers to Self-Test

1. 0 (zero)
2. 1 (return unused inputs to logical 0)
3. inverter
4. truth table
5. DeMorgan's

EXPERIMENT

26

DIGITAL ICs: BINARY ADDITION, HALF AND FULL ADDERS

OBJECTIVES

1. To study binary numbers and learn how they are added
2. To learn how to convert decimal (base 10) numbers into binary (base 2) numbers and how to convert base 2 numbers into decimal numbers
3. To construct a simple full-adder circuit and demonstrate how it performs addition of binary numbers

BASIC INFORMATION

As discussed in previous experiments, digital devices have only two input or output states: 1 or 0. Many items used in digital systems also have two conditions. A hole is either present or absent in a punched card; a relay is either energized or not; a light bulb is ON or OFF; a piece of magnetic material may be magnetized in one direction to represent information or magnetized in the opposite direction to show lack of information. Presence of data on a line may be indicated by a tone of one frequency and absence of data by a tone of another frequency.

The arithmetic of everyday life, decimal or base 10 arithmetic, uses the symbols 0, 1, 2,..., 9 to express all possible numerical values. But the arithmetic of digital devices and computers, base 2 or binary arithmetic, uses only two: 0 and 1. Binary numbers may be indicated in a variety of ways. A hole in a card may represent a 1 and no hole in that same location may represent a 0; magnetic material may be magnetized one way to represent a 1 and the other way to represent a 0; one state of a flip-flop may represent the presence of a 1 and the other state may represent a 0.

A computer must store digits, transmit them from one place to another, and perform logical and numerical operations on them. One of the most basic arithmetic operations a computer must perform is the addition of two or more numbers. Before proceeding to the mechanics of binary addition, let us review binary arithmetic.

Table 26-1 shows the correspondence between the decimal numbers from 1 to 11 and the equivalent binary representation. Notice that the binary and decimal numbers 0 and 1 are identical, but the base 2 representation for decimal "2" looks like "ten."

Actually the same idea is used here as in the base 10 representation of "ten" and "eleven." There is a place value.

TABLE 26-1. Decimal and Binary Numbers

Decimal	*Binary*	*Decimal*	*Binary*
0	0	6	110
1	1	7	111
2	10	8	1000
3	11	9	1001
4	100	10	1010
5	101	11	1011

The base 10 representation of the number 10 means: one 10 and zero 1s. The base 10 representation of the number 11 means: one 10 and one 1. In base 2, the quantity 10 means one 2 and zero 1s. In base 2, the number 1010 means: one 8, zero 4s, one 2, and zero is. The base 10 number 11 (eleven) is expressed in base 2 as 1011: one 8, zero 4s, one 2, and one 1. In other words, in base 2 the place value of the columns reading from the right are 1, 2, 4, 8, etc., and the quantity indicated is the sum of the columns in which there is a 1.

Table 26-2 shows how various decimal numbers may be written in the binary system. To understand how a decimal number, such as 75, may be expressed in binary, or base 2, the first step is to determine the largest multiple of 2 contained in it. Since 64 is the largest multiple of 2 in 75, place a 1 in the 64 column. The

TABLE 26-2. Decimal Numbers and Their Binary Equivalents

Decimal Number	*Binary position* 256	128	64	32	16	8	4	2	1	*Binary Number*	*Decimal Number*
34				1	0	0	0	1	0	100011	34
15						1	1	1	1	1111	15
225		1	1	1	0	0	0	0	1	11100001	225
75			1	0	0	1	0	1	1	1001011	75

remainder is 11 (base 10). What is the largest multiple of 2 in 11? Since 32 is not, nor is 16, place a zero (0) in those columns. But 8 is the largest multiple of 2 in 11, so place a 1 in the 8 column. Subtracting 8 from 11, the new remainder is 3. Now 2 is contained in this remainder, so place a 1 in the 2s column. Since the new remainder is 1, place a 1 in the 1s column. There is no further remainder, and the decimal number 75 may be written as 1001011 in base 2:

$$75 \text{ (base } 10) = 1001011 \text{ (base } 2)$$

$$75 = 64 + 8 + 2 + 1$$

Binary Addition

Although binary arithmetic may be new to the student, it is very simple since very few addition facts must be remembered:

1. $0 + 0 = 0$
2. $0 + 1 = 1$
3. $1 + 0 = 1$
4. $1 + 1 = 0$ with 1 carry to the left column

Table 26-3 shows an example of binary addition. The decimal numbers 225 and 75 are to be added together in both decimal and binary. Beginning at the right-hand column, $1 + 1 = 0$ with 1 carry to the column on the left. In the next column, $1 + 0 = 1 + 1$ (carry) $= 0$ with 1 carried to the next column. This procedure is continued until all columns have been added with each carry taken into account. The sum in binary is 100101100:

$$100101100 \text{ (base } 2) = 256 + 32 + 8 + 4 \text{ (base } 10)$$
$$= 300 \text{ (base } 10)$$

TABLE 26-3. Adding Binary Numbers

Decimal Number	*Binary Value* 256	128	64	32	16	8	4	2	1
	1	1					1	1	
225	0	1	1	1	0	0	0	0	1
+75	0	0	1	0	0	1	0	1	1
300	1	0	0	1	0	1	1	0	0

Subtraction can be performed by adding to another number the "complement" of the number to be subtracted. Multiplication and division are performed as repeated additions or subtractions. Therefore, a circuit which can perform addition on two binary numbers is basic to arithmetic operations in a computer.

The binary half adder, which has two inputs and two outputs, can provide the correct sum of two binary digits and a carry. Its truth table is given in Table 26-4. The boolean algebra expression for the sum and carry is:

$$\text{Sum} = (B \times \overline{A}) \quad or \quad (A \times \overline{B})$$ This is the exclusive OR function.

$$\text{Carry} = A \times B$$

TABLE 26-4. Truth Table for Half Adder

Input		*Output*	
A	*B*	*Sum*	*Carry*
0	0	0	0
0	1	1	0
1	0	1	0
1	1	0	1

Figure 26-1 shows one of several different ways to generate the sum and carry outputs for inputs *A* and *B*. A carry should only occur when *A* and *B* are both logical 1. It is accomplished with a simple AND gate.

The sum should only be true when either *A* or *B* is true but should be false if both are either true or false.

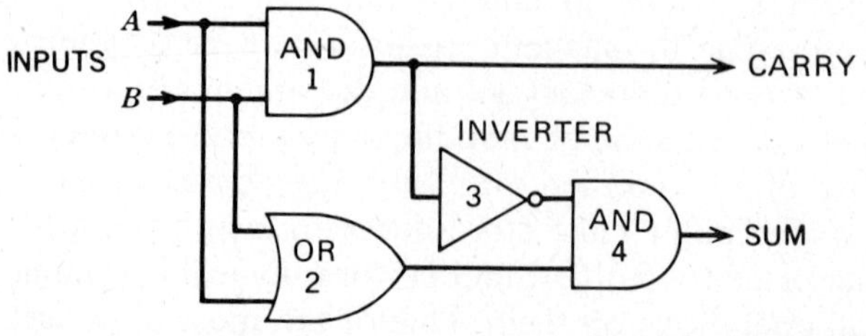

Fig. 26-1. Half adder.

Notice that AND gate 1 will have a 0 output any time either A or $B = 0$. The inverter will invert this 0 to a 1 and cause AND 4 to have a 1 output any time either A or B is true. If A and B are both true, however, the inverted carry output causes the sum output to be 0. The sum output is also 0 if both A and B are 0.

Full Adder

The half-adder circuit produces the correct sum and carry outputs for two binary inputs A and B but does not provide for a carry input. The full adder whose truth table is shown in Table 26-5 will produce the correct sum and carry outputs in response to the three inputs A, B, and carry. Figure 26-2 is a logical block diagram showing one method of providing the full-adder function. To demonstrate that the sum and carry outputs of this circuit are correct, take each of the eight combinations of A, B, and C, one at a time, as inputs to the full adder. For example, for row one, where A, B, and C are all zero, what are the outputs of each logical block?

OR*1* = *0* because its inputs are both *0*

OR*2* = *0* because its inputs are both *0*

AND*3* = *0* because its inputs are both *0*

AND*4* = *0* because its inputs are both *0*

AND*5* = *0* because its inputs are both *0*

OR*6* = *0* because its inputs are both *0*

= CARRY

INV*9* = *1* because its input is *0*

OR*7* = *1* because one input is *1*

AND*8* = *0* because one input is *0*

= SUM

It is left as an exercise to show that the other combinations of A, B, and C yield the remaining sums and carries in Table 26-5.

TABLE 26-5. Truth Table for a Full Adder

Input			*Output*	
A	*B*	*C*	*Sum*	*Carry*
0	0	0	0	0
1	0	0	1	0
0	1	0	1	0
0	0	1	1	0
1	1	0	0	1
1	0	1	0	1
0	1	1	0	1
1	1	1	1	1

Figure 26-3 shows a block diagram of a 5-bit adder made up of five full adders similar to Fig. 26-2. This 5-bit adder will calculate the correct sum of two 5-bit binary numbers. Instead of using small-scale ICs to make individual full adders, today one would probably choose an IC which performs all of the required functions. The TTL 7483, for example, is an IC 4-bit parallel adder which may be cascaded with additional 7483s to perform binary addition on two numbers of any length (number of 1s and 0s).

Experimental Full Adder

As mentioned above, the TTL 7483 is a complete 4-bit full adder on a single IC chip, and, in terms of cost and convenience, would be preferred in an industrial application.

For demonstration purposes, however, it is better to breadboard the full adder using individual logic modules so that the operation of each part of the circuit may be observed. We will use a 7408 quad, 2-input AND gate, a 7432 quad, 2-input OR gate, and a few discrete components to breadboard the full adder. Figures 26-4 and 26-5 show the pin connections of these two ICs. Remember that pin 14 = +5 V and pin 7 = 0 V on both ICs and that the pins are numbered from a reference point on the top. Pin 1 is the one to the immediate left of the reference mark (top view).

The experimental circuit is a 1-bit full adder. To perform the function shown in Fig. 26-3, it would be necessary to duplicate Fig. 26-6 on page 169 five times. With discrete components (resistors, diodes, etc.), it would require many parts and much space. Even with small-scale ICs, it would still be formidable. But with ICs such as the 7483, its just a case of connecting up the power leads and the inputs and outputs.

SUMMARY

1. The binary number system has only two symbols, 1 and 0. In this system, they represent a quantity or numerical value of 1 or 0 (the symbols 1 and 0 are also used in logical systems to indicate the two possible states).

2. The decimal number 123 means 100 + 20 + 3. Or $1 \times 100 + 2 \times 10 + 3 \times 1$. We write the coefficients 1, 2, and 3 in the 100s, 10s, and 1s columns. In base 2 a number such as 1011 means $1 \times 8 + 0 \times 4 + 1 \times 2 + 1 \times 1$. In both cases we write only the coefficients and the place value is understood.

3. In the binary system the place values of the columns starting on the right are multiples of 2: 1, 2, 4, 8, 16, 32, 64, 128, 256, 512, etc., instead of the familiar base 10 values 1, 10, 100, 1000, etc.

4. To convert a binary number to a decimal number, add together the place values of all the columns in

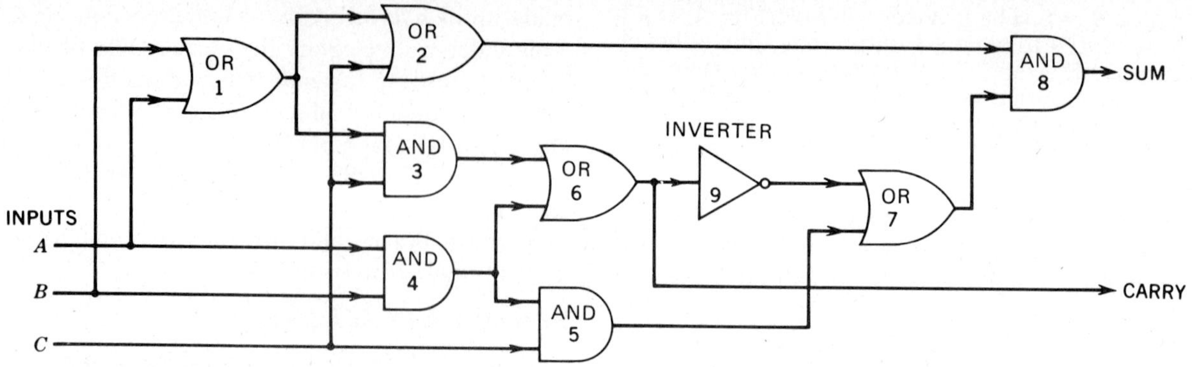

Fig. 26-2. Binary full adder.

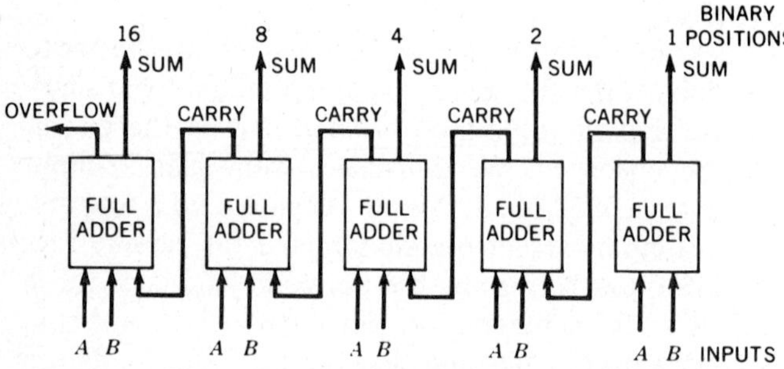

Fig. 26-3. Block diagram of five-position binary adder.

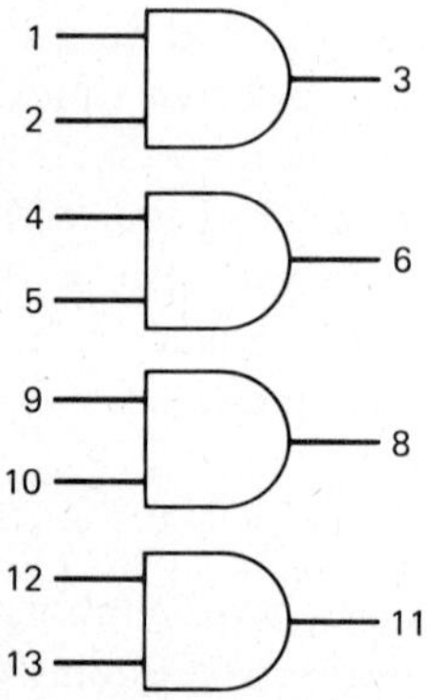

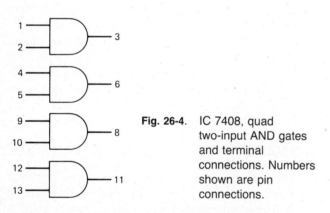

Fig. 26-4. IC 7408, quad two-input AND gates and terminal connections. Numbers shown are pin connections.

which the binary number has a 1. For example, convert 10110 (base 2) to a decimal number:

32 16 8 4 2 1	place value
0 1 0 1 1 0	binary number
16 + 4 + 2 = 22	base 10

5. Binary addition facts:
$0 + 0 = 0$ $0 + 1 = 1$ $1 + 0 = 1$ $1 + 1 = 0$ with *1* carry

(Note that the + sign means "plus" here. In logic circuits, it means logical OR.)

6. Digital computers use binary arithmetic because there are many two-state devices which make it convenient.

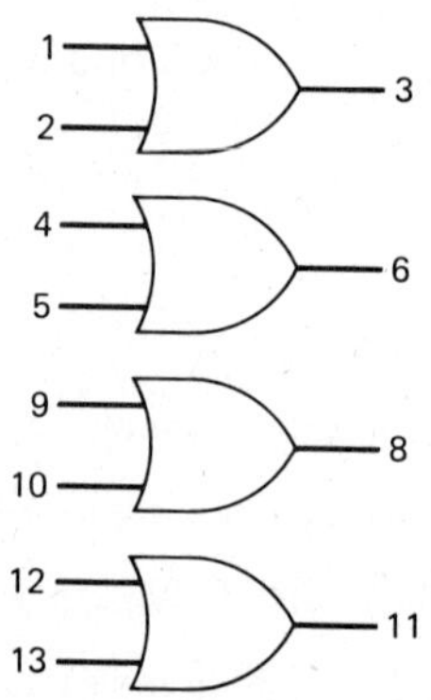

Fig. 26-5. IC 7432, quad two-input OR gates and terminal connections. Numbers shown are pin connections.

7. A full binary adder produces the correct sum and carry for two binary inputs and a carry input. A half adder is similar but has no provision for a carry input.

8. A carry is the result from a previous addition.

SELF-TEST

Check your understanding by answering these questions.

1. The number 201 is not a valid number in base 2. __________ (true/false)

2. The binary number 0111 represents the base 10 number __________ .

3. If 1011011 is added to 1101001, the answer is __________ in base 2.

4. Express the answer from question 3 in base 10: __________

5. If the sum output of a half adder is inverted, the inverted sum output will be logical 1 when __________ .

6. In binary form, 100 (base 10) is __________ .

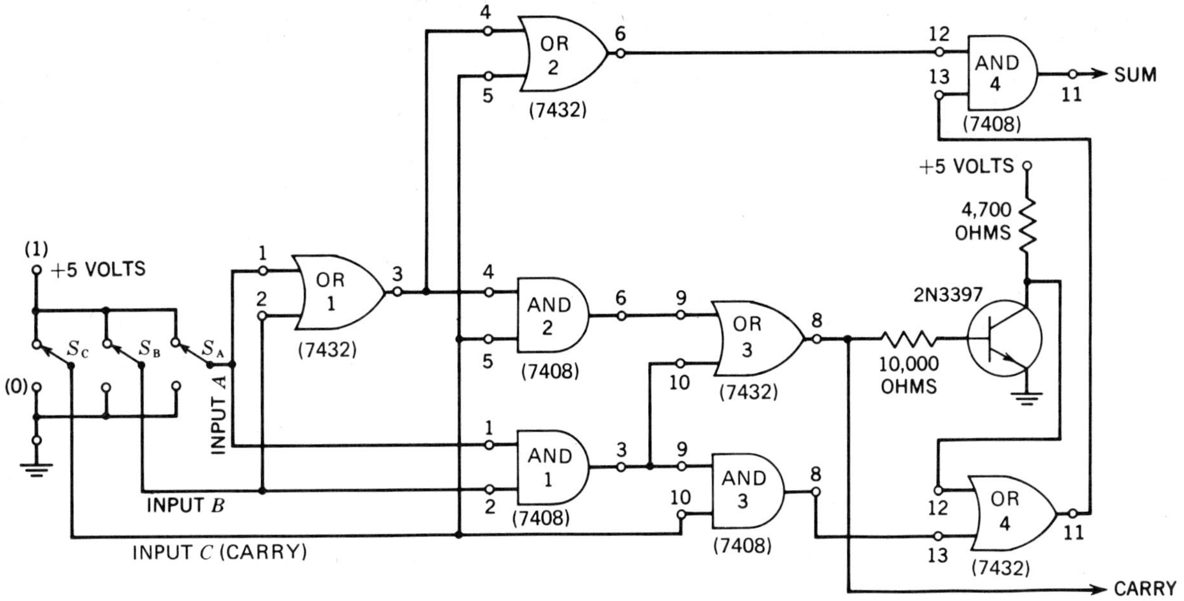

Fig. 26-6. Full-adder experimental circuit.

MATERIALS REQUIRED

- Power supply: Variable regulated low-voltage dc
- Equipment: EVM or VOM
- Resistors: ½-W 4700-, 10,000-Ω
- Integrated circuits: 7408 (AND gate); 7432 (OR gate)
- Transistor: 2N3397 or equivalent
- Miscellaneous: Three SPDT toggle switches

PROCEDURE

1. Connect the full-adder circuit (Fig. 26-6). The AND gates are contained in IC 7408 and the OR gates in IC 7432. The inverter is the 2N3397. Connect pin 14 of each of the ICs to +5.0 V of the supply and pin 7 to ground. The other pins are interconnected as shown.

2. The level on each of the three input lines is controlled by the respective SPDT switch, S_A, S_B, or S_C. When the switch is closed on the +5.0-V side, a binary 1 is present on the input line. When the switch is closed on the 0-V side, a DOWN level representing a binary 0 is present.

Using the notations UP and DOWN, set the input switches for the combinations shown in Table 26-6 and record the conditions for the sum and carry output lines. When a positive level exists on an output line, record UP in the proper position in the table. For a negative output level, record DOWN.

Do the output results recorded in the chart agree with the outputs in Table 26-5?

3. Refer to Fig. 26-6 and assume the following conditions:

Input A = binary *1*
Input B = binary *1*
Input C = binary *1*

Terminal 13 in AND block 4 is grounded. What is the level on the sum line under these conditions?

TABLE 26-6. Logic of a Full Adder

Input			*Output*	
A	*B*	*C*	*Sum*	*Carry*
DOWN	DOWN	DOWN		
UP	DOWN	DOWN		
DOWN	UP	DOWN		
DOWN	DOWN	UP		
UP	UP	DOWN		
UP	DOWN	UP		
DOWN	UP	UP		
UP	UP	UP		

Is this level correct for the input conditions given? Check your answer by operating your laboratory circuit with the short circuit and with the input switches set for the conditions shown. Remove the short circuit when finished.

4. Refer to Fig. 26-6 and assume the following conditions:

   Input A = binary *1*
   Input B = binary *1*
   Input C = binary *0*

   Base-to-emitter short-circuits in the inverter. Determine the levels of the output lines by analyzing the circuit operation under the conditions stated. Is output correct for the input conditions used? Check your answer by operating your laboratory circuit with the short. Remove the short circuit when the test is completed.

5. Assume that there is no inverter stage in Fig. 26-6 and that the output of OR block 3 (terminal 8) is connected directly to terminal 12 of OR block 4. Fill in the blanks in Table 26-7 for the input combinations shown. Check your answers by operating your laboratory circuit with the inverter removed and terminal 8 of OR block 3 shorted to terminal 12 of OR block 4. Observe the outputs for the various input combinations and compare the results with your answers in Table 26-7.

TABLE 26-7. Defective Full Adder

Input			*Output*	
A	*B*	*C*	*Sum*	*Carry*
0	0	0		
1	0	0		
0	1	0		
0	0	1		
1	1	0		
1	0	1		
0	1	1		
1	1	1		

QUESTIONS

1. Computers use base 2 because there are many devices which have two states, such as flip-flops, and thus it is convenient. How many states would be required to do base 10 arithmetic directly? Do such devices exist? If so, name them.

2. What is the main difference between a full adder and a half adder?

3. If the output of an exclusive OR circuit is inverted, we have the exclusive NOR circuit. Why is the exclusive NOR circuit sometimes called a "comparator?"

4. The *least significant bit* (LSB) of a binary number is the 1 or 0 on the extreme right. The addition of the LSBs of two binary numbers can be performed with a half adder. Why? If a full adder is used to add the LSBs of two binary numbers, what should be done with the carry input?

Extra Credit

Obtain a 7483 TTL 4-bit parallel full adder and demonstrate its operation. LEDs may be used to indicate the status of the two 4-bit input words as well as the 4-bit output and carry. Use inverters to turn the LEDs ON when the outputs go high.

Answers to Self-Test

1. true
2. 7
3. 11000100
4. 196
5. they are the same: $A = B$
6. 1100100

EXPERIMENT

27

DIGITAL ICs: THE SET-RESET FLIP-FLOP (TRIGGER) AND THE *JK* FLIP-FLOP

OBJECTIVES

1. To construct a flip-flop using IC NOR gates and experimentally determine its logic characteristics
2. To trigger a *JK* flip-flop

BASIC INFORMATION

Bistable multivibrators, or "flip-flops," were introduced in Experiment 22. Because they will remain in either of two stable states until triggered, they are used extensively in computers and other digital systems. The flip-flop may be used as a temporary storage device to retain information, as in the main memory of a computer. A group of flip-flops may be used as a register to temporarily store data or to perform operations on it. Flip-flops are used to count pulses in binary or decimal counters.

Flip-Flop Logic

The flip-flop may be considered as two dc coupled inverters in a positive feedback circuit as shown in Fig. 27-1. Regardless of the logic type or voltage levels, the basic circuit has two stable conditions. The logical conditions assumed in Fig. 27-1*a* are consistent with the operation of each inverter: a 1 at an input produces a 0 at the output and a 0 at an input produces a 1 at the output.

If the other set of conditions is assumed, as in Fig. 27-1*b*, the input-output conditions of the two inverters are also satisfied. At this point we have a simple latch which is the building block for more useful circuits.

The elementary circuit has two complementary outputs, usually labeled Q and $\overline{Q}$, or the 1 and 0 sides of the flip-flop. Adding a means to trigger the flip-flop between its two states and logical steering at the inputs transforms the basic latching circuit into the set-reset, *D*-type, *T*-type, and *JK* flip-flops.

In applications where a visual indication of the state of the flip-flop is required, a driver circuit may be added to turn ON a light-emitting diode (LED) or light bulb when the flip-flop is in its 1 state. A typical driver circuit is shown in Fig. 27-2. If an NPN driver is used, it should have a high dc current gain (H_{FE}) so that it does not require much current from the flip-flop when it is in the 1 state. (The design of the driver circuit must

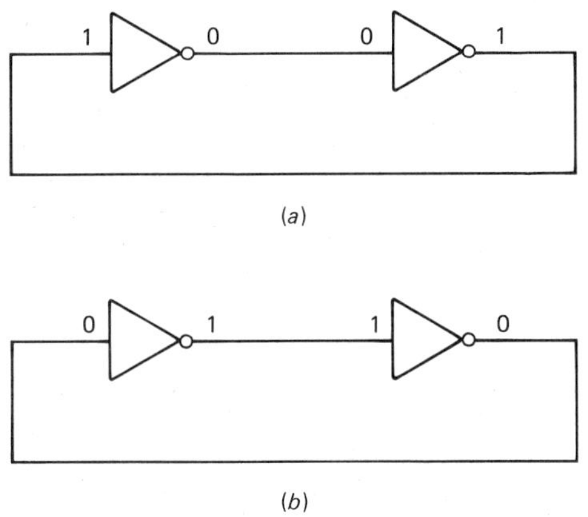

Fig. 27-1. A bistable circuit consisting of two inverters connected in a positive feedback arrangement.

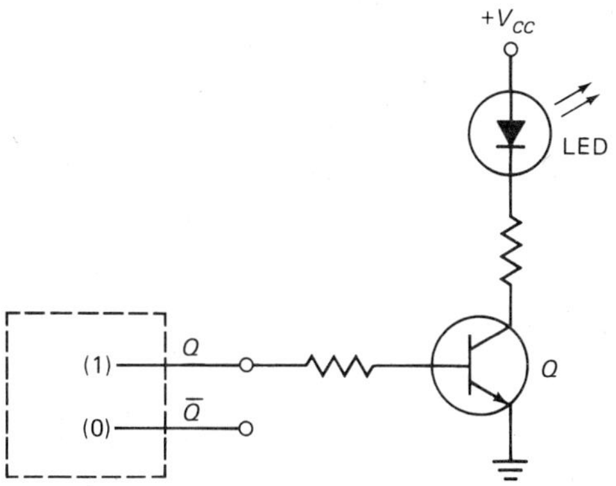

Fig. 27-2. LED driver.

take into account whether the flip-flop is TTL, CMOS, or other logic family and whether it is *current-sinking* or *current-sourcing logic*.)

LEDs are typically used as status indicators because they operate at low values of voltage and current, are available in various colors, and do not have a burn-out mechanism like tungsten filament light bulbs do.

Setting and Clearing the Flip-flop

When the *Q*, or 1 side, of a flip-flop is in the 1 state (in a positive logic system, the more positive voltage is present), it is said to be "set." When the *Q* side is in the 0 state, it is "cleared."

The *RS* Latch

The *RS* latch, the most elementary flip-flop, may be constructed of two NAND gates as in Fig. 27-3. This circuit is useful as a switch debouncer to eliminate multiple inputs from mechanical switches. When such switches close, they tend to bounce several times between the closed and open conditions; a high-speed circuit would consider each of these bounces an "input." Where only one input per switch depression is desired, a debouncer circuit is used.

The block diagram symbol for the *RS* latch is shown in Fig. 27-4*a*, and the truth table which describes the operation of this circuit is given in Fig. 27-4*b*. Note that this circuit, constructed of NAND gates, requires a low, or 0, input on *R* or *S* to cause a change of state, and once set or reset, further inputs to *S* or *R* have no effect. Attempting to set and reset the flip-flop simultaneously will produce an indeterminate or unpredictable state and therefore is illegal (or undefined).

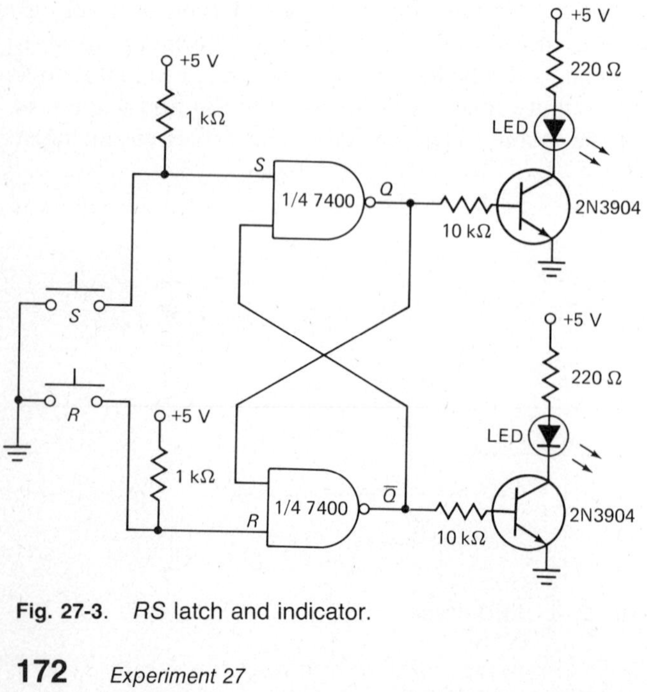

Fig. 27-3. *RS* latch and indicator.

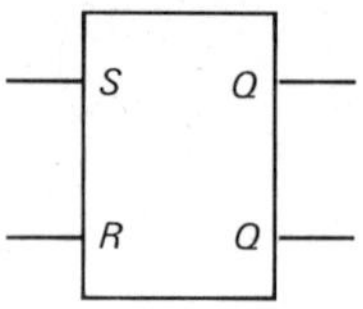

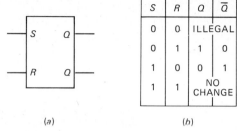

S	*R*	*Q*	$\overline{Q}$
0	0	ILLEGAL	
0	1	1	0
1	0	0	1
1	1	NO CHANGE	

(*a*) (*b*)

Fig. 27-4. *(a)* Block diagram for *RS* latch; *(b)* truth table.

Synchronous and Asynchronous Operation

When the state of a flip-flop can be changed at any time, it is operating *asynchronously* (not synchronized) with respect to any other event. On the other hand, digital systems often are timed by a system clock pulse. By adding gating circuits to the *RS* flip-flop, it can be made to operate synchronously: the *R* and *S* inputs can only change the state of the flip-flop when the "enable" signal is high (1). Due to the added logic in Fig. 27-5, this strobed (or clocked) *RS* latch is set or reset when *R* or *S* is high and the enable signal is high.

The *D*-type Flip-flop

The *D*-type flip-flop, or *D* latch, shown in Fig. 27-6, includes an inverter between the *R* and *S* inputs so that their inputs are always complementary and no illegal state can exist. The *Q* output assumes the same state as the *D* input when the enable signal is high and holds that state after *D* goes low. This circuit is useful for holding information produced by a counter to produce a steady display. The TTL 7475 is a 4-bit *D* latch which is used between a 7490 counter and a 7447 seven-segment decoder/driver in many counting and display applications.

The Binary Counter

With appropriate steering circuitry, a train of trigger pulses can be applied to a flip-flop in such a way that these trigger pulses cause the flip-flop to be set and reset alternately. This produces a square wave at the *Q* output which is at one-half of the trigger, or clock,

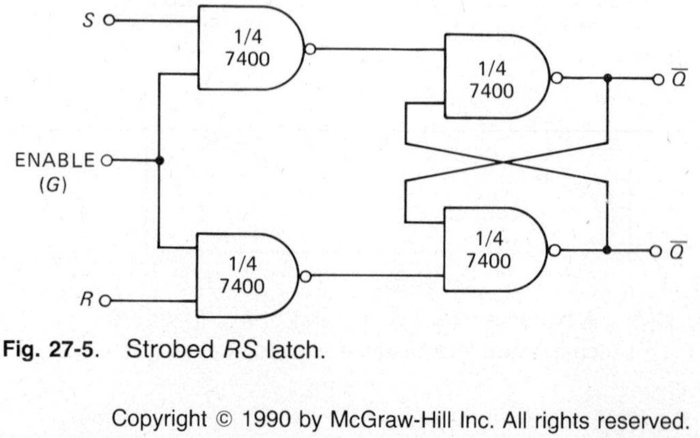

Fig. 27-5. Strobed *RS* latch.

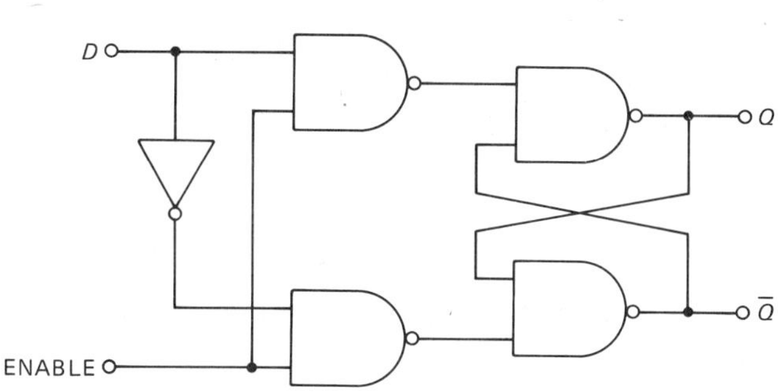

Fig. 27-6. *D*-type latch.

frequency. This concept is shown in Fig. 27-7. The output of one flip-flop can be used as the trigger source for another flip-flop. Groups of interconnected flip-flops are available in IC form as counters, registers, and latches.

The *JK* Flip-flop

The most versatile flip-flop is the *JK* flip-flop. It has no illegal states; the output is always predictable in terms of the input conditions. The logic symbol for the *JK* flip-flop is given in Fig. 27-8 together with its truth table. Two points should be noted: (1) if *J* and *K* are both zero, the trigger pulse will not cause a change of state, and (2) if *J* and *K* are both high, the output will "toggle," that is, it will change to the complement of the condition it had before the trigger pulse arrived. If *Q* was previously low, it will go high. If it was high, it will go low.

SUMMARY

1. A bistable multivibrator, or flip-flop, has two stable states and will remain in either one indefinitely. A brief trigger pulse is adequate to cause it to change state.
2. If a device can change states at any random time, it operates asynchronously. If it can only change states when a clock pulse or other timing event takes place, it operates synchronously with the clock pulse.
3. For positive logic, a flip-flop is set when its *Q* output is high and is reset, or cleared, when its *Q* output is low. The *Q* and $\overline{Q}$ outputs of a flip-flop are complementary. At any given time they have opposite states.
4. A flip-flop requires two trigger pulses to change states from high to low to high and therefore requires two trigger pulses for one complete cycle of its output. It inherently divides the frequency of its trigger pulses by 2.
5. The state of a flip-flop is indicated by the voltage level present at the *Q* and $\overline{Q}$ outputs. This voltage may be measured with a voltmeter or indicated by an LED status indicator driven by the *Q* or $\overline{Q}$ output.

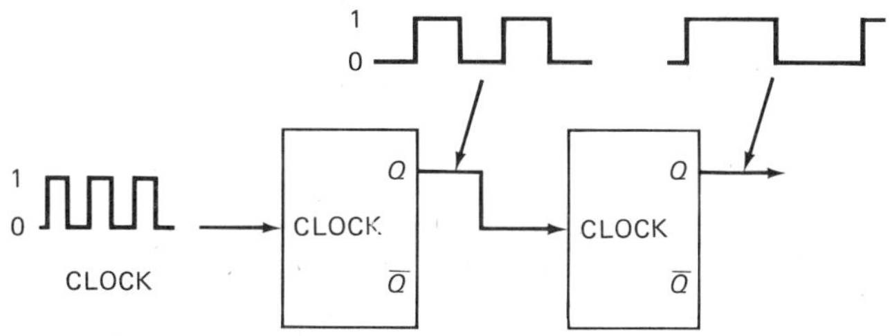

Fig. 27-7. Triggered flip-flops triggering on positive edge of clock waveform.

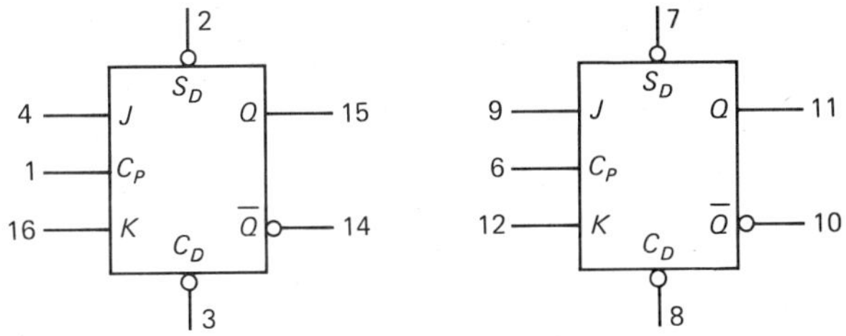

J-K F/F Truth Table

J	*K*	*Q*	$\overline{Q}$	
0	0	NO CHANGE		
0	1	0	1	
1	0	1	0	
1	1	$\overline{Q}$	*Q*	(TOGGLES)

Fig. 27-8. TTL *JK* flip-flop (7476). Asynchronous direct set *Sd* and direct clear *Cd*.

SELF-TEST

Check your understanding by answering these questions.

1. Another name for the flip-flop is the bistable ______ .
2. If the *Q* side of a flip-flop is high, the $\overline{Q}$ side of the flip-flop will be ______ .
3. A flip-flop is said to be set when its *Q* output is logical ______ .
4. A binary flip-flop whose output is low requires ______ trigger pulses to return it to the low state.
5. An LED driver circuit should turn the LED ON when the flip-flop is in the ______ state.
6. A flip-flop which can change states at any time operates ______ .
7. A register is a group of ______ .
8. Flip-flops produce ______ waveforms.

MATERIALS REQUIRED

- Power supply: Variable regulated low-voltage dc source
- Equipment: EVM; oscilloscope
- Resistors: ½-W 150-, two 1000-, 2700, 22,000-, 100,000-Ω
- Capacitors: 0.001-μF, 0.0015-μF, 25-μF 50-V
- Solid-state devices: IC 7427 (triple three-input NOR gate); IC 7476 (dual *JK* flip-flop); IC 555 (timer); 2N6004; LED RL2000 Litronix or equivalent
- Miscellaneous: Two SPDT switches; DPDT switch; two SPST switches; push-button make-break switch

PROCEDURE

Set-Reset Flip-flop

1. Connect the circuit shown in Fig. 27-9, using two of the NOR gates in the IC 7427. Connect terminal 14 of the IC to +5 V dc and terminal 7 to ground.
2. Complete Table 27-1 for the input conditions given, measuring and recording the voltage at Q and $\overline{Q}$. Follow the sequence of switch positions exactly as it is given in the table.
3. Design a switching circuit, using a DPDT switch, to replace S_1 and S_2 in Fig. 27-9 so that in one position of the switch the flip-flop is set and in the other reset.
4. Draw the circuit, connect it, and try it out. Prepare your own table to record the data obtained.

TABLE 27-1. Set-Reset Flip-Flop Characteristics

Switch Position		Trigger Status: Set or Reset	Output, V	
S_1	S_2		Q	$\overline{Q}$
0	+5			
0	0			
+5	0			
0	0			
0	+5			
+5	+5			
+5	0			
+5	+5			
0	+5			

Pulse-Triggering a Flip-flop

NOTE: Mechanical switches deliver a dc input to logic circuits which exhibits holding action on the circuits. Moreover, switch contacts produce noise; they bounce and make repeated contacts when first closed. Pulse circuits produce a more stable triggering action on logic circuits.

The monostable multivibrator (one-shot) discussed in Experiment 23 provides a trigger pulse and acts as a no-bounce switch for the experimental *JK* flip-flop in this experiment. Trigger pulses are supplied by the IC 555 which can be connected as an astable or as a monostable multivibrator. In this experiment it will be connected as a monostable circuit. When +5 V is applied to the IC 555, its output is TTL-compatible.

5. Construct the circuit shown in Fig. 27-10*a*. Terminal 2 is the input to the IC 555. *RC* is the time constant which determines the width of the positive pulse delivered at terminal 3. S_1 is a make-break switch which permits momentary grounding of pin 2 to start the one-shot action. The 100-kΩ resistor *R* and the 25-μF capacitor *C* will gener-

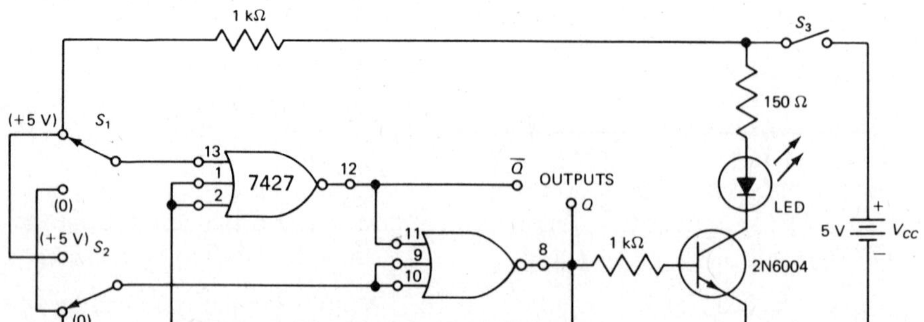

Fig. 27-9. Experimental set-reset flip-flop with LED status indicator.

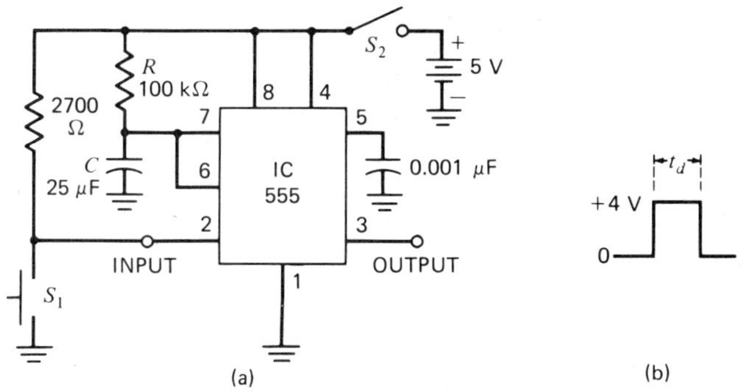

Fig. 27-10. *(a)* One-shot (OS) used to trigger a flip-flop; *(b)* output pulse when triggered.

ate a 3.5-s pulse (approximately) when S_1 is closed and quickly released. Set the output of the power supply at +5 V.

6. Apply power to the circuit by closing S_2. With an EVM, measure the voltage from terminal 3 to ground while S_1 is open. It should be 0 V. Momentarily close and release S_1, monitoring the voltage at terminal 3. The voltage should rise to +4 V, approximately, the instant S_1 is closed, and should remain at +4 V for a time interval t_d, even though S_1 is released. After a time interval t_d determined by the *RC* in Fig. 27-10*a*, the one-shot automatically resets and the output at terminal 3 drops to 0 V. The effect of triggering the one-shot by momentarily grounding terminal 2 is to generate a +4-V pulse of interval t_d, as in Fig. 27-10*b*.

7. Again momentarily close S_1 and release it, permitting the one-shot to run through its cycle. Monitor the voltage at terminal 3. Repeat.

8. Open S_2. Power OFF. Replace *C* by a 0.1-μF capacitor. What should be the effect on t_d as the one-shot is now triggered? ____________

9. Close S_2 Power ON. Repeat step 7. Did the pulse interval t_d increase or decrease? ____________ . Repeat. You are now ready to trigger a flip-flop with the output from the one-shot.

10. Connect one of the flip-flops in the IC 7476 as in Fig. 27-11. Connect pin 5 to +5.0 V and pin 13 to ground. The output of the one-shot, terminal 3 in Fig. 27-13*a*, is connected to terminal 1 on the IC 7476. $C = 0.1$ μF. $R = 100$ kΩ.

11. The flip-flop can assume either of its two stable states. Measure and record in Table 27-2 the voltage at *Q* (15) to ground; at $\overline{Q}$ (14) to ground. Indicate the status of the flip-flop. NOTE: It is ON if the LED indicator is lit.

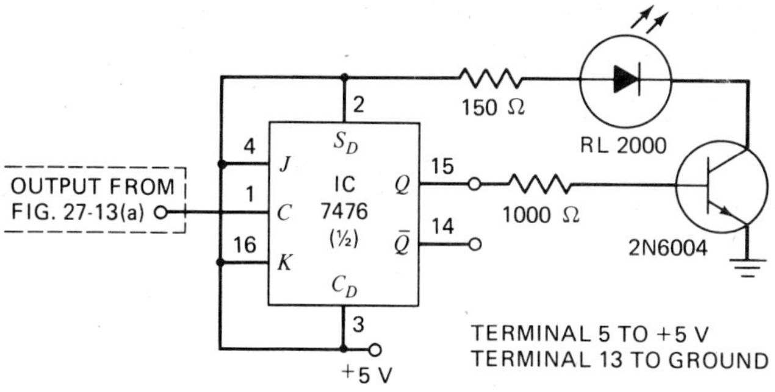

Fig. 27-11. Triggering a *JK* flip-flop with a one-shot.

TABLE 27-2. Pulsing a *J-K* Flip-Flop

	Voltage		*Status*
Step	*Q*	$\overline{Q}$	ON *or* OFF
11			
12			
13			

12. Pulse the flip-flop by momentarily closing and releasing S_1. Measure and record the voltages at *Q* and at $\overline{Q}$ and indicate the status of the flip-flop.

13. Repeat step 12 three more times. What conclusion can we draw about *T* pulsing a flip-flop?

Flip-flop Used as a Frequency Divider

14. Connect the circuit shown in Fig. 27-12. A differentiated square wave is used to *T*-pulse a flip-flop. Set the frequency of the square-wave generator at 1000 Hz, square-wave amplitude at 10 V p-p.

    Externally synchronize the oscilloscope with the square wave from the generator. Observe the input waveform and adjust the sweep controls for four cycles, as in Table 27-3.

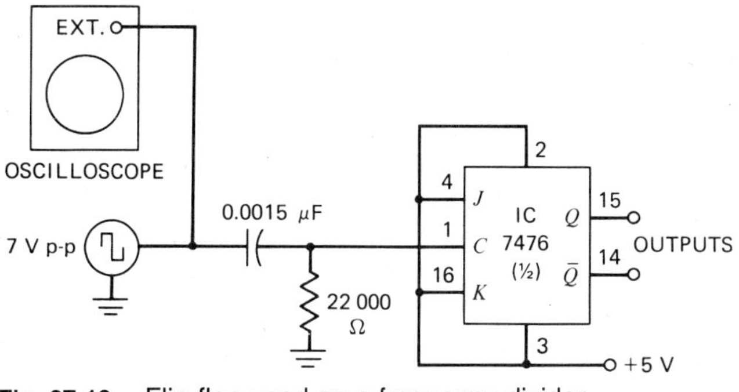

Fig. 27-12. Flip-flop used as a frequency divider.

TABLE 27-3. Flip-Flop as a Frequency Divider

	Waveform	*Frequency,* Hz
Input Waveform	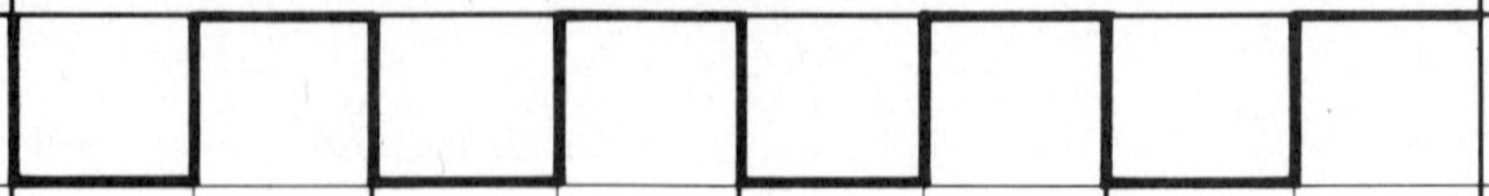	1000
Waveform at Q		
Waveform at $\overline{Q}$		

15. Observe and record the waveform at Q, in proper time phase with the input square wave. Indicate its frequency. Observe also the waveform at $\overline{Q}$ and draw it in proper time phase with the input. Indicate its frequency. NOTE: If you are using a dual-trace oscilloscope, let the input waveform appear on the upper trace, and the waveform at Q on the lower trace. Then let the waveform at Q appear on the upper trace, and that at $\overline{Q}$ on the lower trace.

QUESTIONS

1. How does a flip-flop differ from an astable multivibrator?
2. In the experimental circuit (Fig. 27-11), the flip-flop is set (LED is ON). How can we reset it? How can we then set it again?
3. How can we tell if the flip-flop in Fig. 27-6 is set? If it is set, how can we reset it?
4. What is meant by a binary (T) input to a flip-flop?
5. What effect does each pulse on the T input of a flip-flop have on its status?
6. What is meant by "steering" diodes?
7. Where are steering diodes used?
8. Why can we say that a flip-flop which is T-pulsed divides by 2?
9. What is an LED? When will it glow?
10. Why are LEDs replacing filamentary pilot lamps?

Answers to Self-Test

1. multivibrator
2. low
3. 1
4. 2 (the first produces a 1, the second produces a 0)
5. high or 1
6. asynchronously
7. flip-flops
8. rectangular

EXPERIMENT

28

JK FLIP-FLOPS AND COUNTERS

OBJECTIVES

1. To become more familiar with the *JK* flip-flop
2. To learn how counters are used in computers and other digital systems
3. To demonstrate the use of the *JK* flip-flop in binary and BCD counters

BASIC INFORMATION

Characteristics of a TTL *JK* Flip-flop

The *JK* flip-flop, probably the most versatile member of the flip-flop family, may have as many as five inputs and two complementary outputs. In addition to the *J*, *K*, and clock inputs, there may be a direct (asynchronous) input for clearing the flip-flop and a similar input for setting it. Figure 28-1 gives the block diagram symbol for four- and five-input *JK* flip-flops.

Clocked Inputs

The *J* and *K* inputs establish conditions which determine what state the flip-flop will assume when triggered, or "clocked." This change of state will occur on the negative-going edge of the clock pulse; that is, when the clock input abruptly changes from high to low. This occurs at time t_2 in Fig. 28-2.

The operation of the *JK* flip-flop may be summarized as follows:

1. If both *J* and *K* are positive (1), a train of clock pulses at *C* will "toggle" the flip-flop; i.e., it will alternate between its two states as it receives trigger pulses.
2. If the *J* input is high and the *K* input is low, the next clock pulse will set the flip-flop if it was previously cleared. The *Q* output will go to the high, or 1 state. If it was already high, the clock pulse has no effect.
3. If the *K* input is high and the *J* input is low, the next clock pulse will clear or reset the flip-flop if it was previously set. If it was already cleared, the clock pulse has no effect.
4. If both *J* and *K* are low (0), the clock pulse has no effect: the flip-flop remains as it was. It does not change state.

Direct Inputs

When the direct clear input C_d is low, it immediately clears the flip-flop if it was set, regardless of the status of the other inputs. The *Q* output goes low and the $\overline{Q}$ output goes high.

A digital counter will contain many flip-flops, and

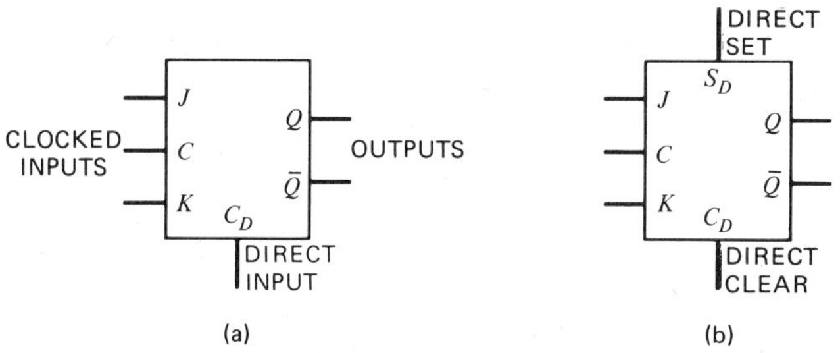

Fig. 28-1. Inputs and outputs in *(a)* four-input *JK* flip-flop; *(b)* five-input *JK* flip-flop.

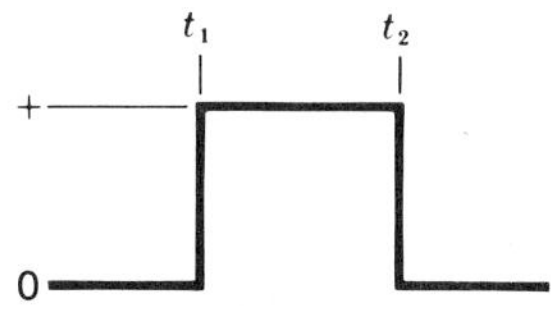

Fig. 28-2. The *C* input flips or flops the *JK* only on a negative transition, that is, on the negative-going edge, t_2

their C_d lines are usually connected together. Thus, before the start of a counting period, for example, they all may be cleared simultaneously so that the count begins from zero.

The direct set input S_d immediately sets the *JK* flip-flop when S_d is brought low. S_d and C_d should never be brought low at the same time since this is a disallowed state and the effect on the flip-flop is undefined.

Counter Applications

Counters are used extensively in modern test equipment with digital readouts. A good example is the digital multimeter, which is replacing, to a large extent, the analog volt-ohm-milliammeter (VOM). Since counting is the basic process in so many digital devices and systems, let us examine some very common types.

BINARY COUNTER

Consider Fig. 28-3. Here four *JK* flip-flops are arranged as a 4-bit counter with the *Q* output of the first acting as the clock input to the second. This is a straight binary counter, and it has 16 unique states, or combinations of outputs.

Notice that the *J* and *K* inputs of all flip-flops are connected to +5 V; they will toggle every time their clock input goes low. Assume that the C_d line is brought low, thereby clearing each flip-flop.

The negative-going edge of the first clock pulse to flip-flop A (FFA) causes its state to change from 0 to 1, and the next clock pulse causes FFA to change from 1 to 0. This negative-going output from FFA causes flip-flop B (FFB) to change states from 0 to 1. Note that it took 2 clock pulses into FFA to cause FFB to change states. It will require a second negative-going output from FFA to cause FFB to change its state back from 1 to 0 and thereby trigger flip-flop C (FFC). Each flip-flop requires two negative-going transitions from the previous flip-flop to change its state from 0 to 1 to 0: each stage divides the frequency of its clock pulses by 2.

If the clock pulses applied to FFA are at 800 Hz, the *Q* output of FFA will be a square wave at 400 Hz; at FFB, 200 Hz; at FFC, 100 Hz; and at flip-flop D (FFD), 50 Hz. This is an example of using the counter as a binary frequency divider.

It will take 16 input pulses to FFA to cause the 16 possible combinations of states of FFA, FFB, FFC, and FFD. These states are shown in Table 28-1.

The counter in Fig. 28-3 used four counters to count to 16. In general, a straight binary counter using *n* flip-flops has how many states?

$$\text{Number of states} = 2^n$$

Notice that the 4-bit counter has 16 states, beginning with zero, and that the highest count which can be indicated is:

$$\text{Highest count indicated} = 2^n - 1$$

where *n* is the number of flip-flops.

Notice that, for each pulse applied to the input of FFA, the entries in Table 28-1 are the binary representations of the decimal number of input pulses. If an indicator, such as an LED, were connected to the *Q* output of each flip-flop, the number of input pulses applied to FFA could be determined by noting which LEDs were ON and adding together the place values of those LED indications. In early computers it was common to have such binary indicators rather than the typical seven-segment decimal readouts which are so common today.

BCD COUNTERS

The straight binary counter discussed above has many applications within digital computers and other devices, but it is inconvenient to use by untrained people, who are accustomed to decimal quantities and notation. The interface between the natural mathematical language of digital computers, the binary system, and the human user can be improved by using a modified binary system called *Binary-coded decimal,* or BCD.

In the BCD system the same patterns of 1s and 0s are used to represent the decimal numbers 0 through 9 as are found in straight binary notation. However, the binary patterns beyond 1001 are not allowed. In other words, a BCD counter counts from 0 (0000) to 9 (1001). It has 10 states instead of 16.

Fig. 28-4 shows how a binary counter can be modified so that it will count by 10 instead of 16. NOR gates *X* and *Y* are connected to the $\overline{Q}$ sides of FFB and FFD

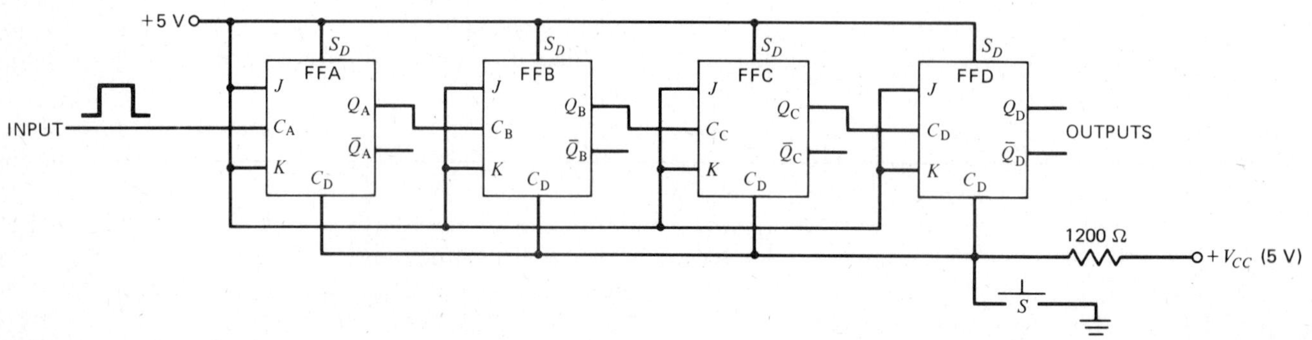

Fig. 28-3. Four-position binary counter, using TTL *JK* flip-flops.

TABLE 28-1. Flip-Flop Status versus Input Count

Count	*1*	*2*	*3*	*4*	*5*	*6*	*7*	*8*	*9*	*10*	*11*	*12*	*13*	*14*	*15*	*16*
Flip-flop A binary 1	*		*		*		*		*		*		*		*	
Flip-flop B binary 2		*	*			*	*			*	*			*	*	
Flip-flop C binary 4				*	*	*	*					*	*	*	*	
Flip-flop D binary 8								*	*	*	*	*	*	*	*	

NOTE: * equals ON status; blank equals OFF status

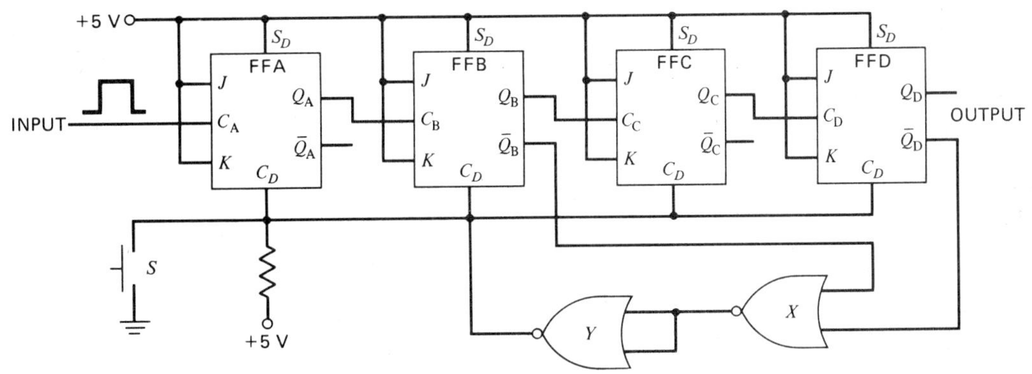

Fig. 28-4. Modified binary counter is a BCD counter.

to recognize the pattern 1010 (decimal 10). When this pattern occurs, the output of NOR gate Y goes low and resets the counter to 0000. NOR gates X and Y constitute an OR gate, and the inputs to this gate are $\overline{Q}_b$ and $\overline{Q}_d$. The boolean algebra expression for the clear signal is:

$$C = \overline{Q}_b + \overline{Q}_d$$

C is true (high, 1) when Q_b is false (low, 0), when Q_d is false, and when they are both false. C is only low when Q_b and Q_d are both high. In other words, the reset signal could also have been obtained by NANDing Q_b and Q_d. (C is low when Q_b and Q_d are both high.)

A BCD counter capable of counting to 999,999 is shown in Fig. 28-5. Four flip-flops are required for each decade, or decimal digit, in the count. Such a counter made with individual flip-flops would require a lot of space but would be preferred over a counter using discrete components. The physical size of a 6-decade counter could be reduced significantly by using larger scale ICs, such as the TTL 7490 BCD counter, which includes all four flip-flops on a single chip. Other IC counter chips are available which include all the flip-flops and other circuitry needed for a complete three- or four-digit decimal counter.

ACCUMULATOR

The BCD counter just discussed is an example of a serial counter. Input pulses occur one at a time, one after the other.

Fig. 28-6 shows a BCD accumulator, a type of adder which accepts "parallel" inputs to each decade. To illustrate its operation, let us assume that every flip-flop, including the carry flip-flops, are cleared. Let us enter the number 325 into the accumulator by pulsing the unit's input five times, the ten's input two times, and the hundred's input three times. The number 325 is now in the accumulator, and no carry flip-flop is set. Now let us enter the quantity 176 into the accumulator by pulsing the unit's input six times, the ten's input seven times, and the hundred's input one time. Because 5 + 6 = 11, the unit's decade sets the carry flip-flop, and the unit's decade indicates 1. Similarly, in the ten's column, 7 + 2 = 9, and the carry flip-flop is not set. In the hundred's column, 3 + 1 = 4, and the carry flip-flop is not set.

When the carry signal goes high, any carry flip-flops which are set produce an output from the AND gates (A1, A2, A3) and inverters (I1, I2, I3), thus triggering the next decade and advancing its count by one where there was a carry. When the carry signal goes low, the

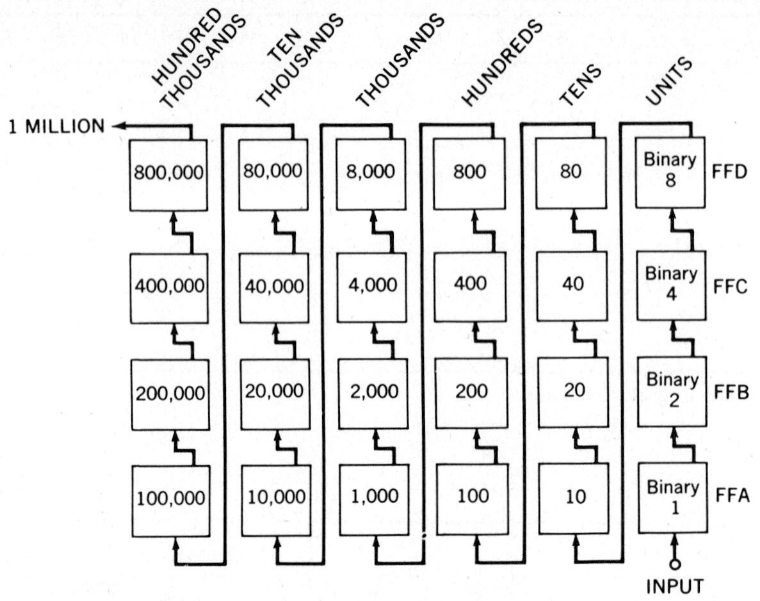

Fig. 28-5. BCD counter using modified binary counter in each decimal position.

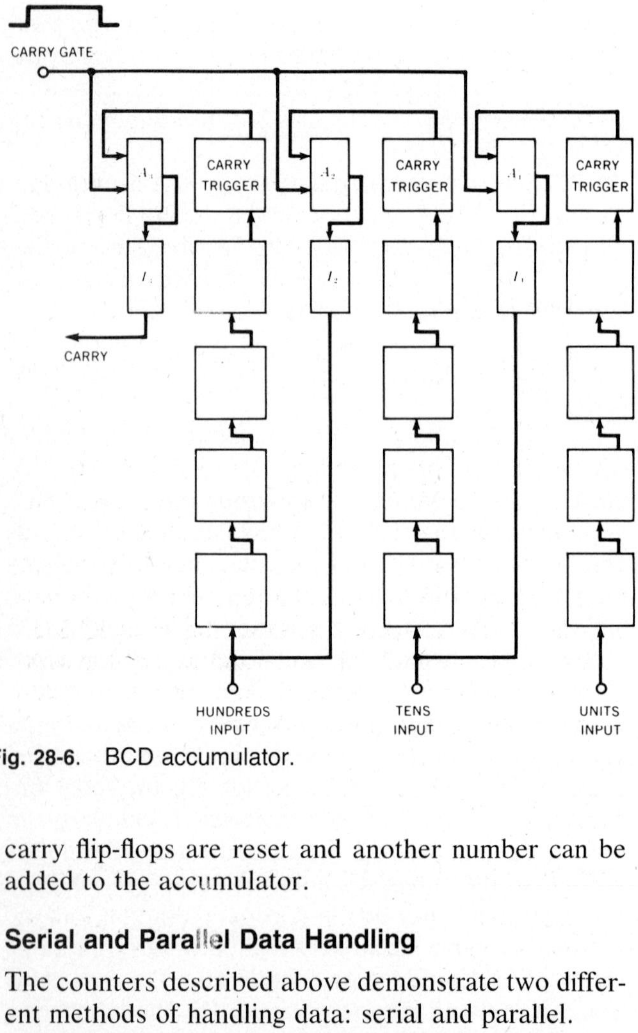

Fig. 28-6. BCD accumulator.

carry flip-flops are reset and another number can be added to the accumulator.

Serial and Parallel Data Handling

The counters described above demonstrate two different methods of handling data: serial and parallel.

Serial processing means that each bit of the data precedes the next bit in time. The bits are strung out in time, one after the other. Serial processing is simple but slow.

Parallel processing means that the bits of data are all available at the same time. The accumulator above had parallel inputs to the 3 decades. The trigger pulses could have been applied to the 3 decades simultaneously. Parallel processing is more complex than serial processing, but it is faster.

SUMMARY

1. A clocked *JK* flip-flop has a clock (or trigger) input, and *J* and *K* inputs which determine which state the flip-flop will assume on the next clock pulse.
2. A *JK* flip-flop may also have direct set and direct clear inputs S_d and C_d, which asynchronously set or clear the flip-flop.
3. The operation of a clocked *JK* flip-flop is described by the following truth table:

J	*K*	Q	*On the next clock pulse:*
0	0	*Q*	(no change)
0	1	1	(if cleared)
1	0	0	(if set)
1	1	$\overline{Q}$	(toggles on each pulse)

4. The *JK* flip-flop is found in binary counters, BCD counters, accumulators, registers, and many other applications.

SELF-TEST

Check your understanding by answering these questions.

1. The *JK* flip-flop described in this chapter changes state only on the 1 to 0 transition, or the ___________ -going edge of the clock pulse.
2. A straight binary counter using five flip-flops would have ___________ states.
3. The highest binary number the counter of question 2 could indicate is ___________ (decimal 31).
4. The binary number 1010 is an allowed number in BCD. ___________ (true/false)
5. When the *J* and *K* inputs are both high, the flip-flop ___________, or changes states on every pulse.
6. To count to 999 using BCD counters, ___________ flip-flops would be required.
7. If a BCD counter is cleared and 10 pulses are applied to the clock input of the first flip-flop, the state of all four flip-flops will be ___________ .

MATERIALS REQUIRED

- Power supply: Variable, regulated dc source
- Equipment: EVM
- Resistors: ½-W, 150-, 220-, 330-, 470-Ω, two 1000-, 2700-, 6800-Ω
- Capacitors: 0.001-μF; 100-μF 10-V
- Semiconductors: ICs: two 7476, 7427, 555; LEDs: four RL2000; or the equivalents
- Miscellaneous: SPST switch; two make-break (doorbell-type) switches

PROCEDURE

Clocked Characteristics of a TTL *JK* Flip-flop

1. Construct the circuit shown in Fig. 28-7. This monostable (one-shot) multivibrator will be used to pulse (single-step) the *JK* flip-flops in this experiment. When make-break switch S_1 is momentarily closed, the output of the one-shot at terminal 3 will rise from 0 to +4 V (approximately) and will remain at this level for a period of time determined by the time constant *RC*.
2. With an EVM set on +dc volts, monitor the output at terminal 3.
3. Momentarily close and release S_1. Observe the voltage change at the output, as indicated by the EVM. Repeat, and measure the pulse width t_d: ___________ s.
4. Leave Fig. 28-7 connected. Connect the circuit shown in Fig. 28-8. The 7476 is a dual *JK* flip-flop. Only one of the flip-flops is used now. Connect terminal 5 to +5 V, and terminal 13 to ground. The direct clear (C_D) and direct set (S_D) terminals are both connected through 1000-Ω resistors to +5 V. Voltmeter V_1 is used to monitor the input to flip-flop A.

   The LED connected in the output of the flip-flop acts as a status indicator. When the flip-flop is set, the LED will light. When the flip-flop is reset, the LED will not light. Clear the flip-flop, if it is set, by momentarily closing make-break switch S_3.
5. Momentarily close S_1, pulsing flip-flop A. Observe V_1 and the LED and determine whether the flip-flop sets on the negative or positive edge of the triggering pulse. ___________ Clear flip-flop A.
6. Pulse and set flip-flop A. Remove the *J* input (terminal 4) from +5 V, and ground terminal 4. All the other flip-flop connections are undisturbed. The LED should remain lit, showing that flip-flop A is still set. Pulse flip-flop A. Does this clear it? Indicate in Table 28-2. Measure and record the voltages at Q and $\overline{Q}$. Pulse again and indicate in Table 28-2 flip-flop A status and the voltages at Q and $\overline{Q}$.
7. Repeat step 6 for each of the *JK* input conditions in Table 28-2 and record your results.

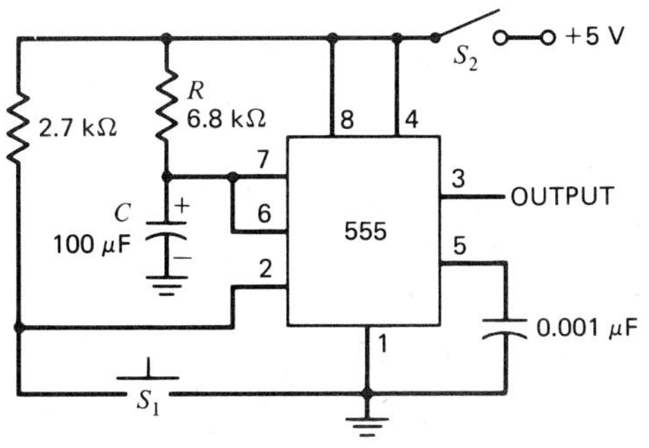

Fig. 28-7. The IC 555 connected as a one-shot pulse generator.

Direct (C_D, S_D) Characteristics of a *JK* Flip-flop

8. Reconnect the *JK* flip-flop as in Fig. 28-8. The output of the one-shot (Fig. 28-7) remains connected to terminal 1 of the *JK* flip-flop. The *J* and *K* inputs are connected to +5 V. The C_D and S_D terminals are connected through 1000-Ω resistors to +5 V. Power ON.

TABLE 28-2. Clocked Characteristics of a *J-K* Flip-Flop

Flip-Flop Status at Start	*Input Conditions*		*Flip-Flop Status After Pulsing*	*Voltage at*	
	J	*K*	ON *or* OFF	Q	$\overline{Q}$
Set	0	+5 V			
Clear	0	+5 V			
Clear	+5 V	0			
Set	+5 V	0			
Set	+5 V	+5 V			
Clear	+5 V	+5 V			
Set	0	0			
Set	0	0			

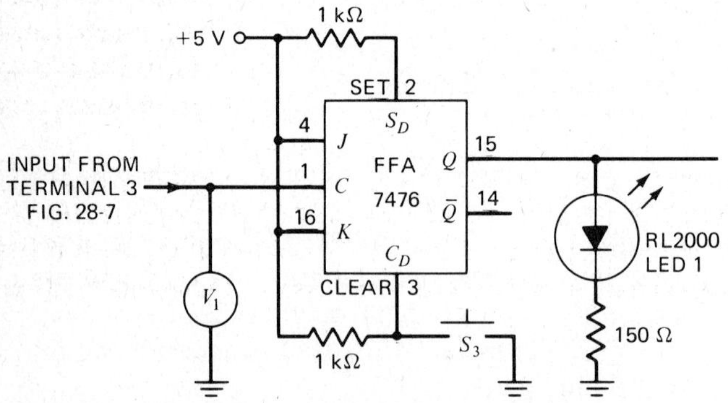

Fig. 28-8. Experimental *JK* flip-flop is pulsed by the monostable multivibrator from Fig. 28-7.

9. If the *JK* flip-flop is set, clear it by pulsing.
10. Pulse the *JK* flip-flop and indicate its status in Table 28-3, *after* pulsing. Also measure and record the voltages at Q and $\overline{Q}$.
11. The flip-flop should be set. *Leave it set.* Ground terminal 3, C_D. Leave terminal 2, S_D, connected through its 1000-Ω resistor to +5 V. Observe and record the flip-flop status for these input conditions.
12. Pulse the flip-flop and indicate its status after pulsing. Measure and record the voltages at Q and $\overline{Q}$.

TABLE 28-3. Direct Characteristics of a *J-K* Flip-Flop

Flip-Flop Status at Start	*Input Conditions*		*Flip-Flop Status After*		*Voltage at*	
	C_D	S_D	*Direct Input*	*Pulsing*	Q	$\overline{Q}$
Clear	+5 V	+5 V	X			
Set	+0	+5 V				
Clear	0	+5 V	X			
Clear	+5 V	0				
Set	+5 V	0	X			
Set	0	0				
Set	0	0	X			

13. Complete Table 28-3 for the input conditions shown.

Binary Counter

14. Connect the circuit of Fig. 28-9. The *JK* flip-flops are contained in two 7476s. The toggle (pulse) output of the one-shot (Fig. 28-7), remains connected to terminal 1 (*C* input) of flip-flop A. Terminals 5 of the two 7476s are connected to +5 V, with terminals 13 to ground. Switch S_3 is used to clear directly all the flip-flops. LEDs 1 through 4 indicate respectively the status of flip-flop A through flip-flop D.

15. Preclear the flip-flops by momentarily closing S_3. All the flip-flops should be OFF. Pulse flip-flop A. Indicate in Table 28-4 the status of each flip-flop. *Check* the space for each flip-flop that is set, and leave blank if the flip-flop is clear. NOTE: In the procedure which follows you will be required to pulse the counter sequentially. Observe voltmeter V_1 in the input to flip-flop A. After each pulse, wait until V_1 shows that the pulse is complete, that is, until the voltage has dropped to 0 V. Pulse the counter sequentially to the count of 16, entering your results in Table 28-4. NOTE: If you lose count, or if the flip-flops trigger falsely, reset the flip-flops and start over again.

16. Repeat step 15. Do not record again the status of the flip-flops, but check your results with those in Table 28-4 to see that they are consistent.

BCD Counter

17. Modify Fig. 28-9 by adding the circuit in Fig. 28-10, as shown. *X* and *Y* are NOR gates from the 7427. The inputs to NOR gate *X* are the $\overline{Q}$ terminals 10 from flip-flop B and flip-flop D. Connect terminal 7 of the 7427 to ground and terminal 14 to +5 V. The LEDs should light normally on toggle pulses 1 through 9. On count 10, however, all the LEDs should go out, showing that the added circuit causes all the flip-flops to reset at the tenth count, as would be expected of a BCD counter.

18. Power ON. Clear the flip-flops if any of the LEDs are lit, by momentarily closing S_3.

19. Pulse the counter. Indicate in Table 28-5 which binary flip-flops are ON. Leave blank the spaces for the flip-flops which are OFF.

TABLE 28-4. Flip-Flop Status versus Input in Experimental Binary Counter

Count	*1*	*2*	*3*	*4*	*5*	*6*	*7*	*8*	*9*	*10*	*11*	*12*	*13*	*14*	*15*	*16*	
Flip-Flop A binary 1																	
Flip-Flop B binary 2																	
Flip-Flop C binary 4																	
Flip-Flop D binary 8																	

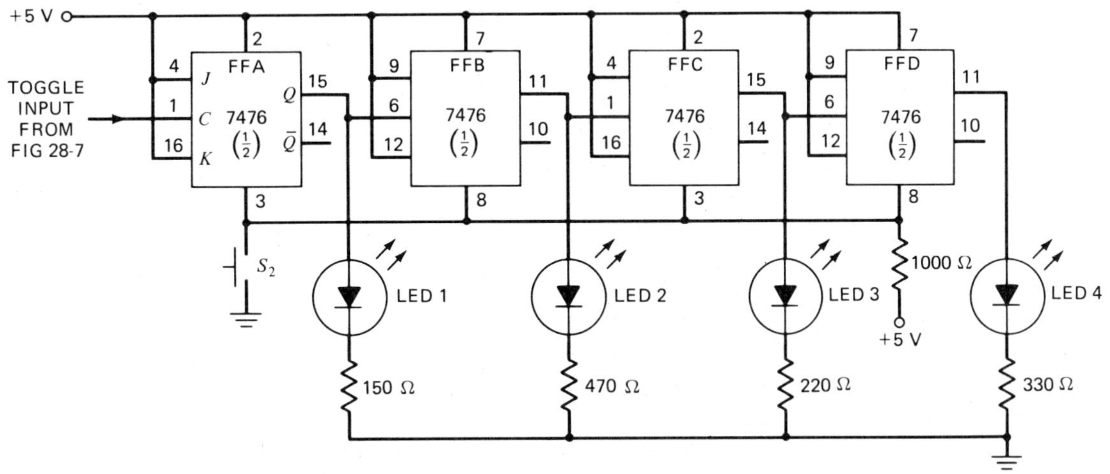

Fig. 28-9. Experimental counter with indicators.

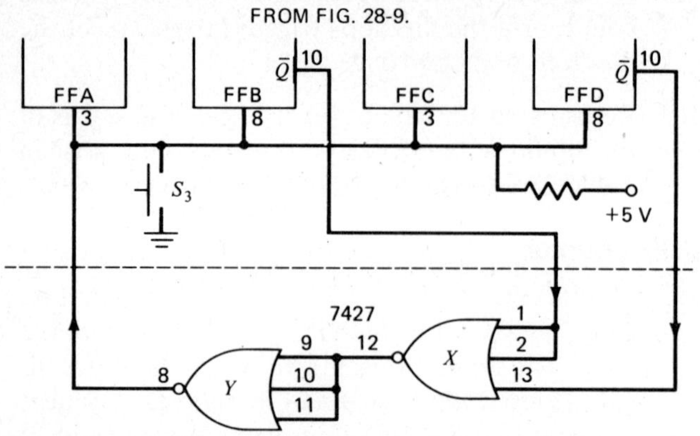

Fig. 28-10. Modifications which convert binary counter of Fig. 28-9 into a BCD counter. The portion above the dotted line is from Fig. 28-9.

20. Pulse the counter sequentially up to and including the count of 10, entering your results in Table 28-5. If you lose count or if the flip-flops trigger falsely, start over again.

21. Repeat steps 19 and 20, comparing your results with those in Table 28-5. They should be the same. NOTE: Leave pulser and BCD counter connected, as it will be used in the next experiment.

TABLE 28-5. Flip-Flop Status versus Input in BCD Counter

Count	*1*	*2*	*3*	*4*	*5*	*6*	*7*	*8*	*9*	*10*
Flip-Flop A binary 1										
Flip-Flop B binary 2										
Flip-Flop C binary 4										
Flip-Flop D binary 8										

QUESTIONS

1. Which of the inputs of the *JK* flip-flop have prior control over the status of the circuit? Support your answer by referring to your data.
2. For what connections of *J* and *K* will a *JK* flip-flop refuse to respond to a toggle input? Refer to your data.
3. What is meant by: A *JK* flip-flop responds to the negative transition on the toggle (*C*) input?
4. What modification is required to make the binary counter in Fig. 28-3 count to 64?
5. In question 4, what will be the status of each flip-flop at the count of 63?
6. Two counters are required: (1) a straight binary counter and (2) a BCD counter. Each must be able to count to at least 999. How many flip-flops are required for each counter?
7. What is the main advantage of a BCD counter over a straight binary counter?
8. In the BCD counter in Fig. 28-4, what is the input to *X* on the seventh count, assuming the counter was clear at the start? What is the output of *Y* on the same seventh count?

Answers to Self-Test

1. negative
2. 32
3. 11111
4. false
5. toggles
6. 12
7. 0000

EXPERIMENT 29

DECODING COUNTERS AND DIGITAL DISPLAYS

OBJECTIVES

1. To learn how the individual states of a counter may be recognized, or "decoded"
2. To show how the states of a BCD counter may be decoded
3. To introduce seven-segment displays and NIXIE tubes
4. To demonstrate the operation of seven-segment LEDs, IC counters, and decoder/drivers

BASIC INFORMATION

Decoding Counters

To decode a count means to recognize when it, alone, of many other patterns of 1s and 0s exists.

For example, rather than present the user with a binary indication of lights ON and OFF, recognizable decimal indications are usually provided today: 9 is more readily understood by untrained users than 1001.

The controller of a machine tool may include a clock pulse generator, counter, and decoders to define time intervals. On the counts of 2 and 3, a part may be heated while it is held in position during counts 1, 2, 3, and 4. During later intervals of the sequence, other operations may be performed on the part before the sequence is repeated. Electronic sequencers of this type replace older methods involving motors, cams, and microswitches. (A programmable controller is an even more recent method for controlling machines.)

Logical Decoding

The state of a counter is decoded using logic. For example, if the four $\overline{Q}$ outputs (Fig. 29-1) from a BCD counter are connected to the inputs of a four-input AND gate, its output will only be high on the count of zero, when all four $\overline{Q}$ outputs are high. A light bulb or LED driven by the output of this AND gate would be ON only on the count of zero and could be labeled 0.

Similarly, if the inputs to another four-input AND gate were connected to the Q_a, Q_b, Q_c, and Q_d outputs as shown in Fig. 29-1, its output would be high only on the count of 1. With 10 four-input AND gates, therefore, the 10 states of a BCD counter could be decoded and a unique indication provided for each count.

Using the method above, the user would be presented with 10 indicators labeled 0, 1, . . . , 9. The indicator lighting up shows the count. This method is adequate for some purposes, but a better and more modern method uses a seven-segment LED display.

Seven-Segment Displays

The basis for the seven-segment display is shown in Fig. 29-2. Each of the segments is a light source which can be turned ON individually. In small displays, the segments may be LEDs, and in physically larger displays, they may be incandescent light bulbs or other light sources.

By activating various combinations of these segments, the digits through 9 may be formed as well as a few letters. Figure 29-3 shows the appearance of the 10 digits formed by a seven-segment display.

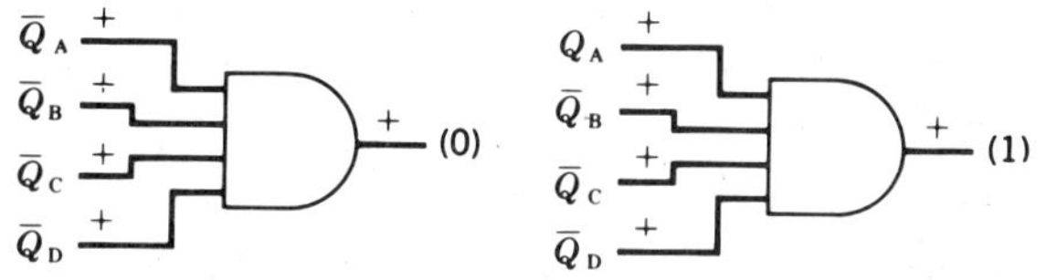

$\overline{Q}_A, Q_B, \overline{Q}_C, \overline{Q}_D = 2$
$Q_A, Q_B, \overline{Q}_C, \overline{Q}_D = 3$
$\overline{Q}_A, \overline{Q}_B, Q_C, \overline{Q}_D = 4$
$Q_A, \overline{Q}_B, Q_C, \overline{Q}_D = 5$

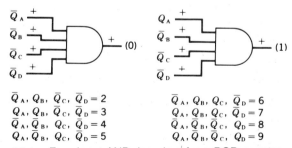

$\overline{Q}_A, Q_B, Q_C, \overline{Q}_D = 6$
$Q_A, Q_B, Q_C, \overline{Q}_D = 7$
$\overline{Q}_A, \overline{Q}_B, \overline{Q}_C, Q_D = 8$
$Q_A, \overline{Q}_B, \overline{Q}_C, Q_D = 9$

Fig. 29-1. Four-input AND decoders for a BCD counter. Each of the outputs of the AND gates activates a unique numerical driver, 0 through 9.

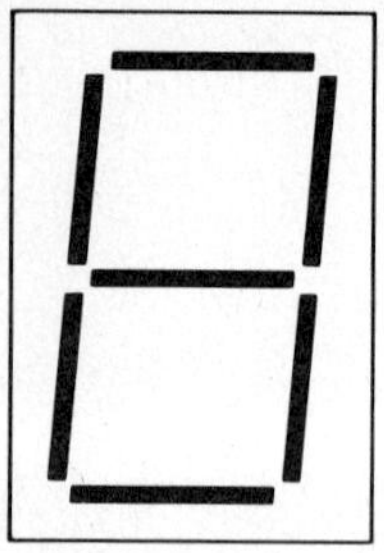

Fig. 29-2. Seven-segment LED display.

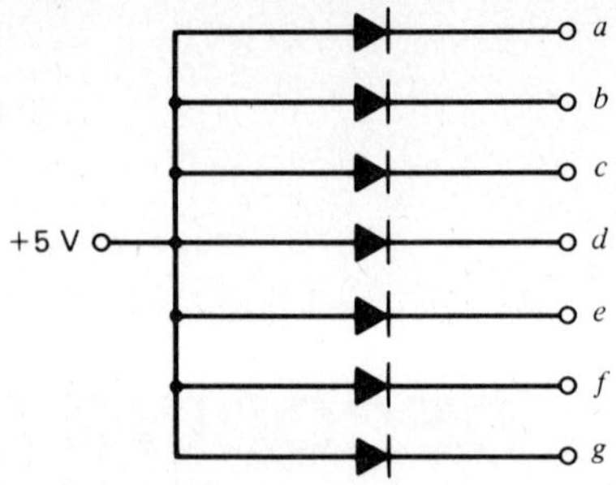

Fig. 29-4. Schematic diagram of a common-anode, seven-segment LED.

An LED seven-segment display consists of seven diodes having either a common-anode or a common-cathode connection. (NOTE: The two types are *not* interchangeable and require different drivers.) Figure 29-4 shows the equivalent circuit of a common-anode, seven-segment LED display.

Figure 29-5 shows how to test the individual segments of a seven-segment LED. Since the voltage drop across a conducting LED is about 1.7 V, the current-limiting resistor R should be 330 Ω if the current through the LED is to be 10 mA.

$$R = \frac{5 - 1.7}{0.01} = \frac{3.3}{0.01} = 330\Omega$$

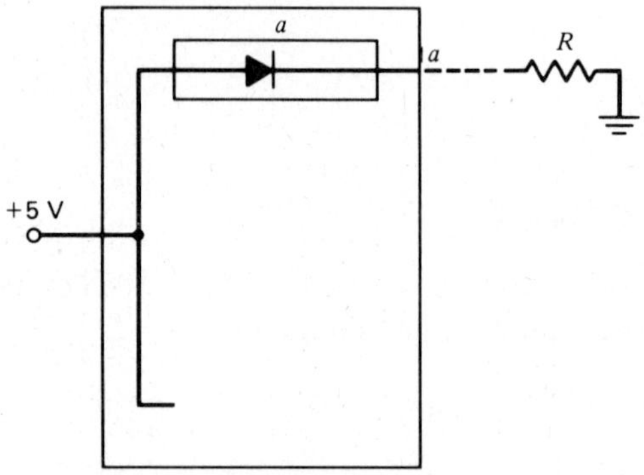

Fig. 29-5. Testing line segment *a* of a seven-segment LED display.

Decoder/Driver

A special IC device is available which will ground the appropriate cathodes through a current-limiting resistor so that the desired digit is seen. For the common-anode, seven-segment LED display, this decoder/driver is the TTL 7447. It accepts as inputs the four outputs from a BCD counter, decodes the 10 states, and causes the required segments to light up.

Digital Counter/LED Display

Figure 29-6 shows how a BCD counter, decoder/driver, and LED indicator may be combined to show the number of times a pulse has been received. Manually generated pulses may be used to demonstrate the operation of this circuit, but it will just as easily count and display pulses occurring millions of times per second.

Alphanumeric Displays

LED displays are low-voltage devices and are inherently compatible with ICs. They have limitations, how-

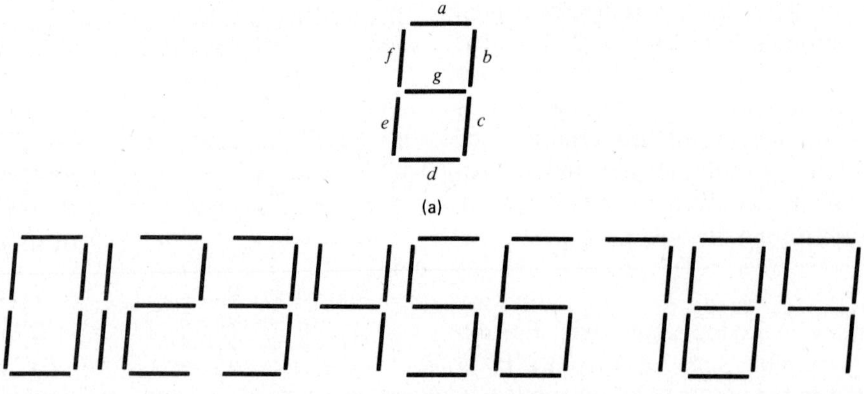

Fig. 29-3. *(a)* Lettered LED segments; *(b)* numerals which can be formed, 0 through 9, when the proper combination of segments is lit.

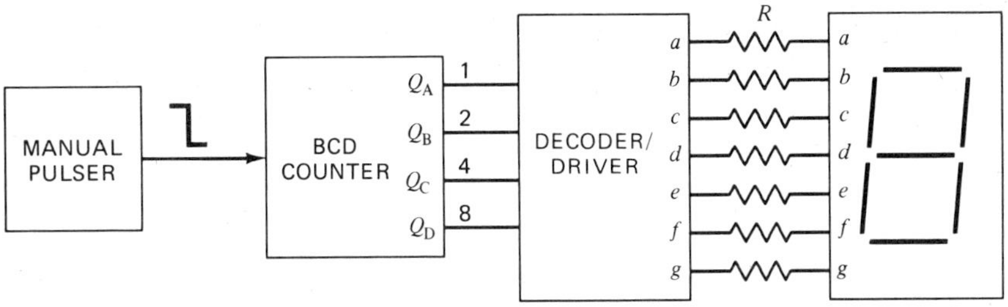

Fig. 29-6. Block diagram of a simple digital display system.

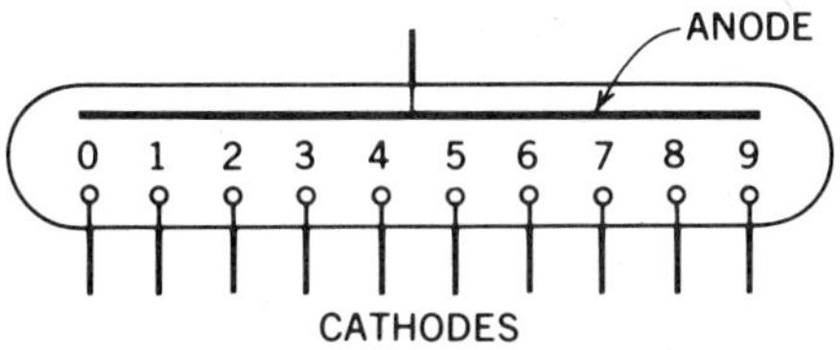

Fig. 29-7. Symbol for a 10-cathode NIXIE tube displaying the numerals 0 through 9.

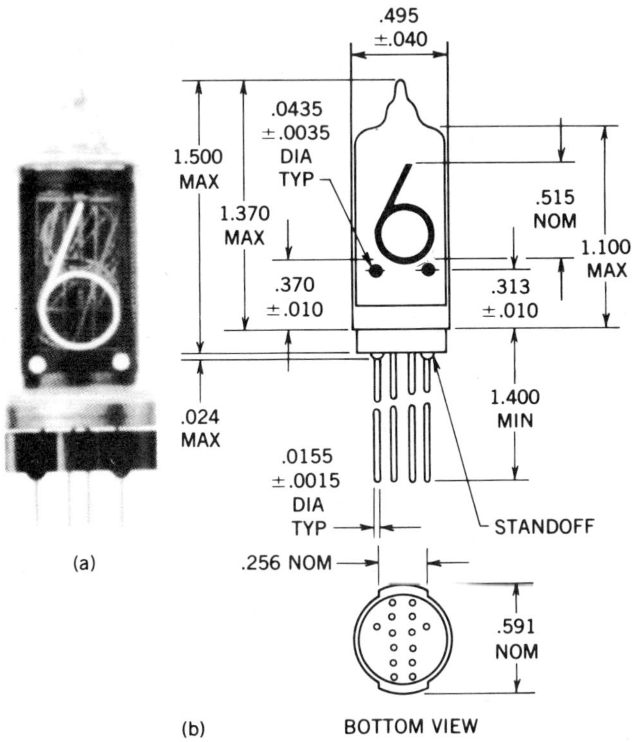

Fig. 29-8. *(a)* NIXIE tube displaying the numeral 6 and two decimal points; other numerically shaped cathodes are shown in the background; *(b)* outline drawing (dimensions are in inches). *(Burroughs)*

ever; their size is relatively small, their intensity is limited, and sometimes the viewing angle is a problem. Another type of display, the NIXIE tube, has been used in equipment where their larger size, greater brightness, and more fully formed characters are desirable.

A NIXIE tube is a single-anode, multiple-cathode,

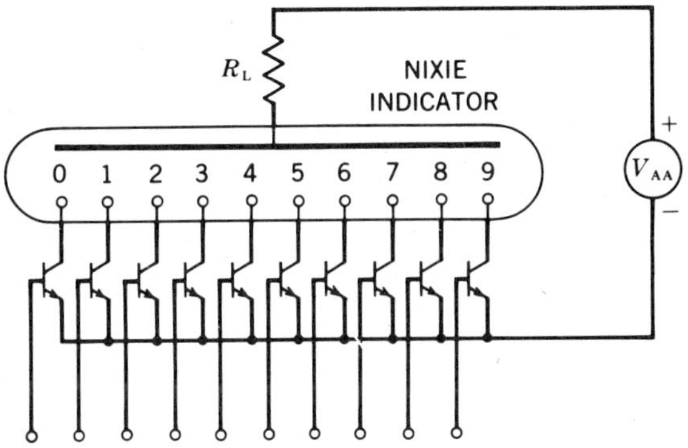

Fig. 29-9. NIXIE tube typical drive circuit. *(Burroughs)*

gas-discharge indicator. Each of its cathodes is formed into the shape of a number (0 through 9) or other symbol. When a cathode is selected, the corresponding character is seen as a visible gaseous discharge glow around the cathode. Figure 29-7 shows the circuit schematic diagram symbol for a 10-cathode NIXIE tube. Figure 29-8 shows a B5750 NIXIE tube which includes decimal points on both sides of the digit.

A disadvantage of the NIXIE tube is that it requires a relatively high voltage to operate. In Fig. 29-9 the cathode driver transistors must have a collector breakdown voltage rating of 60 V or more so that they do not conduct until selected. A BCD "1 of 10 decoder," or NIXIE tube driver, will cause the appropriate cathode to be grounded and the desired digit or symbol to be displayed.

A single IC is available which includes the functions of counting, decoding, and driving a NIXIE tube display. The Burroughs BIP-2610-1 shown in Fig. 29-10 and a NIXIE tube provide a very simple, one-decade, cascadable, decimal counter and display system. Additional decades may be added by connecting the carry output of the first to the count input of the second decade, etc., as in Fig. 29-11.

Frequency Counter

Many digital devices convert a parameter to be measured, such as temperature, voltage, or force, to a stream of pulses whose frequency is proportional to the quantity to be used or measured. For example, to measure the torque in a shaft, a strain-gage bridge might be attached to the shaft. The change in its resistance

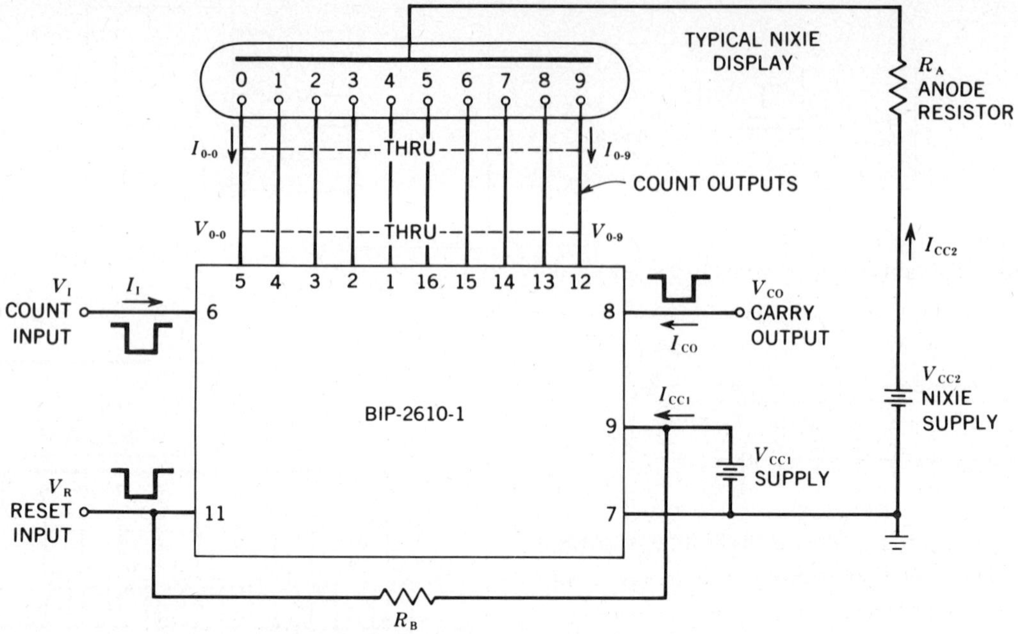

Fig. 29-10. Typical NIXIE tube operational circuit BIP-2610-1. *(Burroughs)*

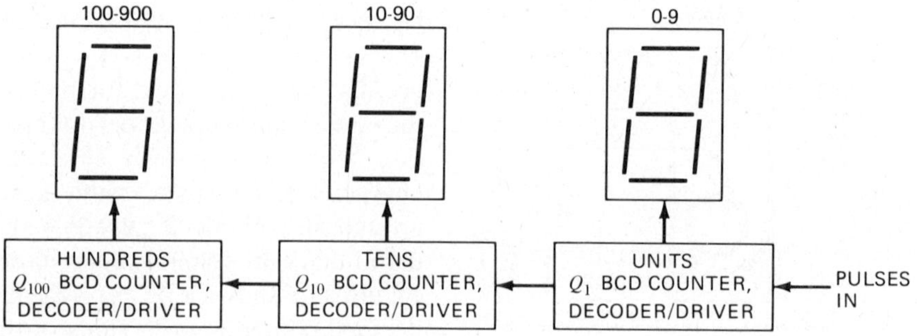

Fig. 29-11. Three-stage, serial, decade counter.

could cause the frequency of an oscillator to change. A frequency counter then could count this frequency and be calibrated in torque.

A counter is made into a frequency counter by gating the input pulses into the counter for a known time, such as 1 s. The number of pulses counted in that interval is the frequency in pulses per second, or Hertz.

Figure 29-12 shows an AND gate providing inputs to a counter and display system. The gate pulse applies a 1 to one input of the AND gate, and the pulses to be counted are applied to the other input. Pulses can only go through the AND gate during the interval T_1T_2. If the counter is reset to zero before T_1, at time T_2 the counter will have counted and will display the number of pulses in the chosen time interval. The counter is periodically reset to zero and again counts the input pulses. A typical sequence of events is:

1. Reset the counter (1-, μs pulse)
2. Start enable gate (T_1 T_2: 1 s)
3. End enable gate, start display (10 s)
4. End display, reset counter; repeat steps 2 and 3

SUMMARY

1. A specific count, or states, of the flip-flops in a counter is decoded using logic gates which have a 1 output only when that state exists.
2. The patterns of 1s and 0s present in a counter may be decoded with special ICs such as the 7441, 7442, 7447, and 7448, which are used with BCD counters and display devices to provide an easily interpreted decimal indication to the human operator.
3. Seven-segment LED displays are used extensively as numerical output devices because they are compatible with TTL logic and do not require a special power supply. They are small, rugged, solid-state, and are adequate for most applications.
4. NIXIE tubes have multiple cathodes and operate from a relatively high voltage supply. They have been used extensively in the past where a larger

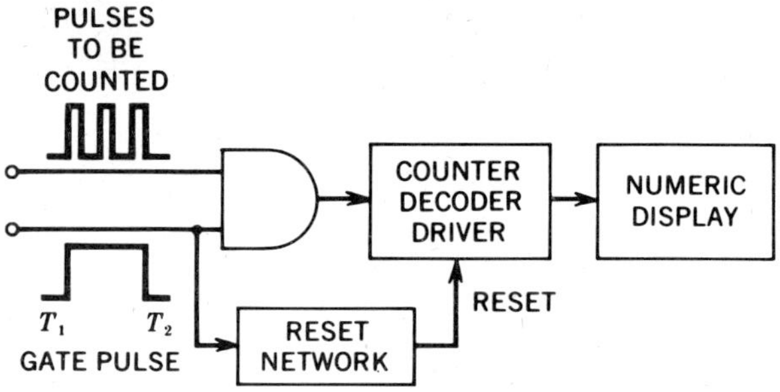

Fig. 29-12. AND gate permits driving pulses to appear at input of counter when they coincide in time with the appearance of an enabling pulse *T1T2*. The counter then adds the pulses passed on by the coincidence (AND) gate during the interval *T1T2* and displays them numerically.

and brighter display was needed than an LED could provide. They are cold-cathode tubes in which the ionization of a gas produces a luminous glow around one of several cathodes.

5. To minimize the number of individual components and connections in a digital display, ICs are available which incorporate the functions of counting pulses, decoding the count, and driving LED or NIXIE tube displays.

SELF-TEST

Check your understanding by answering these questions.

1. The number of four-input AND gates necessary to decode all the states of a four stage binary counter is __________ .

2. The segments of a seven-segment display that would light up if the input to a BCD decoder/driver were 0111 are __________ , __________ , and __________ .

3. A type of gaseous discharge indicator tube is the __________ .

4. The decoder/driver for a seven-segment display has __________ outputs, but the decoder/driver for a NIXIE tube has __________ .

5. To limit the current flow through a segment of a seven-segment LED display when it is ON, there should be a current-limiting __________ in series with the segment.

6. If a counter counts 1000 pulses in one-tenth of a second, the frequency of the pulse train is __________ .

MATERIALS REQUIRED

- Power supply: Variable, low-voltage regulated dc supply
- Equipment: EVM; square-wave generator (low-frequency)
- Resistors: ½-W, seven 330-Ω
- Solid-state devices: ICs 7447, 7490, 7408, or the equivalents; common-anode, seven-segment LED such as FND 360, or DIALCO 745-0017, or the equivalent
- Miscellaneous: The pulser and counter circuits, connected, from Experiment 28 (Figs. 28-7, 28-9, and 28-10); 10-kΩ 2-W potentiometer.

PROCEDURE

Testing Seven-Segment LED

1. Connect the common-anode of the seven-segment LED to the +5-V source and terminal *a* through a 330-Ω resistor to ground, as in Fig. 29-13. The *a* segment should light. In Table 29-1 list the *a* terminal.

2. Test, in turn, each of the line segments and the decimal point and record their respective terminal numbers in Table 29-1.

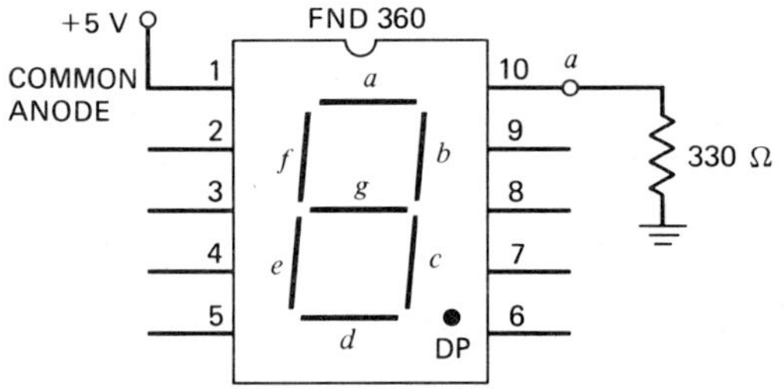

Fig. 29-13. Experimental circuit for testing a common-anode, seven-segment LED.

TABLE 29-1. Seven-Segment LED Terminals

LED type number FND 360

Segment	Anode	*a*	*b*	*c*	*d*	*e*	*f*	*g*	Decimal Point
Pin Number									

Testing Numeric Readouts

3. Determine *experimentally* which segments of the seven-segment LED light for each of the numerals 0 through 9, in turn. Record these segment in Table 29-2. (HINT: To light several segments, connect the cathode of each of these through individual 330-Ω resistors to ground. The common anode is connected to +5 V.)

TABLE 29-2. Displaying Decimal Digits

	Segment						
Numeral	*a*	*b*	*c*	*d*	*e*	*f*	*g*
0							
1							
2							
3							
4							
5							
6							
7							
8							
9							

Counter, Decoder/Driver, Numeric Readout

NOTE: *Figure 29-14 shows the terminal connections of a 7447 BCD decoder/driver.*

4. To the pulser and BCD counter in Figs. 28-7, 28-9, and 28-10, connect the 7447 and the seven-segment LED display, as in Fig. 29-15. If you have substituted other devices for those shown in Fig. 29-15, be certain that you use the proper terminal numbers. The *Q* outputs from the BCD counter serve as the inputs to the decoder/driver. Note that Q_A is the *Q* output from flip-flop A, Q_B from flip-flop B, and so on. Note also that terminals 9 through 15 of the 7447 serve as the inputs to the cathodes of the LEDs, as shown. The 330-Ω resistors hold LED segment current to a safe value. The common anode of the LED display is connected to the +5-V supply.

5. Power ON. One of the numerals on the seven-segment LED may light. Clear the counter. The LED display should now read 0.

6. Pulse the counter. The numeral 1 should appear. Pulse the counter again, and again, for a total of 10 pulses. The numerals 1, 2,..., 9, and finally 0, should appear in sequence.

7. Repeat step 6.

IC Counter, Decoder/Driver

NOTE: *The 7490, whose terminal connections are shown in Fig. 29-16, will replace the BCD counter used in the preceding steps.*

8. Connect the circuit shown in Fig. 29-17 on page 192. This is the equivalent of the circuit in Fig. 29-15, except that the 7490 has replaced the counter in Figs. 28-9 and 28-10, and that the square-wave generator has replaced the pulser. If necessary, reset the counter to 0 by momentarily lifting terminals 2 and 3 of the 7490 off ground, and immediately reconnecting them to ground, while the square-wave generator is OFF.

9. Set the square-wave generator frequency at 1 Hz, 3 V p-p output. When the square-wave generator is turned on, the input signal will cause the output of the 7490 and 7447 to advance by 1 for each square-wave cycle of input. As a result, the decimal numbers on the seven-segment LED will advance by 1 from 0 through 9 to zero and will then start over again.

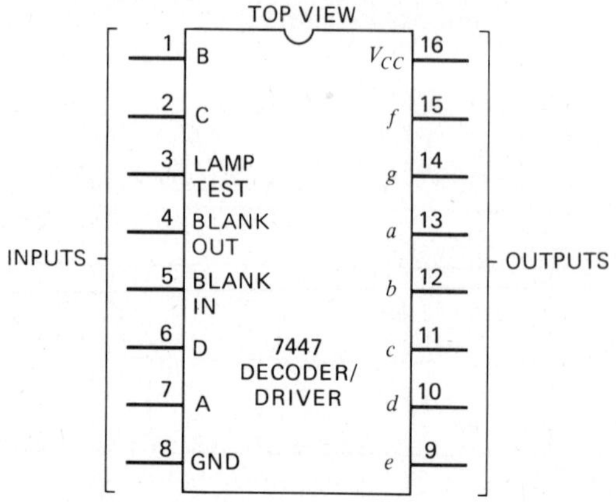

INPUT DECIMAL VALUES
A = 1
B = 2
C = 4
D = 8

LAMP TEST CONDITIONS
Normally high (+5 V)
When low (0 V), all outputs go low.

BLANKING CONDITIONS
a) BLANKING INPUTS: low, will extinguish only character *O*.
b) BLANKING OUTPUTS: low, will extinguish the display.

Fig. 29-14. Terminal diagram and input conditions for the 7447 BCD decoder/driver.

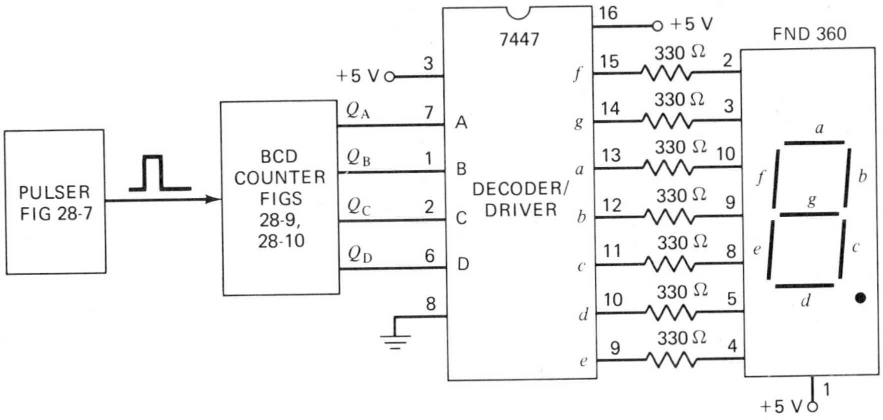

Fig. 29-15. Experimental pulser, BCD counter, decoder/driver, and seven-segment LED display.

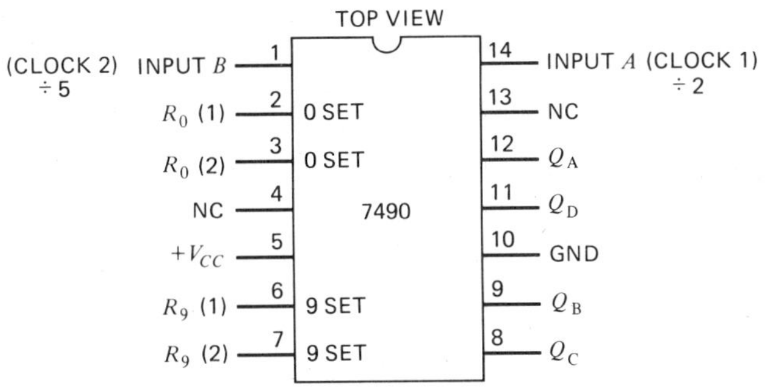

Notes:
1) For a BCD counter connect terminals 1 and 12.
2) Counter may be reset to 0 by bringing either or both 0-set inputs positive.
3) Counter may be preset to 9 by bringing either or both 9-set inputs positive.

Fig. 29-16. Terminal connections for the 7490 BCD counter.

10. Power ON. Check the operation of the counter to see that it is functioning as indicated. Power OFF.

Frequency Counter (One Decade)

NOTE: The circuit in Fig. 29-17 may be modified to operate as a 1-decade frequency counter. This is achieved by adding the one-shot pulser, adjusting its pulse width to 1 s, and feeding its positive pulse output to one leg of an IC 7408 (quad two-input) AND gate. A square wave is fed to the other leg, and the output of the AND gate is then coupled to the input of the counter (decoder/driver), as in Fig. 29-18. When the one-shot is pulsed, pulses derived from the square-wave generator are applied to the counter for a 1-s interval. If the counter is cleared (set at 0) and the square-wave generator is set at 3 Hz, the counter will cycle up to a count of 3. The next time the one-shot is pulsed, the counter will add another 3 and go to 6. The maximum count before overloading is 10, that is, 0 through 9 and back to 0. The 0 therefore represents 0 input, or 10 pulses in, or an integral multiple of 10.

11. Modify the input to the counter by adding the pulser and AND gate, as in Fig. 29-18. Note that a 10,000-Ω potentiometer, connected as a rheostat, replaces the 6.8-kΩ resistor previously used. The rheostat is adjusted to a value of 6.3-kΩ, for a pulse width of 1.0 s, approximately. R may be readjusted, if necessary, to assure a 1.0-s pulse output.

12. Set the square-wave generator at 3 Hz, 3 V p-p. Power ON. Reset the counter to 0 and pulse the one-shot. In Table 29-3 record the count indicated on the seven-segment LED display.

TABLE 29-3. Frequency Counter (One-Decade)

Frequency, Hz	*Pulse Number*	*Count*
3	1	
5	2	
6	3	
8	4	
10	5	
15	6	
25	7	

13. Reset the counter and set the square-wave generator at 5 Hz. Again pulse the one-shot (pulse number 2). Record the count in Table 29-3.

14. Repeat step 13 for each of the frequencies shown in Table 29-3.

Three-Decade Frequency Counter

NOTE: A three-decade frequency counter may be constructed by cascading three decade counters. This will require three counters, decoder/drivers, and three seven-segment LEDs.

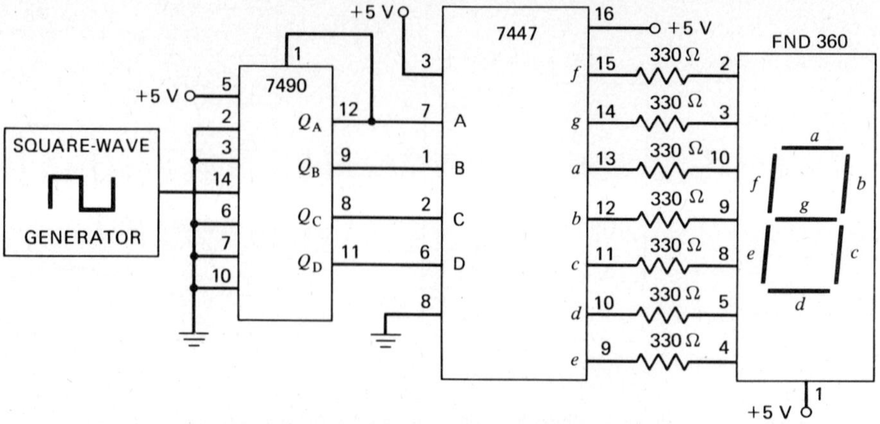

Fig. 29-17. Experimental circuit counts pulses delivered by square-wave generator.

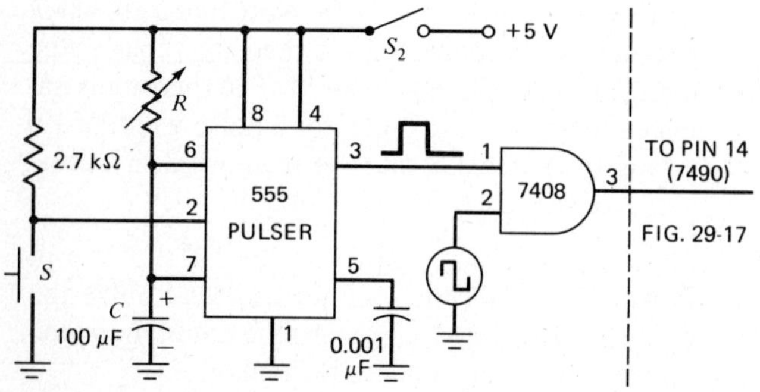

Fig. 29-18. Converting counter in Fig. 29-17 into a one-decade frequency counter.

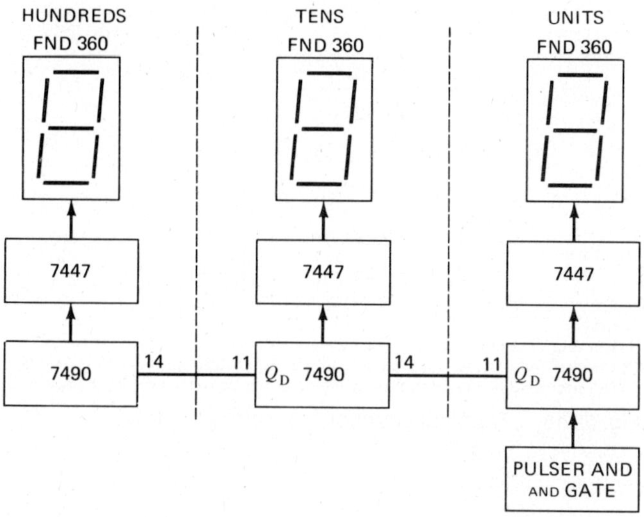

Fig. 29-19. Block diagram of a three-decade frequency counter.

15. Leave the decade frequency counter units connected as they were in steps 11 through 14, but disconnect the pulsers and the square-wave generators from two of the units. Interconnect the units as in Fig. 29-19, using Fig. 29-18 as the unit's counter. The ten's counter requires its input from Q_D in the 7490 of the unit's counter, and so on.
16. Reset the counter at 0–0–0. Set the output of the square-wave generator at 5 Hz and pulse the one-shot. Record the reading in Table 29-4.
17. Repeat step 16 for each of the frequencies listed in Table 29-4.

TABLE 29-4. Frequency Counter (Three-Decade)

Frequency, Hz	*Pulse Number*	*Count*
5	1	
12	2	
55	3	
90	4	
125	5	
250	6	
650	7	
950	8	
1050	9	

QUESTIONS

1. In your experimental procedure, which of the segments of the seven-segment LED lit up for the following? (*a*) 0 (*b*) 2 (*c*) 4 (*d*) 5
2. In the experimental procedure in step 5, how did you *clear* the counter? What did the LED read when the counter was cleared?
3. Operationally, how does the frequency counter in Fig. 29-18 differ from the counter in Fig. 29-17?
4. What are the limitations, if any, of a one-decade frequency counter?
5. What are the advantages of a three-decade frequency counter over those of a one-decade counter?
6. What procedure did you follow to determine that the width of the output pulses of the 555, in Fig. 29-18, was 1.0 s?
7. Assuming that the frequency counter and square-wave generator were accurately calibrated, what should your readings have been for each of the frequencies in Table 29-4?
8. How did your actual readings compare with the predicted values? Explain any discrepancies.
9. Why is a decoder necessary in driving a numeric indicator for a BCD counter?
10. What is the maximum number that a two-decade BCD counter can indicate?

Answers to Self-Test

1. 16
2. *a; b; c* (7)
3. NIXIE tube
4. 7; 10
5. resistor
6. 10,000 Hz

EXPERIMENT

30

COMPUTERS AND MICROPROCESSORS: INTRODUCTION

OBJECTIVES

1. To study the basic principles of digital computers
2. To learn about the hexadecimal number system
3. To become familiar with the architecture of a microprocessor
4. To use the Heath ET3400 microprocessor trainer to demonstrate some elementary microprocessor operations

BASIC INFORMATION

The ability to manufacture large-scale integrated circuits (ICs) inexpensively has led to the use of digital computers or computerlike devices in countless ways. For a small investment, powerful personal computers are available with capabilities which were beyond imagination just a short time ago. The smallest business can afford to "computerize" its inventory and other records.

In industry, the programmable controller, a special-purpose computer, can replace hard-wired logic, relays, timers, and a wide array of control circuits.

Computer Basics

A digital computer system consists of four basic sections: input, processor, memory, and output. Sometimes the memory function is combined with the processing function and the abbreviation *IPO* is used to describe these functions: *I*nput, *P*rocessing, and *O*utput.

A block diagram of the parts of a digital computer is given in Fig. 30-1*a*, and a photograph of an IBM 360 installation is given in Fig. 30-1*b*.

The input section receives data from the outside world. At one time data was entered by punched cards, but that method is being phased out. Today data is entered by manual keyboards, from tape, over telephone lines, and in other ways.

The *central processing unit* (CPU) is the heart of the computer. It contains circuits which can perform the logic and arithmetic functions which are designed into the computer's *instruction set*. It also includes *registers* (groups of flip-flops) which are used for temporary storage of data, addresses, and other information. A register which is important to input and output functions is the *accumulator* (there may be more than one).

There are two types of storage in a computer: internal and external. Internal storage, or memory, is where programs and data reside. Early computers had small internal memories and could only run relatively short programs.

It is common today, however, to have internal memory [random access memory (RAM)] which is very large. A modest personal computer may have as much as 512,000 bytes of memory. Large internal memory allows long and complex programs to be run.

External memory includes disks, magnetic tape, and other media. Access to data stored in internal memory is rapid. Access to externally stored data is slower.

The output section of a computer consists of those devices which record the output of the computer: cathode-ray tubes, printers, plotters, magnetic tape drives, and other devices.

Stored Program Concept

Internally a computer is designed to perform a number of basic operations. These operations are called the instruction set of the computer and are relatively simple. For example, it is often useful to be able to add 1 to a stored number. This is called *incrementing*, and one of the basic instructions referred to above may be to "increment the accumulator."

To perform a task, the computer must be instructed as to which of its basic operations it should perform,

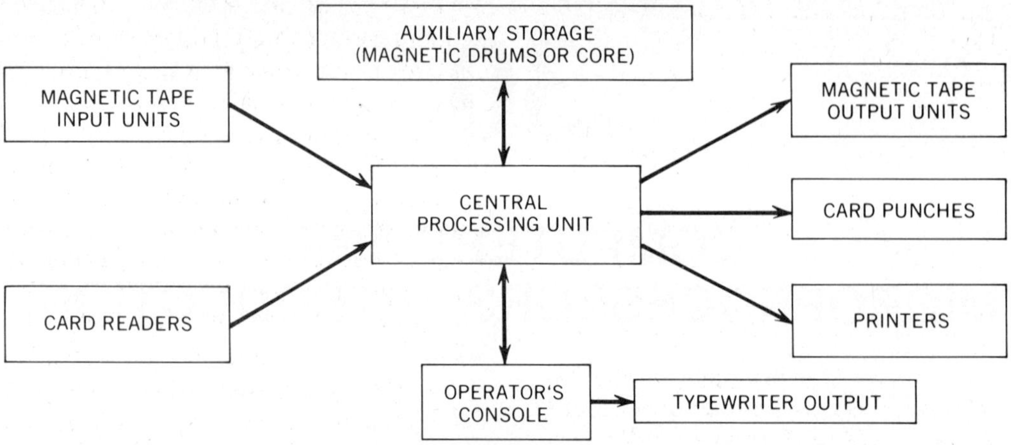

(a)

(b)

Fig. 30-1. *(a)* Computer system block diagram; *(b)* IBM computer system. *(IBM)*

and in what order they should be performed. This is called *programming* the computer. If the computer programmer writes a program in terms of the most basic operations the computer can perform, it is termed *machine language* programming. This program will be stored in the internal memory of the computer in 8-bit bytes of 1s and 0s. When the program is run, these instructions (and other data) will be examined, one by one, by the *arithmetic-logic unit* (ALU) of the computer to determine what is to be done. Although the individual operations performed may be simple, they are performed in microseconds and can be tirelessly repeated as many times as required.

Let us illustrate the operation of a hypothetical computer which includes the following instruction set to perform the addition of 7 and 8:

What to do	Operation Code	Mnemonic
Clear Accumulator	50	CLA
Load Accum from Memory#	51	LDA
Add to Accum from Memory#	52	ADD
Store Accum in Memory#	53	STA
End program	128	END

Instructions 51, 52, and 53 will require a memory address to specify where data is stored or to be stored. Therefore, these instructions will occupy two successive locations in memory.

A *mnemonic* is an abbreviation, or memory aid, for use by the person writing a program. For example, it is much easier to remember that STA means "store the accumulator" than to remember the operation code (53), or worse yet, the actual entry in the memory, 00110101. A more sophisticated method of writing a program is to use these mnemonics with a program called an *assembler* which will produce the machine language code from the mnemonics.

The contents of the memory will be:

Address	Contents	Notes
0000 0000	0011 0010	"50" base 10 CLA
0000 0001	0011 0011	51 LDA
0000 0002	0000 0010	10 address of 7
0000 0003	0011 0100	52 ADD
0000 0004	0000 1011	11 address of 8
0000 0005	0011 0101	53 STA
0000 0006	0000 1100	12 address of sum
0000 0007	1000 0000	128 END
0000 0008	xxxxxxxxx	unused
0000 0009	xxxxxxxxx	unused
0000 000A (10)	0000 0111	7 first number
0000 000B (11)	0000 1000	8 second number
0000 000C (12)	0000 1111	15 7+8

When this program is run, the address is set to 0000 0000 initially. The instruction at that location (CLA: set all bits of the accumulator to zero) is performed and the address is incremented to 0000 0001. This instruction (LDA) requires two adjacent memory locations because the computer must be told what to do and where data is located. The second instruction tells it to add to the accumulator the data at the address specified. After performing the other operations, the program ends at step 0000 0007. Because each of the instructions takes only microseconds to perform, the sum (15) will be available in memory location 0000 000C almost instantaneously.

Notice that, in this example, three types of numbers are used: for addresses, hexadecimal; for memory contents, binary; and decimal numbers in the notes. A computer programmer must be familiar with all three types of numbering systems. The hexadecimal system is discussed in a later section.

With this introduction to the parts of a computer, some computer terminology, and a simple program, let us see what has caused a revolution in the computer and related industries.

Inexpensive but Powerful Computers

Whereas in the past a digital computer might have required many printed circuit cards full of discrete components or small-scale ICs to perform the complex functions which take place in a digital computer, today the heart of a computer, the central processor, is usually on a single IC chip. Furthermore, the physical size and cost of memory chips is small. The power supply and disk drives tend to occupy more space than the heart of the computer. Let us see how this has been accomplished.

Microprocessor Defined

The past decade has witnessed a remarkable growth in IC technology. The development of large-scale integrated circuits (LSICs) and very large scale integrated circuits (VLSICs) coincided with the design and the manufacture of highly sophisticated devices called *microprocessors*. The microprocessor unit (MPU) is a digital IC component which contains almost all the circuits which make up a digital microcomputer. Thus a microprocessor can do arithmetic operations, logic, and control operations. All the circuitry of a microprocessor is contained within a single IC chip. Figure 30-2 shows the 6800 microprocessor chip which we will use in this and the next three experiments.

Microcomputer Defined

Microprocessors are used in microcomputers. They make up the CPU of a computer. In addition to the microprocessor, a computer contains other components and circuits which permit it to carry on its operations. Thus microcomputers contain *memory* units to store information and instructions (program), I/O (input/output) circuits to drive peripheral devices such as teletypewriter keyboards or mechanical switches for entering information, and printers or cathode-ray tubes for reading out information. Microcomputers also contain internal clocks for timing and synchronization of all the computer circuits.

Some of the more recent microprocessors are actu-

Fig. 30-2. The 6800 microprocessor. (*Motorola*)

ally self-contained microcomputers (not including the peripheral devices) and are used as computers by designers and hobbyists, and in many modern appliances such as microwave ovens, washing machines, and automobiles.

Microprocessors versus Hard-Wired Logic Devices

A microprocessor is a logic device. What is so unique about this device? We have studied other logic circuits, such as AND, OR, NOT, NAND, and NOR gates. As we have seen, these are *simple* logic devices. What distinguishes the more complex microprocessor unit from these simple devices? Well, the operation of the simple devices cannot be changed. That is, the data is presented to their input, and the output is the result of the "logical" action of these devices. Again, the operation of these simple logic devices cannot be changed. The logic of these simple devices is called *hard-wired* logic. On the other hand, the logic operation of the microprocessor *can* be changed by the instructions (program) which are presented to it. Thus a microprocessor can be made to perform a multiplicity of logical operations by simply *changing* its program.

The design of a microprocessor, to make certain that each of its circuits is operating properly alone and in conjunction with the other circuits, may take a team of engineers and technicians many months to accomplish, at a very high cost. To repay this expenditure of development effort, hundreds of thousands of units of this microprocessor have to be manufactured and sold. The versatility of the micro in responding to a variety of different instruction combinations (programs) makes it possible to design different systems employing a particular micro and, therefore, makes it economically feasible to design and manufacture the microprocessor.

In the next experiment we will learn how to communicate with (that is, to program) the microprocessor which we will be using.

All microprocessors are *not* designed and built in exactly the same manner. Indeed not! There are many different types and families of microprocessors. The early ones were much simpler than the units being manufactured now. Micros contain thousands of transistors, diodes, and gates. As we mentioned, some even contain memories and other circuits which make them suitable for use as microcomputers. Of course, the more sophisticated the circuitry of a microprocessor, the more highly complex control functions it can be made to perform.

Bits, Nibbles, Bytes, Words, and Long Words

Microprocessors and microcomputers operate on binary numbers. A single binary digit is called a *bit*. When 4 bits are combined in a binary number (that is, when a binary number has four places, for example, the number 0111), the resultant number is called a *nibble*. An 8-bit binary number is a *byte,* while a 16-bit binary number is a *word*. A *long word* is a 32-bit binary number.

Microprocessors are designed to work on a specific-length binary number. The more advanced units are 16-bit processors; that is, they work on words. Some work on long words, others on bytes, and still others on nibbles. The 6800 is an 8-bit processor, operating on bytes.

Microprocessor and Computer Terminology

In your study of microprocessors you will encounter many terms which are widely used in microprocessor and computer terminology. We will define some of these terms here. These are arranged in alphabetical sequence, not necessarily in the order of their importance.

- *Address* is a code that identifies the location or the destination of data.
- *ALU* is the abbreviation for arithmetic-logic unit. This is one of the basic logic units in a micro. The ALU adds, subtracts, and performs digital logic (e.g., AND operation).
- *Assembly language* consists of a set of symbols describing computer operations, with address and memory locations.
- *Buffer* refers to a circuit placed between other circuit components to prevent interaction between them. A buffer is also the term used for temporary storage, where data is assembled during data transfer.

- *CPU* is the abbreviation for central processing unit. This section of a computer controls the interpretation and the execution of the instructions (program) fed to the computer.
- *Compiler* is a program which changes high-level computer code into machine language code. The microprocessor and the computer respond only to machine language.
- *Data* comprises the information that can be processed or produced by a computer.
- *Decode* means to interpret a coded instruction.
- *Direct memory access* (DMA) is a means of gaining direct entry to a memory location in order to store or retrieve data.
- *EPROM* is the abbreviation for erasable programmable read-only memory. This is a programmable read-only memory (*PROM*) which can be erased and used again. Most EPROMs can be erased by ultraviolet light.
- *Input/output* (I/O) refers to the peripheral equipment used to communicate with a computer. It also refers to the data used in the communication.
- *Least-significant digit* (LSD) is the extreme right-hand digit in a number. This digit carries the least numerical weight—it is the unit's digit.
- *Machine language* is the system of codes, in binary form, used to represent instructions and data in a particular computer or microprocessor.
- *Memory* is the component in a digital system where information which may be recalled is stored.
- *Most-significant digit* (MSD) is the extreme left-hand digit in a number. This digit carries the most numerical weight.
- *OP code* refers to the machine language instruction for the operation which is to be performed. Thus + and − in a computer instruction are OP codes.
- *Program* constitutes the set of instructions, arranged in sequence, which orders a microprocessor or a computer to do a certain job. The program details every step to be taken and the order in which it is performed.
- *RAM* is the abbreviation for random access memory. This is a storage device which can hold and produce on demand any data entered in it. A RAM can be read from and written to.
- *ROM* is the acronym for read-only memory. A ROM is a device where data is permanently stored. It is the basic program memory and can only be read from.
- *Register* is a temporary storage device which holds one word or one byte of data, depending on the micro.
- *Software* refers to programs, languages, and procedures for a computer system. *Hardware* refers to the physical components (machine) of the computer system.
- *Vocabulary* is a list of operating codes or instructions which a programmer may use in writing a program for a particular computer.

Architecture of the 6800 Microprocessor

Figure 30-3 is a block diagram of the 6800. For ease of understanding, only the major circuit groupings (blocks) are shown. These blocks constitute the architecture, that is, internal construction, of this processor. We note that the 6800 contains the following major circuits:

1. ALU
2. Accumulators
3. Various registers, including the data and address registers
4. Data bus; address bus
5. Program counter
6. Instruction decoder and control
7. Stack pointer
8. Condition code register

Note that this processor by itself does not constitute a microcomputer; external program memory, an external clock, and I/O must be added to make it operate as a microcomputer.

Let us examine the functions of each of the major blocks of the 6800.

1. *ALU* This unit performs arithmetic and logic operations on the sets of eight-digit binary numbers that are presented to it by the addressed data memory (RAM) and by the internal accumulators. The ALU can be made to add, in various ways, the input binary numbers. What the ALU does and the sequence in which it does it are determined by the control signals which are fed to it by the instruction decoder. The control and timing signals are generated within the microprocessor according to the programmed instructions, based on the pulses supplied by an external clock.
2. *Accumulator* This is a special register. To speed up the performance of operations and to simplify the internal construction of the microprocessor, many instructions can be performed on the data in a particular register. This register is called the *ac-*

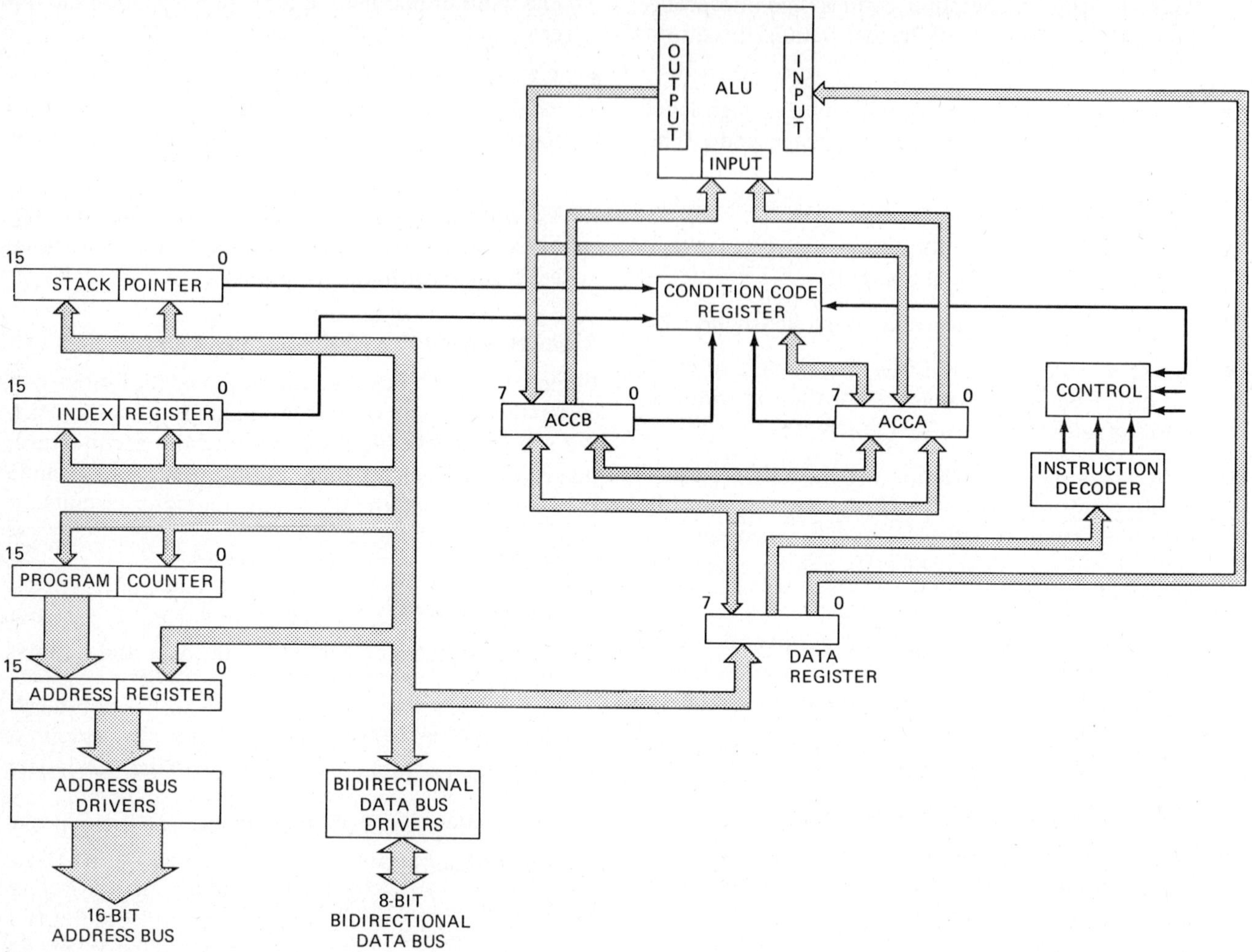

Fig. 30-3. Block diagram of the 6800. *(Heath)*

cumulator because, among its other functions, it will hold the sum or the accumulation of the bits after an addition. The 6800 contains two accumulators.

3. *Registers* These are temporary memory devices where information in the form of bits can be entered and stored as long as the unit is not shut off. When desired, the information in these registers can be "read" and then used in the calculations performed in the ALU. These RAM devices are also known as *scratch pad* memory since they serve the same purpose as a scratch pad used in hand calculations.

4. *Data bus, address bus* A bus consists of a group of conductors on which digital information is transferred. In the 6800 there are the 8-conductor data bus and the 16-conductor address bus. The data bus is bidirectional; that is, information can be sent along its lines from the 6800 into an external memory device (the 6810 in our trainer), or the information can be sent into the 6800 from the external memory. The address bus, which transfers only address information into the external memory, is unidirectional.

5. *Program counter* The instructions for the micro come from an external ROM, which was initially programmed when the microprocessor trainer used in our experiments was being designed. The program counter keeps track of the address at which the next instruction is found. Normally the next consecutive address calls for the next instruction. But under program control the order of instructions can be changed in two ways. In the first, some instructions *branch* to other address locations depending on the outcome of some test. The second possible exception is a special instruction which loads a specified address into the program counter. In this way the program can be changed when required.

6. *Instruction decoder/control* The machine code sent to the micro by the ROM must be interpreted, and the micro must then take the proper action. This function is entirely internal to the micro. After interpreting the instruction, this section (instruc-

tion decoder/control) sends out the necessary signals which assure that the instruction is carried out as ordered.

7. *Stack pointer (SP)* This is a special register which is used to identify the addresses in the *stack* where information is or can be stored. The stack is a special set of locations in the internal 6800 memory which are set aside as operating registers. The stack pointer contains the address of the next available empty location in the stack. Data is normally entered into or taken from the stack one location after another. The stack pointer makes possible the use of subroutines in a program. Subroutines are consecutive program steps which achieve a specific result, such as the routine for lighting up a specific digit in our display. A long program may require the use of a specific subroutine many times. The stack pointer and the stack make this possible.

8. *Condition code (CC) register* This special register keeps track of special conditions in coded form, as the name implies. There are 6 bits in the CC:

   Bit 1 is called o*v*erflow, abbreviated V

   Bit 2 is called *z*ero, abbreviated Z

   Bit 3 is called *n*egative, abbreviated N

   Bit 4 is called *i*nterrupt, abbreviated I

   Bit 5 is called *h*alf carry, abbreviated H

   Bit 0 is called *c*arry, abbreviated C

In various steps of addition or subtraction, these condition indicators are very important. These indicators are often called *flags,* since they are used to flag or warn the operator of possible problems. The condition code is important because it permits the operator to change the program flow depending on the result of some operation. For example, we know that division by zero is not permissible. Thus, before calling subroutine *divide,* we check the zero (Z) flag. The instruction *BEQ* checks the Z flag and branches to the address which is the start of the instruction if the flag shows that the dividend is zero. Otherwise, the program will continue to divide.

Timing the 6800 (Clock)

The program is carried out in an orderly fashion by the micro. This is assured by a *clock* which is external to the 6800. The clock pulses are used to synchronize the entire system. Thus the rise of clock pulses may signal that a new instruction is to be obtained. The fall of the pulse may signal that the data obtained is now valid. The next rising pulse may start the execution of an instruction, and so on. Most instructions require a number of clock cycles to be completed. One measure of the efficiency of a micro is the speed of the clock. Another is how few clock cycles are required to complete most instructions.

I/O Ports

Microprocessors normally communicate with the outside world by way of input/output ports. Specific pins on the micro are arranged in groups so that they may be accessed together. A group of pins used for input is called an *input port,* while a group used for output is called an *output port.* In the 6800 system, the micro does not contain I/O ports. Communication with the outside world is accomplished by way of the peripheral interface adapter (PIA).

The ET3400 Trainer

The hardware we have selected to facilitate our study of the operation of microprocessors is the ET3400 (Fig. 30-4). This versatile device uses the 6800 microprocessor in conjunction with external memory ICs, a clock, a display unit (six 7-segment LEDs), a 17-key keyboard, a series of buffers, test sockets, and so on.

The keyboard is used for entering data and programs. Each of the 15 keys (1 through F) has a dual function. Note that above each key designation (digit or letter) is printed a group of small-type letters, such as ACCA above the 1, ACCB above the 2, and so on. With power ON, when the operator presses the 1 key, either the 1 will appear somewhere in the display, or the letters ACCA will be shown. Which one appears will depend on what key the operator has pressed just before pressing the 1. For example, if the operator has just pressed the RESET key and then presses the 1, the trainer per-

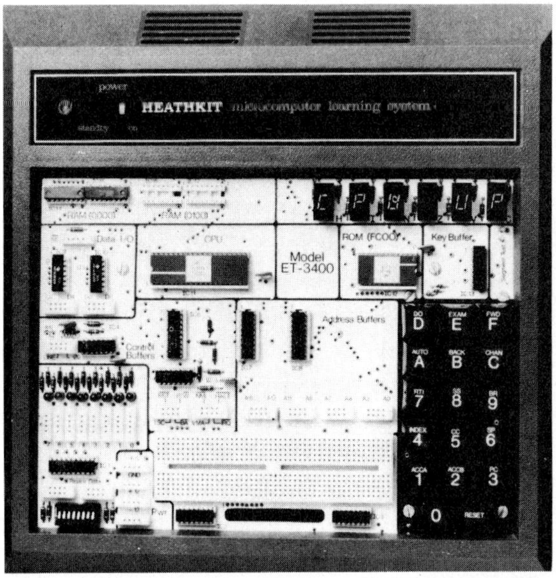

Fig. 30-4. ET3400 microcomputer/microprocessor trainer. *(Heath)*

forms the ACCA function and ACCA will appear in the display. This is true of the other dual-purpose keys. After RESET is pressed, and a dual-purpose key is pressed next, the trainer performs the function specified on the pressed key. The functions of immediate concern to us are the following:

- *ACCA* shows the contents in accumulator A. The display will show ACCA, followed by two hexadecimal digits. The hexadecimal system of representing numbers will be discussed later in the experiment.
- *ACCB* shows the contents of accumulator B in the same manner as above. Note, however, that the B appears in the LED display as a lower-case letter, b that is, as a 6 without the bar on top. The purpose is to prevent confusing an upper case B in the display with the number 8. So in reading hex numbers on the trainer, do not confuse 6 and B.
- *PC* causes the system to display the contents in the program counter. The display will be PC followed by the four digits stored in the program counter. The trainer uses the hexadecimal numbering system (base 16). In this system the numbers above 9 are represented by letters: A = 10 in the decimal system; B = 11; C = 12; D = 13; E = 14; F = 15. The program counter can go to the hex number FFFF, which is the decimal number 65,535. This is the highest number that the 6800 can address.
- *CC* is the display of the condition code. This uses the six LED displays to indicate the value of each of the condition code bits. The letter next to each display LED (lower left corner) shows the letter (H, I, N, Z, V, C) to which each bit corresponds.
- *SP* is the stack pointer. Its display is SP followed by the four digit number in the stack pointer.
- *RTI* stands for the instruction *Ret*urn from *I*nterrupt. It is a command to the micro to resume operation where last stopped. We will become acquainted with some of its uses later.
- *SS* is also a command to the micro. It stands for *single-step*. It causes the program to advance a step at a time following the instructions to the branches or jumps, or just the next instruction, depending on the program.
- *BR* is the acronym for *br*eakpoint and forces the micro to stop at any address we choose. The display is _ _ _ _ b r when we first press the key. We can now enter any address by pressing the digits of that address. Remember that the address is in hex. When the program is running, it will stop at the address we set as breakpoint. This permits us to find problems that we may have in our program.
- *AUTO* is used to enter program information. The display which appears on the trainer when this key is first pressed is _ _ _ _ A d. We can again enter any memory address we choose. This will be the starting address of the program we wish to enter. Suppose that we wish to start at 0000; we then enter the digits 0000 and our display changes to 0000 _ _ . If we now enter the two digits corresponding to the machine code for the first instruction, the address will automatically change to 0001 and the trainer will wait for us to enter the next instruction or operand as required by the program. This arrangement is a great help in entering programs quickly. The only way to stop entering more program steps is to press the RESET key.
- *EXAM* stands for examine memory. Pressing this key again asks us to enter an address by displaying _ _ _ _ A d. As soon as we enter an address of four digits, the last two digits change to show the contents of the memory at that address. We now have a number of options; if we press E again, we can enter a new address we may wish to examine. Pressing *FWD* will advance the address to the next location (increment the address) and display the memory value stored at that location. Pressing *BACK* will decrement the address and display the memory contents of the preceding address. Pressing *CHAN* will change the display by replacing the contents with _ _ . We can now enter any value we wish into the displayed memory location. This arrangement makes it possible to check our program and correct any errors we may have made when it was first entered in the AUTO mode.
- *CHAN* can also be used to change any of the registers we displayed before. Suppose that we are displaying the ACCA register and our display reads ACCA32. If we wish the contents of accumulator A to be 23 (hex), we press CHAN, getting ACCA _ _ . Then we press, in turn, keys 2 then 3, and we have entered 23 into accumulator A and the display will show ACCA23.
- *INDEX*, when pressed, will display the contents of the index register Inxxxx (x stands for a digit of any value) in a similar manner to that of the program counter display.
- *DO* is an instruction to the micro to perform the program in memory, starting at the address we enter. For example, suppose that we wish the micro to start carrying out the instructions starting at address 0001. We press DO and the display that appears is _ _ _ _ do. We now enter the address 0001, and the trainer will start executing the program beginning at the address shown.

Hexadecimal System

We are acquainted with two ways of representing quantities. The first is the familiar decimal system, in which 10 digits (0, 1 through 9) are used to indicate numerical values. The second is the binary system, which we studied in a previous experiment. In the binary system only two digits (0 and 1) are used to represent a number. Binary numbers are used in the machine code of digital computers because they can be easily represented on a digital machine, because of the two-state nature of computer circuits. Thus a gate is either open or closed; a line is either UP or DOWN; a switch is either ON or OFF. Digital computers do not recognize any numbers other than digital numbers. If another number system is used to enter information or instructions into a digital machine, these numbers must be converted to binary quantities by a program which has been built into the machine.

Other systems are used to represent numerical values, as, for example, octal and hexadecimal numbers. So we must identify the number system we are using if any system other than decimal is being employed. We do this by affixing a subscript to the number. Thus 234_{10} is the number 234 in the decimal system, 1101_2 is the decimal number 13 in the binary system, 27_8 is the decimal number 23 in the octal system, and $A2_{16}$ is the decimal number 162 in the hexadecimal system.

Let us see how the hexadecimal system is set up. There are 16 digits in this system 0, 1 through 9, and A through F. The decimal value of these digits is as follows:

Hex Digit	*Decimal Equivalent*	*Hex Digit*	*Decimal Equivalent*
0	0	8	8
1	1	9	9
2	2	A	10
3	3	B	11
4	4	C	12
5	5	D	13
6	6	E	14
7	7	F	15

The value of the decimal number 234_{10} is calculated as follows:

$$234_{10} = 2(10)^2 + 3(10)^1 + 4(10)^0$$

Similarly, the decimal value of the binary number 1101_2 is:

$$1101_2 = 1(2)^3 + 1(2)^2 + 0(2)^1 + 1(2)^0 = 13_{10}$$

and the decimal value of the hexadecimal or hex number $A2_{16}$ is:

$$A2_{16} = 10(16)^1 + 2(16)^0 = 162_{10}$$

Now, what is the decimal value of the hex number $C342_{16}$? It is:

$$C342_{16} = 12(16)^3 + 3(16)^2 + 4(16)^1 + 2(16)^0 = 49{,}986_{10}$$

where

$$16^0 = 1 \qquad 16^1 = 16 \qquad 16^2 = 256$$
$$16^3 = 4096 \qquad 16^4 = 65536 \qquad \text{etc.}$$

Data and instructions in the register of a binary computer are in the form of binary numbers, that is, in machine code. Programming instruction codes are given in hex code, for hexadecimal machines, because the hex code is easier to remember than the binary code. The keyboard on the ET3400 trainer is a hex keyboard.

SUMMARY

1. A microprocessor is a digital IC which can perform arithmetic, logic, and control operations. It acts as the CPU in a digital microcomputer.
2. In addition to a CPU, a computer contains the following major circuits: memory, I/O circuits and devices, and a clock.
3. The most sophisticated microprocessors are, in fact, microcomputers.
4. A microprocessor is a logic device whose operation and output may be changed by the instructions (program) presented to it.
5. A microprocessor operates on instructions presented to it in binary code.
6. A microprocessor is designed to operate on a specific-length binary number. The 6800 is an 8-bit processor; that is, it operates on 1 byte.
7. In its internal construction (architecture) the 6800 contains, among other circuits, the following: ALU, two accumulators, data and address registers, data and address buses, a program counter, an instruction decoder control, a stack pointer, and a condition-code register.
8. The ALU performs arithmetic and logic operations.
9. Registers are temporary storage devices.
10. The accumulator is a special register which holds the data on which arithmetic operations are performed. It also holds the final result of these operations.
11. Digital information is transferred over a group of conductors called a *bus*.
12. In the execution of a program, the program counter keeps a record of the address at which the next instruction is found.

13. The machine code instructions which a micro receives are interpreted and decoded by the instruction decoder/control.

14. Timing or synchronization for the 6800 circuits is provided by an external clock.

15. The hexadecimal number system utilizes 16 digits (0 through 9 and A through F) to represent numerical (symbols) values. It is used in programming the ET3400 trainer.

SELF-TEST

Check your understanding by answering these questions.

1. The 6800 is a self-contained microcomputer. ___________ (true/false)
2. A microprocessor is a hard-wired logic device. ___________ (true/false)
3. A ___________ is a 16-bit binary number.
4. The term ___________ , as used in microprocessor terminology, identifies the location or the destination of data.
5. The section of a computer which controls the interpretation and execution of a program is the ___________ .
6. The set of instructions which orders a micro to do a certain job is called a ___________ .
7. Program instructions for the ET3400 trainer are written in ___________ code.
8. The 6800 micro ___________ (does/does not) contain an internal clock.
9. $11011_2 = 27_{16}$ ___________ (true/false)
10. $123_{16} = 291_{10}$ ___________ (true/false)

MATERIALS REQUIRED

- Equipment: Microprocessor-microcomputer trainer, Heathkit ET3400, plus operating instructions

PROCEDURE

Checking the Trainer

1. Read the operating instructions for the ET3400.
2. Plug the trainer into the 120-V 60-Hz power line. The power indicator LED should light. Turn power ON the trainer. The display should read CPU UP.
3. Perform the checks required in the ET3400 wiring manual. If all the checks show normal operation, proceed to step 4; otherwise, inform your instructor.
4. With power ON, press the RESET key. Note and record display in Table 30-1.

TABLE 30-1. Checking the Trainer

Step	*Key Pressed*	*Trainer Display*
4	RESET	
5	ACCA	
6	RESET	
7		
8		

5. Now press the ACCA key. Note and record display in Table 30-1.
6. Press RESET. Note and record trainer display in Table 30-1.
7. Now display the contents of accumulator B by pressing the proper key. Record in Table 30-1 which key you pressed. Also record the contents of the display at this time.
8. In Table 30-1 list the steps you would take to display the contents of the program counter. Perform the steps and record the trainer display.

Familiarization with Keyboard Operation

9. To determine the operation of the keys shown in Table 30-2, press RESET before pressing each key.

TABLE 30-2. Keyboard Familiarization

*Key Pressed**	*Trainer Display*
CC	
SP	
RTI	
SS	
BR	
AUTO	
EXAM	
FWD†	
BACK†	
CHAN†	
INDEX	

* You must press RESET each time before pressing any one of the *function* keys in this table.
† These keys should be pressed after pressing RESET, EXAM, and 0001.

Indicate in the right column of Table 30-2 the trainer display after each of the keys (other than RESET) is pressed. NOTE: Checking the keyboard operation gives us one indication of whether the trainer is operating normally. We have other techniques to determine if trainer operation is normal. One such technique is to enter a program whose results we already know and observe if the micro and its attendant circuitry in the trainer run this program properly. We will perform such a check by entering a very short program, using the built-in functions of the trainer.

Checking the Trainer with Subroutine OUTCH

*NOTE: At address FE3A (hex) there is a subroutine built into the trainer which the designer calls OUTCH. This humorous acronym stands for the words "output character to display." (It is common practice to put together initials or word beginnings into a word as a label for a subroutine, OUTCH for out*put *ch*aracter.) The character that will be outputted as a result of this subroutine will be the leftmost LED in the display, and the segments that will be lit will depend on the contents of accumulator A.

10. First instruct the micro that we wish to use subroutine OUTCH. The program we enter to achieve this is: (*a*) jump to subroutine FE3A and (*b*) stop. The OP code for jump to subroutine is BD, and it is followed by the address in the next two words. The Stop instruction is identified by OP code 3E (it is really a Wait for Interrupt instruction). Assume that we wish to start the program at address 0000. Proceed as follows:

Press		*Display*	
RESET		C P U U P	
AUTO		_ _ _ _ A d	
0		0 _ _ _ A d	
0		0 0 _ _ A d	
0		0 0 0 _ A d	
0		0 0 0 0 _ _	
B		0 0 0 0 b _	
D		0 0 0 0 b d	(while pressed)
	on release	0 0 0 1 _ _	
F		0 0 0 1 F _	
E		0 0 0 1 F E	(while pressed)
	on release	0 0 0 2 _ _	
3		0 0 0 2 3 _	
A		0 0 0 2 3 A	(while pressed)
	on release	0 0 0 3 _ _	
3		0 0 0 3 3 _	
E		0 0 0 3 3 E	(while pressed)
	on release	0 0 0 4 _ _	
RESET		C P U U P	

11. We must now check to determine if we entered the four-word program (addresses 0000 through 0003) properly, since it is easy to enter errors. To check, do the following:

Press	*Display*
EXAM	_ _ _ _ A d
0	0 _ _ _ A d
0	0 0 _ _ A d
0	0 0 0 _ A d
0	0 0 0 0 b d
FWD	0 0 0 1 F E
FWD	0 0 0 2 3 A
FWD	0 0 0 3 3 E
RESET	C P U U P

NOTE: The display showed that at address 0000 we have bd. This is the instruction to jump to the address that follows. At address 0001 is FE, the most-significant digits of the address we wish to jump to. At address 0002 is 3A, the least-significant digits of the address to which we wish to jump. At 0003 is 3E, the instruction to stop.

12. Now enter into the accumulator those digits which will identify the segments of the leftmost LED (in the display) we wish to light. Suppose that we wish to light all segments, that is, that we wish to light the number 8. Proceed as follows:

Press		*Display*	
ACCA		A c c a x x	(x is a random digit)
CHAN		A c c a _ _	
F		A c c a F _	
F		A c c a F F	
DO		_ _ _ _ d o	
0		0 _ _ _ d o	
0		0 0 _ _ d o	
0		0 0 0 _ d o	
0		0 0 0 0 d o	(while pressed)
	on release	8	
RESET		C P U U P	

NOTE: The program has been executed; displays have appeared as expected. This is another confirmation that operation of the trainer is normal. Pressing the RESET key at the end creates an interrupt and restores the trainer to normal function.

TABLE 30-3. Display of Digits in Leftmost LED

Digit Displayed	*Contents of ACCA*
1	
2	
3	
4	
5	
6	
7	
9	

13. Determine experimentally what digits must be entered into accumulator A to light up the 0 in the leftmost LED. Remember that to enter any two digits into accumulator A, you must press ACCA; CHAN; XX, where X stands for the required digit. To obtain the display of these digits, you must press DO then 0000. Record these digits in Table 30-3. Experimentally determine the digits that must be entered in accumulator A to cause each of the following digits to light in the leftmost LED: 1, 2, 3, 4, 5, 6, 7, and 9.

QUESTIONS

1. What is the difference between a microprocessor and a microcomputer?
2. Which of the following more closely approximates a microcomputer, the 6800 or the ET3400? Why?
3. What steps must you perform on the ET3400 to display the contents of accumulator B?
4. How would you change the last two digits in the final display in question 3? Write the instructions to achieve this.
5. What does executing the OUTCH subroutine accomplish?
6. What key must you press to display CPU UP?
7. When CPU UP is displayed on the trainer, what would be the effect of pressing ACCB? Write the display which you would see.
8. The display shows Accb67. What will happen when you next press CHAN? Indicate the new display.
9. After performing the instruction in question 8, what must you do to cause the display to read Accb23?
10. Why is the keyboard in the ET3400 called a hex keyboard?

Answers to Self-Test

1. false
2. false
3. word
4. address
5. CPU
6. program
7. hex
8. does not
9. false
10. true

EXPERIMENT

31

PROGRAMMING THE MICROPROCESSOR—PART 1

OBJECTIVES

1. To become acquainted with the components of the ET3400 system
2. To examine some of its built-in functions
3. To write, enter, and execute a simple program

BASIC INFORMATION

Block Diagram ET3400

To help understand microprocessor programming, let us look more closely at the hardware we are using, the ET3400 trainer. Figure 31-1 is a block diagram of the fundamental components which make up this micro-

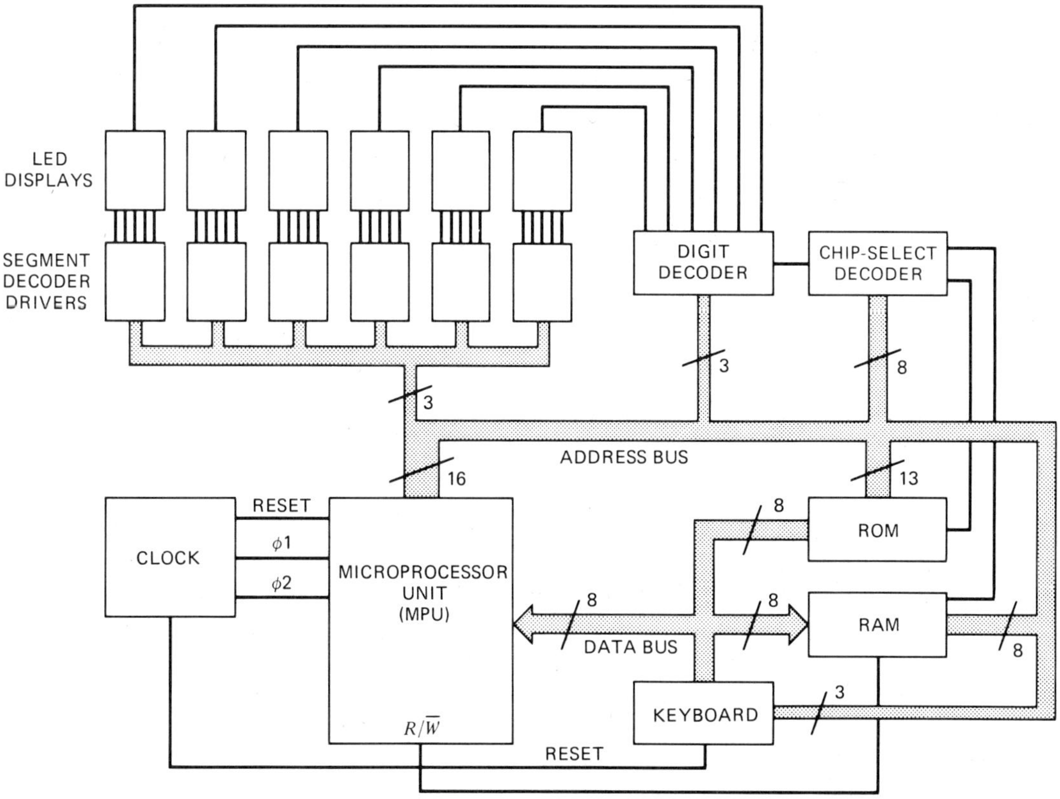

Fig. 31-1. Block diagram of the ET3400 trainer.

processor trainer. These include the 6800 microprocessor unit (MPU), the heart of our system; a clock; a keyboard; a RAM memory; a ROM memory; six seven-segment LED display units; LED-segment decoder drivers; a digit decoder; a chip-select decoder; an address bus; and a data bus.

MPU AND CLOCK

The MPU performs all the logic and arithmetic functions of our system. It also acts to process and keep track of our data and program instructions. To operate the MPU, an external clock is required. The clock generates two timing signals equal in amplitude and frequency, but 180° out of phase. The micro starts and stops various actions depending on the transitions, that is, polarity changes of the two signals.

ADDRESS AND DATA BUSES

The micro connects to the rest of the circuits through two buses. We will not concern ourselves with other circuit details, for example, buffers, inserted at various points to isolate circuits and/or to reduce signal losses. The width of the bus, that is, the number of connecting lines in each bus, is shown in Fig. 31-1 by a slash line and a number. Thus /8 on the data bus indicates that there are eight lines of data going from the micro to the RAM, the ROM, and the keyboard. On the other hand, the address bus coming out of the micro is 16 lines wide. In addition, 3 lines go to the segment decoder/drivers, 3 others go to the digit decoder, 8 lines go to the chip select decoding circuit, 13 address lines go to the ROM, and 8 address lines go to the RAM.

The number of address lines determines the number of different addresses that can be written to a device. For the ROM, for example, the 10 address lines select any of 2^{10} or 1024 addresses. The other three address lines are used to select the ROM chip. It should be noted, however, that all the devices external to the MPU are on standby; that is, they do not read the addresses and do not put out data unless they are selected. By exercising this control, the micro can keep operation on the bus and on the system orderly and meaningful.

RAM AND ROM

The arrows on the data bus indicate that data can go from the ROM to the micro, from the keyboard to the micro, and both from the RAM to the micro *and* from the micro to the RAM. That is, the data bus is bidirectional between RAM and micro. A RAM memory, as was indicated in Exp. 30, can be both written into and read from. A ROM memory, on the other hand, can be only read from. It cannot be written into. This is clearly evident from the bidirectional data bus connecting the MPU and the RAM and the unidirectional data bus connecting the MPU and the ROM.

To ensure proper direction of data flow between the micro and the RAM, a $R/\overline{W}$ line is connected from the MPU to the RAM. $R/\overline{W}$ stands for read, <u>not</u> write. The circuit is so designed that when the $R/\overline{W}$ line is *high*, the RAM signal is read, and information passes from the RAM to the micro. When the line is *low*, the opposite happens; that is, the micro is writing to the RAM. The micro automatically sets the $R/\overline{W}$ line high or low, depending on the nature of the instruction.

KEYBOARD, DISPLAYS, AND UTILITY PROGRAMS

The keyboard generates the signals required to communicate with the micro. But these signals by themselves would have no meaning. Individual programs are required to tell the micro what each of the keyboard signals means so that the micro can initiate the action needed to implement the program. These programs, called *utilities*, are permanently entered into the ROM by the system designer and cannot be changed by operating the keyboard. These programs constitute part of the built-in functions of the ET3400 system.

There are other utility programs entered in the ROM, for example, a program which makes the LED displays light up under instruction from the micro. The LED display is one of the features of the trainer which makes it possible for the operator to observe the results of his or her work. The utility program which tells the seven-segment LEDs when and how to light is also permanently contained in the ROM, as are the routines which define the actions of the keys.

DECODERS

Figure 31-1 shows three different decoders, the chip-select decoder, the digit decoder, and the segment decoder. The address bus connects to each of them. Each has a unique function, which will be examined.

The memory addresses coming from the micro may be intended for the ROM or the RAM or I/O units. Some of the address lines are processed by the chip-select decoder and act to enable the specific device being addressed.

The results of the logical and arithmetical actions of the micro in our system are shown on the LED displays. Each LED unit not only has seven segments which form its characters, but also a decimal point. So a complete byte (8 bits) of information is required to identify any of the segments and the decimal point in any one LED. Since there are six LEDs, a total of 48 addresses are required to select all the segments and decimal points. In our system, approximately twice that many addresses are reserved in the memory map for that purpose.

The digit decoder determines which LED display is to be lit. The segment decoder decides which segment of an LED is to be lit. Note that the LEDs are not connected directly to the address bus. They do go directly to the segment and digit decoders which receive their input from the address bus.

Mapping the Memory

ROM ADDRESSES

The utility programs and their addresses were permanently entered in the ROM in manufacture. The ROM contains the utility programs in addresses FC00 through FFFF (see Table 31-1). This is a total of 400_{16} addresses, which is the equivalent of 1024_{10} addresses, and is therefore known as a 1K memory.

We can examine the contents of any of the addresses in ROM. For example, suppose we wished to see what the ROM has entered at address FC02. We would press the RESET key and then EXAM and would then enter the desired address, namely, FC02. The LED display would then read FC02Eb. The two digits that appear in the display after the address, namely, Eb, are the contents in ROM for that address.

Remember that the ROM will not accept anything that we write into its memory. As proof, if we tried writing into any address in the ROM by way of the keyboard and then checked the *new* contents of that address, we would find the *old* contents still there, not the new.

RAM ADDRESSES

There are altogether 256_{10} memory locations in the RAM in the ET3400 system. Addresses for these memory locations are 0000 to 00FF. The ET3400 trainer permits the operator to write many different programs for the micro to execute, by using the RAM as a program memory. (Normally the ROM acts as the program memory.) RAM programs are entered via the keyboard, are temporarily stored, and disappear from memory when power is disconnected from the trainer, unlike the ROM, whose contents remain permanently in memory. You will recall that the ROM memory contains the utility programs which were entered at the time of manufacture.

In our programming we will use the RAM memory for entering *instructions and data,* and we will specify those addresses reserved for instructions and those for data, a process known as *mapping the memory.* We will arbitrarily set aside for instruction memory 144 locations with addresses from 0000 to 008F. We will write all data into the RAM from addresses 0090 to 00C4. Henceforth we must not confuse these two locations (instructions versus data) in memory when programming. For example, we must be careful not to use the program memory locations of the RAM for entering data, because we would in the process, erase any instructions at those addresses.

The 59 locations in the RAM from 00C5 to 00FF are reserved for the stack, interrupt vectors, and other predetermined functions.

The memory area from 0100 to C000 is not implemented in our system; that is, there are no devices in our system which respond to these addresses. However, the ET3400 makes provision for adding two additional RAM units which would add 256 bytes of RAM, responding to the addresses 0100 to 01FF. This optional feature would double the size of the RAM memory.

TABLE 31-1. Memory Map, ET3400 System (Heath)

Addresses	*Purpose*
0000–00C4	197 bytes of user RAM
00C5–00FF	59 bytes of RAM, reserved for monitor
0100–01FF	Optional 256 bytes of user RAM
C000–C002	Not usable
C003–C00E*	Keyboard
C00F–C10F	Not usable
C110–C16F*	Displays
C170–C1FF	Not usable
FC00–FFFF	Monitor ROM

* Not fully decoded.

KEYBOARD AND DISPLAY ADDRESSES

Refer to Table 31-1. The program area from C003 to C00E has been set aside for the keyboard. The area from C10F to C16F is reserved for the displays.

Microprocessor Unit (MPU)

In order to write a program efficiently, we should know the steps that the micro takes in executing an instruction. Figure 31-2 shows the internal architecture of a micro, that is, those features involved in executing a program step. It should be noted at this point that the architecture of different micros differs widely. For example, the length of a word can differ from 4 to 32 bits. The 6800 uses an 8-bit byte. The number of internal registers can differ widely. Some micros contain both RAM and ROM memories inside the MPU, and these devices are frequently called *microcomputers.*

After a program has been entered, each MPU acts to carry out the details of the program. To that end the various MPUs are similarly organized. Each has a program counter (PC) as in Fig. 31-2, which keeps track of the address for the next instruction. At a given time in the clock cycle, the micro is ready to fetch the next instruction. At that time the address in the PC is transferred to the address register and placed on the address bus. The ROM is signaled to accept that address, and it places the instruction stored at that address on the data bus. Next the MPU transfers this instruction to the instruction decoder. The instruction decoder de-

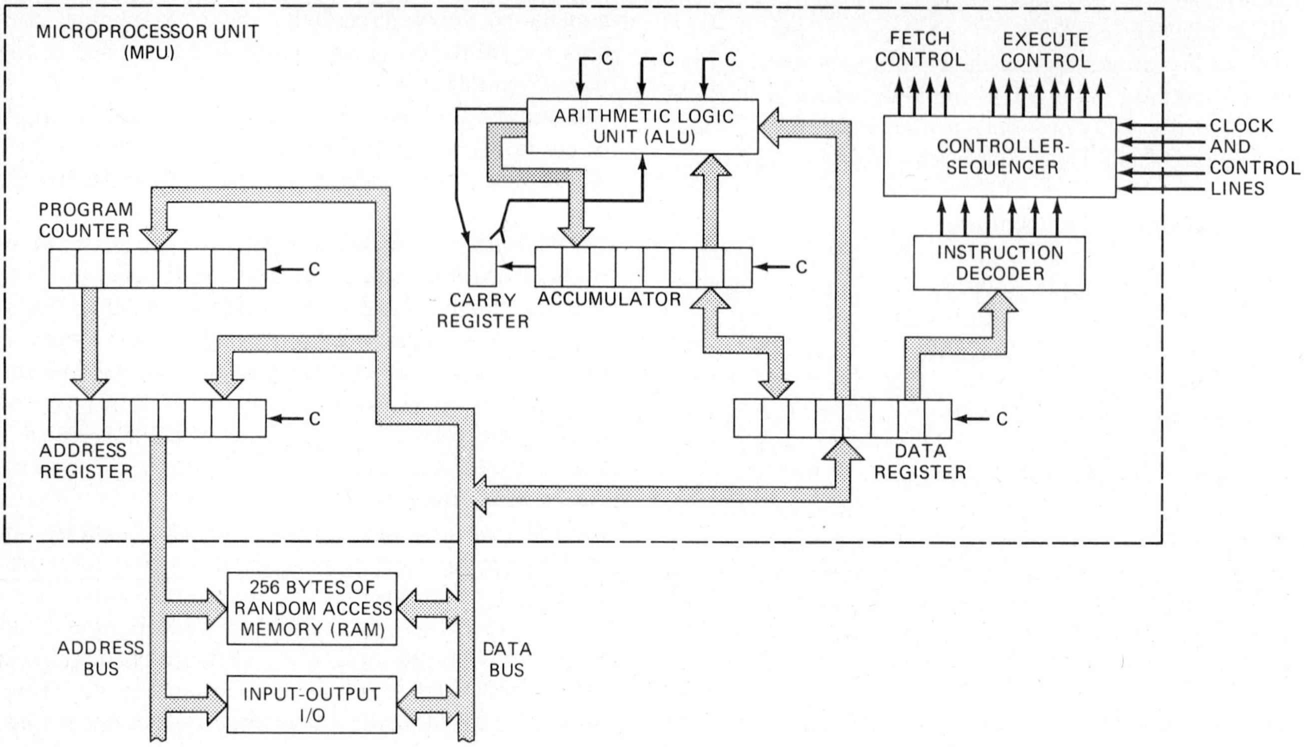

Fig. 31-2. Internal architecture of an elementary microprocessor. *(Heath)*

cides what action the MPU must perform to execute the instruction.

Execution of the instruction is the next phase in the micro operation. The nature of the execution depends on the instruction. Consider the fact that some instructions are single-byte, and others multibyte. For example, the single-byte instruction NOP, meaning *no* *op*eration, just tells the MPU to do nothing except increment the program counter during execution. After NOP the next instruction will be fetched as soon as the clock is at the proper point in its cycle. On the other hand, the JSR (*j*ump to *s*ub*r*outine) instruction, to which the 6800 can respond, requires 3 bytes, including 2 bytes of address for the operand following the instruction, so that the micro will know the address to which it must jump. The JSR instruction requires many steps from the micro. These are:

1. The present address must be placed on the stack. To accomplish this, (*a*) the address in the stack pointer must be placed on the address bus, (*b*) R/$\overline{W}$ must be pulled low to tell the RAM that the MPU is writing, (*c*) present address must be placed on the data bus, (*d*) and stack pointer must be decremented.
2. The next byte in ROM must be fetched and held in the data register because it is part of the new address to be placed in the program counter.
3. The program counter must be incremented again.
4. The address is placed in the address register.
5. The data register information is transferred to the most-significant byte of the program counter.
6. The least-significant byte of the new address is obtained from the next ROM location and transferred through the data register to the lower byte of the program counter. Now the MPU is ready to fetch the next instruction from the new program location.

MPU Instruction Set

Every instruction must be interpreted by the MPU and executed. The instruction set is the means of communication between the programmer and the MPU. The 6800 has 72 instructions, many of which can be executed with different forms of addressing. Built-in operational codes (OP codes) for each addressing mode add to 140 OP codes that the MPU can interpret. Appendix A lists the 6800 Instruction Set and the operating modes for each instruction.

In order to program the micro efficiently, each of the instructions must be understood by the programmer. In this experiment only those instructions used in the introductory programs which follow are discussed in detail. The student can refer to the explanations in Appendix A for any additional instructions if desired.

The instruction set can be considered a language. The difference between our spoken language and a micro or any computer language is primarily the degree of exactness. A statement like "This is a nice picture" will mean different things to different people. A computer cannot use such vague statements. A computer instruction will be executed exactly the same way every time it is presented to a computer. The biggest difficulty that a beginning programmer has is to think concisely and to formulate programs accurately. To help in developing a concise and logically sound program, it is necessary to define exactly what the program should accomplish. The next step is to draw a flowchart. This is a graphic representation of the steps required to accomplish the desired goal. Next, this flowchart is translated step by step into assembly language.

The computer "language" is a set of binary numbers. It is extremely difficult to remember binary numbers, so an assembly language is used to help the programmer. The first step is to associate a specific instruction with a mnemonic. A mnemonic consists of a few letters, usually the first letter(s) of the word(s) of the required operation. The relationship of the letters to the words of the instruction makes the mnemonic easier to remember than the binary code of the operation. For example, there may be times when you wish the micro to do nothing. This is called "no operation," and the mnemonic for that instruction is "NOP." The machine code in binary for that instruction is 0000 0001. In hex it is 01. The 6800 is set up for the hex code, and so we will use hex code in communicating with the MPU. Note that the instruction set listed in Appendix A is given in hex code.

Another example of a mnemonic is the 6800 instruction JSR. This mnemonic stands for *j*ump to *s*ub*r*outine. The machine code for the JSR instruction is either 1010 1101 (AD in hex) or 1011 1101 (BD in hex). Which code to use will be determined by the addressing mode to be used.

Addressing Modes

In analyzing the 6800 Instruction Set listed in Appendix A, it is at once apparent that many instructions can use a variety of addressing modes. Altogether five addressing modes are listed. What does the term "addressing mode" mean? An addressing mode refers to the way that an instruction interprets its operand(s). Thus the instruction Add (see Table 31-2), whose mnemonic is either ADDA or ADDB, depending on which accumulator is used, has four addressing modes. These are IMMEDIATE, DIRECT, INDEXED, and EXTENDED. Another instruction, Add Accumulators, with mnemonic ABA, can use only one addressing mode, INHERENT. What is the difference between these modes, and when would each be used?

Before considering this question, refer again to Table 31-2, instruction ADDA. Note that in the IMMEDIATE mode the OP code for this instruction is 8B, the number of MPU cycles (~) the micro needs to fetch and execute it is 2, and the number of bytes of memory (#) required to store this instruction and its operand is also 2.

The ABA (add accumulators) instruction uses the INHERENT mode of addressing. Its OP code is 1B. It needs two MPU cycles to fetch and execute it and 1 byte of memory.

One fact becomes apparent from this comparison. The INHERENT mode is more economical of mem-

TABLE 31-2. Some 6800 Instructions

Instruction	Mnemonic	*Addressing Modes*				
		IMMEDIATE	*DIRECT*	*INDEXED*	*EXTENDED*	*INHERENT*
		OP ~ #	*OP ~ #*	*OP ~ #*	*OP ~ #*	*OP ~ #*
Add	ADDA	8B 2 2	9B 3 2	AB 5 2	88 4 3	
	ADDB	CB 2 2	DB 3 2	EB 5 2	FB 4 3	
Add Accumulators	ABA					1B 2 1
Load Accumulators	LDAA	86 2 2	96 3 2	A6 5 2	B6 4 3	
	LDAB	C6 2 2	D6 3 2	E6 5 2	F6 4 3	
Store Accumulators	STAA		97 4 2	A7 6 2	B7 5 3	
	STAB		D7 4 2	F7 5 3		
Transfer Accumulators	TAB					16 2 1
	TBA					17 2 1
And	ANDA	84 2 2	94 3 2	A4 5 2	B4 4 3	
	ANDB	C4 2 2	D4 3 2	E4 5 2	F4 4 3	

ory space than the IMMEDIATE mode. There are other characteristics of addressing modes to consider.

IMMEDIATE ADDRESSING

In IMMEDIATE addressing, generally 2 bytes of memory are used in the 6800. Of these, the first is the OP code. The next word in memory *immediately* following the OP code is the *operand*. IMMEDIATE addressing is particularly suitable when the operand addressed remains the same throughout the program.

INHERENT (OR IMPLIED) ADDRESSING

In INHERENT addressing, in the 6800, 1 byte of memory is used. The OP code for this type of instruction has no operand, but the operand is implied. For example, the instruction add accumulators implies accumulator A and accumulator B.

The advantages of the IMMEDIATE and INHERENT modes of addressing, namely, the use of little memory space and short execution time, commend their use whenever possible. However, there are situations when it is not possible to use these modes. That is why microprocessors provide other modes of addressing.

DIRECT ADDRESSING

DIRECT addressing is useful when the data can change or is to be operated on by many different instructions. In this mode, an instruction also requires 2 bytes of memory. The first is the OP code. The second is *not* the data, but the *address* of the data.

Refer again to Table 31-2. In the DIRECT mode, the instruction *load accumulator A (LDAA)* is identified by the OP code 96. It requires three MPU cycles to fetch and execute and 2 bytes of memory. The instruction LDAA 38 in the DIRECT mode would mean load the data at memory address 38 into accumulator A. In the IMMEDIATE mode, the same instruction would mean load the number 38 into accumulator A. In an 8-bit byte, the addresses for 00 to FF_{16} can be entered. Therefore, in the DIRECT mode the data must be in the first 256_{10} bytes of memory.

Table 31-2 shows that the OP codes for the same instruction LDAA in the DIRECT mode is 96, while in the IMMEDIATE mode it is 86. The reason for their difference is to make certain that the micro will know exactly what it is to do.

EXTENDED ADDRESSING

This mode is like DIRECT addressing, except that it permits the programmer to use addresses beyond the first 256_{10} bytes of memory. An instruction that uses EXTENDED addressing, therefore, requires 3 bytes of memory. The first is the OP code. The second and third bytes are two operands for a 16-bit address. The data is at that address. The byte immediately following the OP code (the second byte) is the most significant part of the address, and the third byte, is the least significant part of the address. Altogether $65{,}536_{10}$ memory locations, 0000 to FFFF, can be addressed in this mode. If the address of the data is in the first 256_{10} memory locations, it is possible to use either EXTENDED or DIRECT addressing. In that case, DIRECT addressing is preferred because it saves 1 byte of memory.

In Table 31-2, note the instruction LDAA. Its OP code in the EXTENDED mode is B6. It requires four MPU cycles to fetch and execute and 3 bytes of memory.

INDEXED ADDRESSING

The 6800 has a 16-bit index register. The instructions associated with this register permit it to be loaded from memory or permit us to transfer the contents of this register into memory. The index register can also be incremented or decremented. This register can therefore be used as a 16-bit counter and as an address pointer, since it can point to any address in memory. The index register can be used in connection with tables of numbers stored in memory. For example, by use of INDEXED addressing, we can program by adding a long table of numbers in fewer steps than we could by using any other form of addressing.

INDEXED addressing uses 2 bytes of memory. The first is the instruction OP code, and the second is for an *offset address*. The offset address, added to the number in the index register, is the address of the operand.

In programming, INDEXED addressing is specified by adding an X after the instruction mnemonic. Thus LDAA,X means load accumulator A by INDEXED addressing. Refer to Table 31-2, and note that the OP code for LDAA,X is A6. It takes five MPU cycles to fetch and execute and 2 memory bytes.

RELATIVE ADDRESSING

This mode of addressing is also available in the 6800. RELATIVE addressing is used for program branching. Two bytes make up the instruction: the OP code and the RELATIVE address. The RELATIVE address, added to the program count, makes up the absolute address. A method to determine the RELATIVE address will be given later in the experiment in the discussion on Program E.

Built-In Functions

Sixteen of the 17 keys of the keyboard are associated with direct built-in functions of the ET3400 system. These functions serve as aids to the programmer for they speed up the process of entering instructions and data and make the contents of certain registers available for examination. In a previous experiment we had a brief look at some of these functions. Let us examine them from an operational standpoint. In the discussion which follows, the legend used to identify each key will be associated with its function.

1. *RESET* This key starts the system and causes the trainer to display the legend CPU UP. Pressing RESET resets the stack pointer and permits the programmer to start addresses from scratch. Pressing RESET also removes any breakpoints in a program. We do not always list RESET when we are entering a program, because some previously entered information, which may be needed, may be lost.

2. *ACCA* Pressing ACCA will display Acca and the contents of accumulator A. If any information was entered in this accumulator, it will appear as two hex digits, for example, Acca23. The digits 23 are the data in the accumulator. If no information was entered in the accumulator, two random digits will appear. The contents of this register can be changed by pressing the CHAN key. Then the display will appear as A c c a _ _ . Now you enter the two hex digits you want with two key strokes.

3. *ACCB* This key operates exactly as does ACCA, except contents of accumulator B are displayed.

4. *PC* Pressing this key will display the contents of the program counter. The LED display will light up as follows: Pc followed by four hex digits, as, for example, Pc3054. The digits 3054 constitute the address of the next instruction to be fetched. You can change the address by pressing CHAN and entering the new four-digit hex number desired.

5. *INDEX* We display the contents of the index register by pressing this key. The display will read In, followed by four hex digits, as, for example, In3C00. This is the reference register for indexed addressing which we discussed earlier. You can change the contents of this register by pressing CHAN and entering the new hex number.

6. *CC* Pressing this key displays the contents of the condition code register. Six digits, each of which is 0 or 1, will appear on the LED display. Next to each LED, at the lower right hand, are the letters H, I, N, Z, V, and C. These letters correspond to the condition codes of the Instruction Set shown in Appendix A. This register cannot be changed by pressing the CHAN key.

7. *SP* When this key is pressed, the contents of the stack pointer are displayed, as, for example, SP00d2. The four digits in the example, namely, 00d2, constitute the address of the last data entered in the stack. The contents of this register *cannot* be changed by pressing CHAN.

8. *RTI* These letters stand for *Ret*urn from *I*nterrupt. Pressing this key allows you to return to the next instruction in a program before an interrupt occurred. A program can be interrupted by enabling the breakpoint (BR) routine, or the single-step (SS) routine, or the WAI instruction.

9. *AUTO* This key is used to enter a program quickly. It eliminates the need to advance the address manually at each step. When AUTO is pressed, the display _ _ _ _ A d appears. The address at which you wish to start is then entered. For example, let the starting address be 0000. After it is entered, the display changes to 0 0 0 0 . A two-hex digit instruction or operand is then entered. The trainer automatically advances the address after the second digit is entered. It then waits for the next instruction to be entered at the new address.

10. *EXAM* This function is used to examine the data at any address by pressing EXAM and entering the address. After the fourth digit of the address is entered, the data at that address appears in the display. For example, pressing EXAM gives rise to the display A d. Suppose that the address we enter is 0000. If the data at that address is Ab, for instance, the display will read 0000Ab. We can change the data by pressing CHAN and entering the data desired.

11. *CHAN* This key is associated with the following keys: EXAM, ACCA, ACCB, INDEX, PC, BACK, FWD, and SS. The CHAN key may be used to enter a new value, or to correct a value previously entered by mistake. It should be noted that a program may be entered by using the EXAM and CHAN function or the AUTO function.

12. *BACK* When an address and its data are displayed, pressing BACK will cause the previous address and its data to be displayed. We can change the data by pressing CHAN and entering the new data.

13. *FWD* When an address and its data are displayed, pressing the FWD key will cause the next address and its data to be displayed. We can change the data by pressing CHAN and entering the new data.

14. *SS* This key is used to single step a program; that is, the microprocessor will execute only one instruction of a program at a time. The instruction to be performed comes from the address which is in the program counter. After the step is completed, the next instruction and its address are displayed when SS is pressed again. The displayed instruction may be changed by pressing CHAN and entering the new data.

15. *BR* This key is used to enter breakpoints in testing a program. A breakpoint is a point where a

program is halted to examine the micro's register or memory. A breakpoint is entered by pressing the BR key, and then entering the four hex digits of the address where the breakpoint is to occur. The address must be one of an *operational* code in your program and that code must be in RAM, *not* in ROM. Four breakpoints are possible in any one program. Pressing the RESET key clears all breakpoints.

16. *DO* After the DO key is pressed, the display reads _ _ _ _ d o. Enter the starting address of your program, assuming that your program has already been entered in the trainer. Execution of the program starts from this address. The program will then run until a display is required, or until it comes to a breakpoint, or until the RESET key is pressed.

As was previously noted, there are other built-in functions in ROM. These are software programs and can be activated by jumping to the desired subroutine in ROM. We will refer to some of these functions as they are needed.

INTRODUCTION TO PROGRAMMING

Programming Steps

In order to write a program, the programmer must follow a number of steps. These are:

1. *Define the requirements* (problem definition) for the system, that is, decide exactly what the system is to accomplish.
2. *Draw a flowchart* showing what the input to the system must be, the steps the system must take to accomplish the objective, and finally what the system must do after the objective is achieved.
3. *Write the mnemonic* for each step on the flowchart.
4. *Translate the mnemonics into the machine code* which the system recognizes. In the case of the ET3400, the machine code is written in hex.
5. *Enter the machine code into memory.*
6. *Check the program.*
7. *Run the program.*

Let us briefly examine these programming steps. First, determining the requirements, that is, defining the problem, is the task of the system engineer or the person in charge. Generally, the definition supplied by the person in charge is not sufficiently exact or detailed to be used immediately by the programmer, so a programmer must develop an exact definition and add the necessary details. Then the programmer must translate this definition into a flowchart. Next, the programmer has to convert the flowchart steps into programming mnemonics.

Let us pause a moment to observe that in the "real world" where a micro system has been designed to control a process, the program needed to achieve the degree and the type of control wanted may be very complex indeed. Fortunately, certain aids are available to the programmer to make the job easier. For example, to translate the flowchart into mnemonics, the programmer can use a *high-level* language. This is a language that is closer to the spoken language than the mnemonics of the micro. But this aid requires the use of a *development system,* which is, in fact, a large-scale computer especially designed to develop programs for microprocessors. Any one of several high-level languages may be employed, such as Pascal, Basic, PL/M, or others. For our part, we will simply use the direct machine mnemonics of the 6800 in writing programs for our system.

In the "real world" the translation of mnemonics into machine code is done automatically by assemblers of a development system, and entering the machine code into memory is done by a machine, which frequently is also part of the development system. For our part, we will use the charts of mnemonics and their machine code equivalents, supplied for the 6800, and we will enter by hand the hex code of each program step via the keyboard.

Checking the Program—SS and BR

After a program has been entered, the next step is to check it for typographical errors, logic errors, and hardware errors which may sometimes occur. Errors in a program are called *bugs,* and the program must be *debugged* if it is to be useful. In commercial practice, debugging is aided by a development system. The ET3400 has two functions that are used for debugging. One is the SS, or the single-step function. In this operating mode we can advance the program one instruction at a time, so that we can examine the contents of memory, accumulator, or other registers at each step. Since we know what the program is supposed to do at each step, we can locate the bug because that is the point at which the program either does not perform the expected operation or performs it improperly.

An obvious problem of single-stepping through a program is the time required. This is especially true when a section of the program is repeated many times. If we know that this portion of the program works correctly, we would like to run it at normal speed and stop at a point in the program which has not been checked yet. For this purpose, a second function on the trainer, called BR, is used in debugging. BR stands for breakpoint. We can set an address as a breakpoint, and the program will stop at that point when it is started with the DO command. Then we can single-step from that point on.

After a program has been checked and debugged, it can be run so that the desired result may be achieved.

Programming with Hexadecimal Numbers

In the programming examples that follow, hexadecimal numbers will be used since the trainer can use these numbers directly. Of course, the MPU and the memory devices respond to and store only binary numbers. You will recall, however, that the ET3400 has a built-in function which reads the keys of the keyboard and permits hex numbers to be entered and then interpreted as binary digits.

Table 31-3 is a restatement of hex digit numbers, their binary equivalents, and finally their decimal values. You may wish to refer to this table in the problems that follow.

Each hex digit corresponds to a 4-bit binary number. A two-digit hex number is represented by two 4-bit binary numbers written side by side. For example, the number $2A_{16}$ is written in binary as 0010 1010. The first 4 bits, 0010, is obviously the hex number 2, while the last 4 bits, 1010, is A_{16}. The hex number 49 in binary form is 0100 1001. It is evident from Fig. 31-1 that the eight-line data bus will accommodate a two-digit hex number, while the 16-line address bus can hold a four-digit hex number.

A Sample Program—Adding Two Hexadecimal Numbers

Let a requirement for the first program be the statement: "Add two hex numbers." Is this statement an adequate definition for the programmer? No. The programmer must know what the two numbers are, where they are, or where they are to be placed in the ET3400's memory and must also know where the result is to be stored.

TABLE 31-3. Hex, Binary, and Decimal Numbers

Hex	*Binary*	*Decimal*
0	0000	0
1	0001	1
2	0010	2
3	0011	3
4	0100	4
5	0101	5
6	0110	6
7	0111	7
8	1000	8
9	1001	9
A	1010	10
B	1011	11
C	1100	12
D	1101	13
E	1110	14
F	1111	15

You will recall in an earlier discussion that it was decided to use addresses 0090 to 00C4 of the memory map for data memory. Let the two numbers to be added be 21_{16} and 35_{16}; let 21 be stored in memory address 0090 and 35 in memory location 0091, and let the result be stored in memory location 0092. With these details, we can draw a flowchart for the program (see Fig. 31-3). In this chart two terms, augend and addend, must be defined. *Augend* is the mathematical term for a number to which another number, the addend, is to be added.

The flowchart instructs the programmer first to enter the number 21 into address 0090 and then 35 into address 0091. These steps are not really part of the program. They simply supply the data for the MPU. The program starts with the NOP instruction (no operation) and then directs the programmer to write the instructions for loading the augend into accumulator A (AccA), adding the addend to the number in accumulator A, and storing the result in memory address 0092.

It is not necessary to start the program with a NOP instruction. So why do we do it? The reason is that we will want to check how the program works, after it is entered into the trainer. In order to do this, the SS (single-step) function will be used. The first thing that happens when SS is pressed is that the first instruction is executed and the next address and instruction are shown in the display. By making the first instruction an NOP, we will be able to examine the program starting with the first operative instruction which is load accumulator A. It should be noted that the choice of using accumulator A was an arbitrary one. Either A or B could have been used.

We are now ready to write the mnemonics for the instructions in the program. These mnemonics can be found in the Instruction Set for the 6800 in Appendix A.

- NOP (no operation)
- LDAA 90 (load the contents of address 0090 into accumulator A)

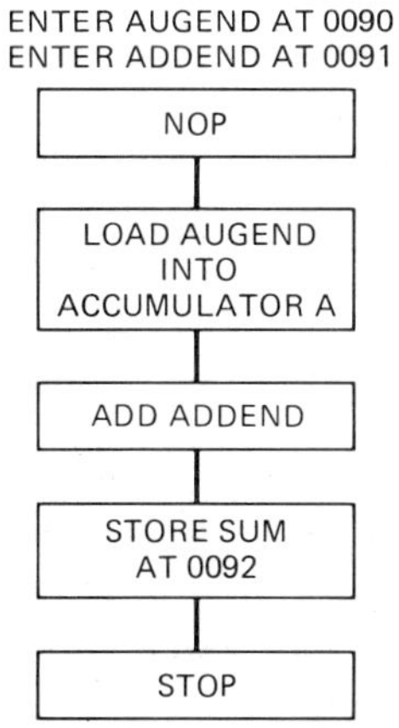
ENTER AUGEND AT 0090
ENTER ADDEND AT 0091
NOP
LOAD AUGEND INTO ACCUMULATOR A
ADD ADDEND
STORE SUM AT 0092
STOP

Fig. 31-3. Flowchart for adding two numbers.

- ADDA 91 (add the number in address 0091 to the number in accumulator A)
- STAA 92 (store the result in address 0092)
- WAI (Wait for Interrupt; we will use this instruction to HALT the program)

Now let us convert the mnemonics into machine code starting with address 0000. The instruction OP codes are found in the Appendix. Since we are loading, adding, and storing the *contents* of the next address following each instruction, we must use DIRECT addressing for the LDAA, ADDA, and STAA instructions. The NOP and WAI instructions have only the INHERENT mode of addressing.

After entering the program and its data, the program should be checked to see that no errors were made in entering. The details to accomplish this process will be given in the Procedure section. The program can then be executed by using the SS or DO functions. If the SS function is used, the final display will be 00073E. This is the last step of the program, but where is the result, the sum of 21 and 35? It is stored in address 0092 and can be found by examining the contents of that address, or of accumulator A. If the DO function is used to execute the program, the display will appear blank after execution. By resetting the trainer and checking the contents of address 0092, the sum, the number 56, will be displayed.

OUTBYT

If we wish to display the sum directly at the completion of a program, without examining the contents of the accumulator or the address where the result may be stored, a display subroutine called OUTBYT may be utilized. The subroutine is stored in ROM starting at address FE20. However, to get to address FE20, we must modify Program A at address 0007 and make use of another instruction called JSR. This mnemonic stands for *j*ump to *s*ub*r*outine. Since FE20 is a four-digit address, the EXTENDED mode of addressing must be used for JSR, whose OP code in this mode is BD. The modified program will be the same as Program A for steps 0000 through 0006. It will then proceed as follows:

The method for modifying Program A and entering steps 0007 through 000A will be given in the Procedure section. When Program B is now executed, the sum, 56, will be displayed. Programs A and B will add any two numbers and display their sum accurately if their sum is not greater than FF, that is, whose sum is restricted to *two* hex digits. So if we now wished to add the numbers 38 and 6A, all we need to do is change the contents of address locations 0090 and 0091 to 38 and 6A respectively and run Program B. The result will be A2.

PROGRAM A.

Address	*OP Code*	*Mnemonics and Addressing Mode*
0000	01	NOP (INHERENT)
0001	96	LDAA (DIRECT)
0002	90	operand
0003	9B	ADDA (DIRECT)
0004	91	operand
0005	97	STAA (DIRECT)
0006	92	operand
0007	3E	WAI (INHERENT)

Data: Enter data and addresses as follows:

Address	*Data*
0090	21
0091	35

Using the Carry (C) Condition Code

What happens if the sum of two two-digit numbers is greater than FF, that is, if the sum is a three-digit number? For example, the sum of F3 and 45 is 138. If we were to enter F3 into location 0090 and 45 into 0091 and execute program B, the display would indicate the sum 38. The most significant digit, 1, is missing because Program B does not take into consideration the final carry bit (1). Again we must modify the program if we wish to add numbers whose sum is as large as 1FE. We use the knowledge that the condition code (CC) of the trainer will indicate whether there is a carry (1) when the Add instruction is executed. After executing Program B, a check of the condition code at address 0007 would have revealed 00000*1*. The final 1 is the condition of the *carry flag* and indicates that the carry is *set,* that is, that there is a carry. If the problem had been to find the sum of 23 and 45, the condition code at address 0007 would have been 00000*0*, showing no carry.

The carry flag is therefore an indication as to whether a carry was generated when the two numbers were added. Now how can the program be altered to give the correct result? One method is to output three numbers which make up the sum. The most-significant digit in the sum just cited would be a 1, the carry condition

PROGRAM B.

Address	*OP Code*	*Mnemonics*
0000–0006 same as in Program A		
0007	BD	JSR (EXTENDED)
0008	FE	operand
0009	20	operand
000A	3E	WAI

code. We do this by transferring the condition code to accumulator A. The instruction mnemonic for this action, taken from the Instruction Set for the 6800 (Appendix A) is TPA, which stands for *T*ransfer from *P*rocessor Condition Code Register to *A*ccumulator. This would result in a 1 in the carry bit if there were a carry, and a 0 if there were none. How about the other condition code bits? There are altogether six, H, I, N, Z, V, and C, in that order. The unwanted condition code bits, H through V, can be eliminated regardless of their value (0 or 1) by ANDing the contents of accumulator A after the TPA instruction has been executed with the hex number 01. For example, assume that the condition code is 000001. When this number is ANDed with 01, the result is 01, that is, 000001 · 01 = 01. The reason becomes obvious when the numbers are written in binary. Here are three examples:

CC = *000001   010001   010000*

OPERAND = *00000001 00000001 00000001*

CC · OPERAND = *00000001 00000001 00000000*

because logical AND is performed bit by bit in the MPU. Remember that the only time the output of an AND circuit is HIGH (1), is when both inputs are 1, and that is true only when the carry bit and the 1 appear together in the input. So now the carry bit appears in the accumulator. It is interesting to note that if the carry bit had been 0, a 0 would have been entered in the accumulator as a result of this process. (The carry bit would have been 0 only if the sum of the number is ≥ FF.) In either case, the carry bit, whether a 0 or a 1, appears in the accumulator as a result of executing the preceding instructions.

The problem now is to modify further the program so that it will output the carry and the lower two digits of the sum.

OUTHEX

To output the carry, the ET3400 provides another subroutine called OUTHEX. The programming for this is also contained in ROM, starting at address FE28. After that is written, the program continues with OUTBYT to output the two digits which remain in the sum. The final display will then consist of three digits, the carry and the two digits of our sum.

Figure 31-4*a* is a flowchart which we will use for writing the new program. The program to accomplish this flowchart is:

Enter the two numbers to be added in address locations 0090 and 0091.

The program follows the flowchart. Necessary OP codes are found in Appendix A. Steps 0000 through 0004 are the same as those written in the preceding program. They add the numbers stored at addresses 0090 and 0091. The instruction store accumulator A direct (STAA) into address 0092 is accomplished by steps 0005 and 0006. Step 0007 transfers the condition code into accumulator A. Step 0008 ANDs the contents of accumulator A which contains the condition code, with 01 in step 0009. Step 000A jumps the program to the OUTHEX subroutine (steps 000B and 000C) which outputs the carry bit. Steps 000D and 000E load into

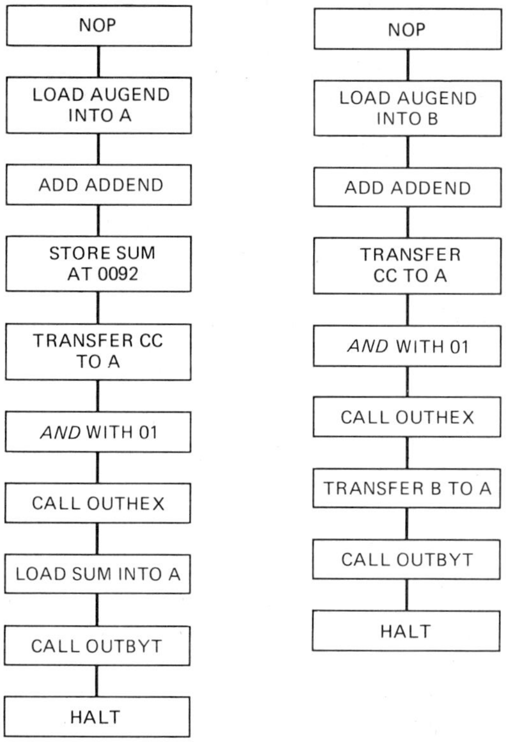

(a) (b)

Fig. 31-4. Flowcharts. *(a)* Program *C*; *(b)* Program *D*.

PROGRAM C.

Address	*OP Code*	*Mnemonic*
0000	01	NOP (INHERENT)
0001	96	LDAA (DIRECT)
0002	90	operand
0003	9B	ADDA (DIRECT)
0004	91	operand
0005	97	STAA (DIRECT)
0006	92	operand
0007	07	TPA (INHERENT)
0008	84	ANDA (IMMEDIATE)
0009	01	operand
000A	BD	JSR (EXTENDED)
000B	FE	operand
000C	28	operand
000D	96	LDAA (DIRECT)
000E	92	operand
000F	BD	JSR (EXTENDED)
0010	FE	operand
0011	20	operand
0012	3E	WAI (INHERENT)

accumulator A the sum stored in address 0092. Step 000F jumps the program to the OUTBYT subroutine which starts at address FE20 (steps 0010 and 0011). Finally, step 0012 halts the program. The data is entered at addresses 0090 and 0091. After the program is entered and executed, the sum appears in the display. Three digits will now indicate the sum. In the case of the sum of F3 and 45, the display will read 138. If the sum is a two-digit number, three digits will still appear in the display, but the leftmost digit will be a 0 because the carry bit was 0. For example, the sum of 23 and 34 will be displayed as 057.

The 19-step Program C is one solution to the problem. Another solution is shown in the flowchart in Fig. 31-4*b*. This will result in a 16-step program, so it is more economical of memory space. In long, complex programs, economy of memory is an important consideration.

In writing the program for Fig. 31-4*a*, it was necessary to store the sum at an address in memory. We chose 0092. The reason we needed to do this is that transferring the condition code into accumulator A would have destroyed the contents of accumulator A (the sum). So it was necessary to store the sum in memory, before the TPA instruction was attempted. Now, knowing that the 6800 has two accumulators, we could use accumulator B to sum the two numbers and leave it there until needed. The rest of the program can be done in accumulator A, and when the sum is finally needed, it can be transferred from accumulator B to accumulator A. This simple change will save three program steps, as is evident from the program for the flowchart in Fig. 31-4*b*, which follows:

Enter the two numbers to be added in memory locations 0090 and 0091.

The program carries out the format in the flowchart in Fig. 31-4*b*. Steps 0000 through 0004 add the two numbers in accumulator B. Step 0005 transfers the condition code to accumulator A. Steps 0006 and 0007 AND the condition code with the number 01 and leave the result, the carry C bit, in accumulator A. Step 0008 jumps the program to the start of the OUTHEX subroutine starting at address FE28 (steps 0009 and 000A). Step 000B transfers the sum of the two numbers from accumulator B to accumulator A. The jump subroutine, in step 000C, takes the program to the OUTBYT subroutine, starting at address FE20 (steps 000D and 000E). Finally, step 000F halts the program with the Wait for Interrupt (WAI) instruction.

You may wonder why the DIRECT addressing mode is used in Programs C and D for the instructions LDAA, LDAB, ADAA, and ADAB. The reason is that this mode makes it possible to add *any* two two-digit numbers, simply by entering the numbers at two address locations in memory. Entering the data is not really part of the program, and so the basic program need not be altered. If the IMMEDIATE mode had been used, it would have required entering the numbers *in* the program in the address immediately following the load and add instructions. The IMMEDIATE mode would have been used if we had wished to write a program just to add two specific numbers. The DIRECT mode can be used to add *any* two numbers.

Another observation, the TPA instruction, can be used only with accumulator A, by system design. It is permanently programmed in ROM. EXTENDED addressing is used for the JSR instruction because this addressing mode can handle a four-digit address. The subroutine is in ROM, and its address is in four digits.

PROGRAM D.

Address	*OP Code*	*Mnemonic*
0000	01	NOP (INHERENT)
0001	D6	LDAB (DIRECT)
0002	90	operand
0003	DB	ADDB (DIRECT)
0004	91	operand
0005	07	TPA (INHERENT)
0006	84	ANDA (IMMEDIATE)
0007	01	operand
0008	BD	JSR (EXTENDED)
0009	FE	operand
000A	28	operand
000B	17	TBA (INHERENT)
000C	BD	JSR (EXTENDED)
000D	FE	operand
000E	20	operand
000F	3E	WAI (INHERENT)

The TPA, TBA, and WAI instructions are in INHERENT addressing mode because their locations are implied.

Branching

Programs C and D are linear (straight-line) programs. Their mode of solution is similar, even though the programming steps are somewhat different. Is there another way of obtaining the result? Yes, there is, and we will examine it. The new method utilizes a technique called *branching*. Linear programs do not branch.

The flowchart in Fig. 31-5 illustrates a solution with branching. The block which calls for branching is the diamond-shaped block labeled "Is Carry Clear." Let us interpret the flowchart.

The first three blocks tell us to add the two numbers in accumulator B. The diamond-shaped block tells us to examine if the carry bit of the condition code is clear. If it is NO, that is, if the carry is 1, we know that the result is a three-digit number. So we will use a program similar to that in Program D. The linear steps following the NO in Fig. 31-5 show this approach. How-

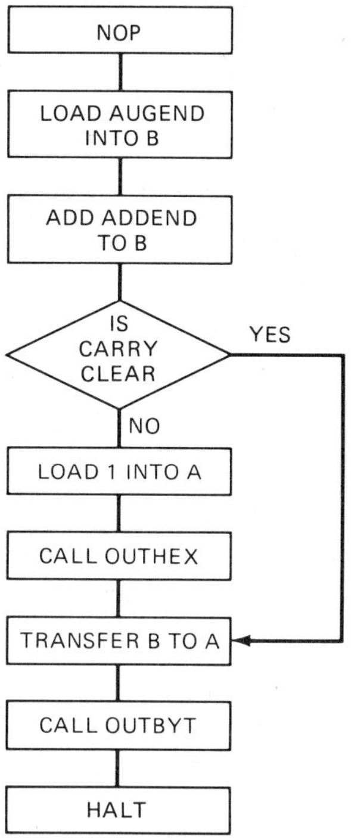

Fig. 31-5. Flowchart demonstrating branching (Program *E*).

ever, if the carry bit is 0, that is, if the carry is clear (YES), we know that the result is a two-digit number. So we bypass the *load 1* and OUTHEX blocks and branch directly to the block labeled "Transfer B to A." We utilize the last three blocks and the first three blocks of the flowchart, in the case of YES.

The instruction in the 6800 which makes the last solution possible is BCC, which means Branch if Carry is Clear. We will write a program utilizing the BCC instruction as required by the flowchart in Fig. 31-5.

As in past programs using direct addressing for the load and add instructions, the two numbers to be added are entered in address locations 0090 and 0091.

Examination of program E will clarify some of the questions you may have. Steps 0000 through 0004, when executed, will load accumulator B with the number at address 0090 and will add the second number in address 0091 to the first. Step 0005 is the BCC (branch if clear) instruction. Reference to Appendix A shows that there is only one addressing mode for this instruction, the RELATIVE mode, and that the OP code for BCC is 24. The address which follows, 0006, should contain a number which will tell the micro how many addresses to skip before executing the TBA instruction if the answer to the branch instruction is YES, that is, if there is no carry (C = 0). For a moment we will leave the space next to 0006 empty and proceed to the remainder of the program.

PROGRAM E.

Address	*OP Code*	*Mnemonic*
0000	01	NOP (INHERENT)
0001	D6	LDAB (DIRECT)
0002	90	operand
0003	DB	ADDB (DIRECT)
0004	91	operand
0005	24	BCC (RELATIVE)
0006	— —	
0007	86	LDAA (IMMEDIATE)
0008	01	operand
0009	BD	JSR (EXTENDED)
000A	FE	operand
000B	28	operand
000C	17	TBA (INHERENT)
000D	BD	JSR (EXTENDED)
000E	FE	operand
000F	20	operand
0010	3E	WAI (INHERENT)

If the answer to BCC is NO, that is, if the carry is not clear and C = 1, we know that the sum of the two numbers to be added is greater than FF, that is, the sum is a three-digit number starting with 1 (the carry). The micro then simply ignores step 0006 and proceeds to execute the instructions at addresses 0007 through 0010. These are identical to steps at addresses 0007 through 000F in program D. It is not necessary to include steps 0005 (TPA) and 0006 (ANDA) from Program D because BCC, the conditional branch instruction, has already established that there is a carry (1) which must be displayed when the sum appears on the LEDs.

Now, what happens if the answer to BCC is YES, that is, if C = 0. In that case the sum of the two numbers is equal to or less than FF. Our answer will be a two-digit number, and the micro can be instructed to skip those steps in Program E which result in outputting the third digit, a 1, on the display. The steps to be skipped therefore are 0007 through 000B, a total of five steps. Now place an 05 in the blank space next to address 0006. This will tell the micro that if the answer is YES, it must proceed from step 0006 to 000C, skipping the five intermediate steps.

The relative mode therefore requires instructing the micro how many steps it must skip. This number is entered in the address immediately following the OP code of the relative mode instruction.

SUMMARY

1. The basic components which make up the ET3400 are: a 6800 micro, a clock, a RAM memory, a ROM memory, a keyboard, six seven-segment

LED display units, LED-segment decoder/drivers, a digit decoder, a chip-select decoder, an address bus, and a data bus.

2. The MPU does all the logic and arithmetic functions.

3. The clock generates timing signals to operate the system.

4. The keyboard is the input device by means of which programs are entered. Its digits are interpreted as hexadecimal digits by the system. The keys also have a second set of functions (Experiment 30).

5. In this system the RAM memory temporarily stores the programs and data entered by way of the keyboard.

6. The ROM is the permanent memory for built-in functions, utility programs, and subroutines entered during system design.

7. The seven-segment LEDs act to display the results of the programmer's input to the system.

8. The various decoders interpret the programmed instructions and turn ON, as programmed, the segments of LEDs, the LED digits, and the various chips (circuits) comprising the system.

9. The address and data buses are the means of communication between the micro and the components of the system and also between the components themselves.

10. Efficient management of the system requires setting aside space in memory for the instructions, data, and so on which are to be entered. This is called *mapping the memory*.

11. Each microprocessor comes with a specified *instruction set,* which details the operations that it can perform. The list of these operations includes a set of initials, called *mnemonics,* for each operation, and an operation code for each operation, in each of its addressing modes.

12. The addressing modes that the 6800 uses are: IMMEDIATE, DIRECT, INDEXED, EXTENDED, INHERENT, and RELATIVE.

13. These steps must be followed in writing and entering a program in the ET3400: (*a*) define the requirements; (*b*) draw a flowchart; (*c*) write the mnemonic for each instruction; (*d*) write the OP code and the operands for each instruction; (*e*) enter the machine code into memory; (*f*) check the program; and (*g*) execute the program.

14. In writing a program for adding two numbers, provision must be made for a carry in the event 1 is generated.

SELF-TEST

Check your understanding by answering these questions.

1. The instruction and address decoders make it possible for the micro to keep operation on the buses and on the system orderly and meaningful. ___________ (true/false)

2. The data bus between the micro and the RAM is ___________ (bidirectional/unidirectional).

3. The keyboard generates ___________ digits.

4. The utility programs are permanently entered in the ___________ by system design.

5. ___________ and ___________ are entered in the RAM memory in the ET3400.

6. The instruction set is the means of communication between the programmer and the ROM, in the ET3400. ___________ (true/false)

7. ___________ is needed before a programmer can determine the actual steps to be taken in writing a program.

8. Computer language is a set of ___________ numbers.

9. The instruction set for the 6800 lists ___________ for each instruction, the addressing modes, and the ___________ for each mode.

10. The DIRECT addressing mode for a particular instruction tells the micro that the next word in memory immediately following is the ___________ of the operand.

11. In IMMEDIATE addressing the next word in memory immediately following the OP code is the ___________ .

12. The subroutines OUTHEX and OUTBYT are built-in functions in ROM in the ET3400. ___________ (true/false)

13. The first step in programming a solution to a problem is to ___________ the ___________ .

14. The instruction ___________ is used to halt or end a program in the ET3400.

MATERIALS REQUIRED

- Equipment: Microprocessor-microcomputer trainer, Heathkit ET3400, plus operating instructions

PROCEDURE

Entering a Program—Method 1

NOTE: *In steps 1 through 9 you will use the EXAM, CHAN, and FWD keys to enter Program A.*

1. Turn power ON the trainer. Press RESET. The display should read CPU UP.
2. Press EXAM. The display should show four blank spaces and the letters Ad; _ _ _ _ Ad. The blank spaces are for the starting address.
3. Enter 0000. The display will now read 0000xx, where x stands for any random number.
4. Press CHAN. The display will then read 0000 _ _ . The two blank spaces are for entering the first instruction OP code, which in Program A is 01.
5. Press, in turn, keys 0 and 1. The display will now read 000001.
6. Press FWD, thus advancing the address by 1. The new display will read 0001xx.
7. Press CHAN, and enter the next instruction OP code, 96. The display will then read 000196.
8. Enter the rest of Program A, using this method.
9. After the program has been entered in addresses 0000 through 0007, press EXAM, and enter the data in address 0090, that is, the number 21. Then press FWD and enter the number 35 in address 0091. This completes entering Program A and the two numbers to be added.
10. You are now ready to check whether the program was accurately *entered.* Do this by pressing EXAM and entering the starting address, 0000. The display should read 000001.
11. Now check the rest of the program by pressing the FWD key. As the key advances the address by 1, the contents at that address (OP code or operands) in our program should appear. For example, when FWD is pressed again, the display should read 000196, and so on. Enter in Table 31-4 the program steps displayed on the trainer as the FWD key advances the program through step 0007. Check also the data at addresses 0090 and 0091. NOTE: The program has been entered, and the accuracy with which it was entered has been checked. It has not been executed yet.

Executing the Program, Automatically

12. Press RESET.
13. Press DO. (This will automatically execute the program after the starting address is entered.) The display now reads do. The four blank spaces are for the starting address. So enter 0000. The trainer then executes the program and blanks out, since there was no instruction in the program to display anything.

TABLE 31-4. Checking Program A

Address	*Display*
0000	000001
0001	000196
0002	
0003	
0004	
0005	
0006	
0007	
0090	
0091	

14. Press RESET.
15. Press ACCA. This displays Acca56, where 56 is the sum of the two numbers we were to add. Check also the contents of address 0092 and again 56. *The program was successful.*

Entering Program A—Method 2

NOTE: *A simpler and faster way for entering a program is by means of the AUTO key.*

16. Press RESET.
17. Press AUTO. The display that appears is _ _ _ _ Ad. The four blank spaces are for the starting address.
18. Enter the address 0000. The display now reads 0000 _ _ . The two blank spaces are for the OP code of the first instruction.
19. Enter 01 from Program A.
20. When your finger is off the 1 key, the address automatically increments by 1. The display now reads 0001 _ _ . Enter the next instruction OP code, 96.
21. In similar fashion enter the remainder of Program A.
22. When you are finished entering the instruction in step 0007 (the last program step), press RESET and then enter the data into addresses 0090 and 0091 by the *EXAM, CHAN,* and *FWD* method. NOTE: It was

not really necessary to enter the program or the data again, since we had not removed power from the trainer and both the data and program were still in memory. The purpose here was to practice the AUTO method of entering a program.

Executing and Checking the Program: SS Method

NOTE: *Another method of executing a program while checking it at the same time is the single-step (SS) method.*

23. Press the PC (program counter) key. The display will be Pcxxxx.
24. Press CHAN. The display now is Pc _ _ _ _ .
25. Set the program counter to 0000 by pressing 0 four times. The display now is Pc0000. We are now ready to execute the program starting with the first instruction.
26. Now press SS. The display which appears is 000196. This action has executed the NOP instruction and has set our program counter to 0001, where the LDAA OP code (96) can be seen.
27. Press SS again. The display which appears is now 00039b. The LDAA instruction was executed, and we are ready to execute the ADDA instruction. NOTE: The SS subroutine will not show any steps involving operands (e.g., 000290 did not appear). But we can check to see if the operand (21) was entered properly in step 0002, by pressing ACCA and observing the contents of accumulator A.
28. So press ACCA. The display should show Acca21.
29. Press SS again. The display now shows 000597. The ADDA instruction has been executed. Again the address of the operand, 0004, was skipped.
30. We can now check accumulator A again. Press ACCA. The display is Acca56. The 56 is the sum of the two numbers, 21 and 35.
31. Press SS again. The display is 00073E. If we press SS again, nothing happens because we have come to the end of our program; that is, the program is halted by instruction 3E at address 0007. Check the contents of address 0092 ____________ . NOTE: Again, our program was properly entered, checked, and executed. Entering a program automatically, and executing it by single-stepping it is the fastest way to enter, check and execute a program in the ET3400.

Correcting an Incorrect Entry

If an entry were incorrectly made, it could be changed by the *EXAM and CHAN* keys. Suppose that in the process of checking program A we find that we made a false entry at address 0004, namely, 92 instead of 91. As a result, our sum was incorrect. We can correct our error as follows.

32. Press RESET, then EXAM, and then 0004. If the error actually were made, we would read 000492. So press CHAN and enter 91. Now press DO, enter 0000, and our program will be correctly executed.

Displaying the Sum—Program B

To see the result of program A, we had to check accumulator A or address location 0092. If we wish to display it directly when the program is executed, we modify Program A with Program B as follows.

33. Press RESET, and EXAM and enter 0007.
34. Press CHAN and enter BD from Program B.
35. Complete entering Program B by the FWD, CHAN method.
36. Execute the new program automatically (press DO and enter 0000). The result is the sum 56 displayed on the LEDs.

Sum of Any Two Numbers—Sum Is Less than FF

37. We wish to add the two numbers 38 and 6A using Program B, which is still entered in memory. Explain the method you will use and execute the program. Record the displayed sum in Table 31-5.

TABLE 31-5. Sum of Two Numbers—Program B

Numbers	*Sum*
38 and 6A	
99 and 28	

38. We wish to add the numbers 99 and 28 with Program B. Explain the method you will use and execute the program. Record the sum in Table 31-5.
39. Try adding other two number combinations whose sum is less than FF.

Sum of Two Numbers with Carry—Program C

We wish to add the two numbers F3 and 45, whose sum we know to be 138 and greater than FF. Can we use Program B? Let us try.

40. Enter F3 in address 0090, and 45 in address 0091.
41. Execute the program. The display reads 38, not 138. Why?

42. Check the condition code by pressing RESET and then CC. The display reads 000001. So we have a carry, and we must write a program which takes the carry into account, Program C will do this.

Program C

43. Enter Program C, using the AUTO key. Enter F3 in address 0090 and 45 in address 0091. Check the program by using the EXAM, CHAN, and FWD keys.

44. Execute the program automatically. The result should be 138.

45. Now enter the numbers 21 in address 0090 and 35 in address 0091 and execute Program C. The result is __________ . What does this show?

Sum of Two Numbers—Program D

46. We wish to add the two numbers A3 and B7 using Program D. Enter Program D into trainer, and enter numbers A3 and B7 into addresses 0090 and 0091, respectively.

47. Check that the program was entered properly by the EXAM, CHAN, and FWD method.

48. Execute the program automatically. The display reads __________ . Is this correct? __________ .

49. Now using Program D, add the two numbers 34 and 53. The display reads __________ . Is this correct? __________ .

Sum of Two Numbers—BRANCHING—Program E

50. We wish to add the two numbers B4 and F4 using Program E. Enter Program E into memory, and enter also numbers B4 and F4 into addresses 0090 and 0091, respectively.

51. What number must be entered in the address 0006? __________ Why? __________ Enter it.

52. Check that the program was entered properly.

53. Execute the program. The display reads __________ . Is it correct? __________ .

54. Now add the numbers 11 and 12 using Program E. The display reads __________ . Is this correct? __________ .

55. Try adding other two number combinations.

QUESTIONS

1. What advantage, if any, is there in entering a program using the AUTO key, over the EXAM, CHAN, and FWD method?
2. What advantage, if any, is there in executing a program by the SS method over the DO method? What disadvantage?
3. In checking Program B, you find that the operand in step 0090 was entered incorrectly. How would you correct this error?
4. What is the advantage of Program B over Program A?
5. What limitations, if any, does Program B have in adding two numbers?
6. Which programs overcame the limitations of Program B? How?
7. What does the carry bit in the condition code (step 42) tell us?
8. Why is the information given by the carry bit of importance?
9. Why are the steps at addresses 0007, 0008, and 0009 in Program C necessary?
10. What is the advantage, if any, of Program D over C? Of Program E over D?

Answers to Self-Test

1. true
2. bidirectional
3. hexadecimal
4. ROM
5. Instructions; operands (data)
6. false
7. flowchart
8. binary
9. mnemonics; OP codes
10. address
11. operand
12. true
13. determine; requirements
14. WAI

EXPERIMENT

32

PROGRAMMING THE MICROPROCESSOR—PART 2

OBJECTIVES

1. To learn how negative binary numbers may be represented by the 2's complement code
2. To write a program for subtracting two numbers

BASIC INFORMATION

Negative Numbers

The microprocessor is frequently required to process both *negative* and *positive* numbers in arithmetic problems. It was therefore necessary to develop a system of representing negative numbers to inform micro circuits when negative numbers were being used. Several systems were tried and discarded. The system finally adopted, called the 2's *complement,* leads to the simplest electronic circuits which process and recognize *signed* numbers.

The 6800 is an 8-bit micro. We will therefore confine our analysis to 8-bit binary digits.

If we were dealing only with positive numbers, an 8-bit micro could accommodate binary numbers in the range 00000000 to 11111111. This, of course, is the equivalent of 00 to FF_{16} or 0 to 255_{10} (see Table 32-1). So in that system the first 255 positive decimal numbers and 0 could be treated. What about negative numbers? One answer is to use the most-significant bit (MSB) of an 8-bit number to signify the *sign* of a number, that is, whether it is positive or negative. Thus in the number *0*1100111 the MSB is *0*. Let this 0 represent a positive number. In the number *1*0111000 the MSB is *1*. Let this 1 represent a negative number. This *is* the system that was adopted to represent signed numbers. By examining the MSB, the micro now has a way of determining whether it is operating on a positive (+) or a negative (−) number. For if the MSB is 0, the number is positive; if it is 1, the number is negative.

TABLE 32-1. Bit Patterns and the Values They Represent in Unsigned Binary, and 2's Complement Systems

Bit Pattern	*Unsigned Binary*	*2's Complement*
00000000	0	0
00000001	1	+1
00000010	2	+2
00000011	3	+3
•	•	•
•	•	•
•	•	•
•	•	•
01111100	124	+124
01111101	125	+125
01111110	126	+126
01111111	127	+127
10000000	128	−128
10000001	129	−127
10000010	130	−126
10000011	131	−125
•	•	•
•	•	•
•	•	•
•	•	•
11111100	252	−4
11111101	253	−3
11111110	254	−2
11111111	255	−1

However, some problems result from the use of this system. These will be treated as the need arises.

In adopting this system, we can represent in an 8-bit pattern, of which the MSB is the *sign* flag, the positive numbers 00000000 to 01111111. This represents the numbers 0 to $+127_{10}$, a total of 128_{10} numbers, and the negative numbers -128_{10} to -1_{10} (see Table 32-1).

2's Complement

It is not necessary for the programmer to memorize Table 32-1 to determine the decimal values of either the positive or negative numbers. From our knowledge of the binary number system, positive values (where the MSB is 0) can be calculated using powers of 2. Negative numbers require another technique.

The number 01111101 = $+125_{10}$. Now for every binary 1 substitute a 0, and for every 0 substitute a 1 in the original number. The result is 10000010. This number is the 1's complement of the original number. To convert the 1's complement to the 2's complement, simply add a 1 to the LSB. The number then becomes 10000011. Referring to Table 32-1, we observe that the value of this 2's complement is −125. We have therefore converted a positive number, +125, into a negative number, −125, by taking the 2's complement of the positive number. Does this system hold for other numbers in our range? The answer is "yes." Try it.

Therefore, the rule for converting a positive number into a negative number having the same absolute value is to (1) convert the positive number to its equivalent binary form, and (2) take the 2's complement of the binary number; the resulting 2's complement is a negative binary of the original positive number. We will see that the micro uses the 2's complement in subtracting two numbers.

Decimal Value of 2's Complement

How can you find the absolute (decimal) value of a negative-coded binary number? It is done quite simply by taking the 2's complement of the negative-coded binary and computing the value of the resulting positive binary. For example, find the decimal value of 10000011_2. You will recognize this as the number -125_{10} from the previous discussion. But suppose that we did not know the coded value. Then take the 2's complement of 10000011. It is 01111101. This binary number equals 125_{10}, and the decimal value of our original binary number is therefore −125.

Subtracting Two Numbers

The rules for subtracting binary numbers are:

(1) 0−0 = 0
(2) 1−1 = 0
(3) 1−0 = 1
(4) 0−1 = 1 with a *borrow* of 1.

Problem: From $+127_{10}$, subtract $+124_{10}$, using binary arithmetic. This problem is easily solved by converting the positive numbers to their binary equivalents and applying the rules of binary subtraction. But first let us define the terms in a subtraction. In a subtraction, the *minuend* is the number from which the *subtrahend* is subtracted. So in our problem, +127 is the minuend, and +124 is the subtrahend.

$$\begin{array}{rl} +127_{10} = & 01111111 \\ +124_{10} = & \underline{01111100} \\ \text{Difference } 3_{10} = & 00000011 \end{array}$$

This is one solution. But for the microprocessor to accomplish this, separate circuits would be required for subtraction, just as separate circuits are needed for addition. There is, however, another way to solve this problem. Represent the subtrahend as a negative number by the 2's complement method, and then add the minuend and the 2's complement of the subtrahend. This procedure makes it possible to use the same *add* circuits as in adding two positive numbers. How would this technique work in the subtraction problem above? Well, the 2's complement of 124 is 10000100. So we may write our problem as

$$\begin{array}{rl} +127_{10} = & 01111111 \\ -124_{10} = & \underline{10000100} \\ \text{Sum} \quad 3 = & 00000011 \\ & \uparrow \ \text{carry} \end{array}$$

Note that there is a ninth bit, a 1 just before the MSB (0) of our 8-bit solution. This 1, the ninth bit, represents a carry which we will ignore. The remaining 8 bits represents the number 3_{10}. Since the MSB is 0, we know that the difference is the positive number 3, as expected.

The use of the 2's complement for the subtrahend converts the subtraction problem into an addition problem and permits the micro to use its *add* circuits whether two numbers are being added or subtracted.

N Bit of Condition Code

Now the use of the N bit of the condition code can be readily understood. In this *signed* binary number system used with micros, it is frequently necessary to know whether a number is positive or negative. A 6800 instruction such as ADD, ADC, SUB, and SBC (see Appendix A) resulting in the arithmetic operations of add, subtract, and so on will set the N bit of the CC register if the result is a negative number, that is, N will be made 1. If the result is positive, the N bit is 0.

One use of signed numbers is in programming. If the relative address following a branch instruction is positive, that is, if the MSB of the binary number giving the relative address is 0, then the program branches *forward*. If, however, the MSB of the relative address is 1, then the micro interprets it as a backward branch. A problem illustrating back branching will be discussed in Experiment 33. Forward branching was illustrated in Program E, Experiment 31, and will be used again in this experiment.

V Bit of Condition Code

In working with unsigned numbers, the carry (C) bit indicates if the result of an addition exceeds the eight bits we are working with. Thus the C bit is set (C = 1) if there is an overflow. With *signed* numbers, the V bit plays a similar role; that is, the V bit of the CC register is set (V = 1) if an overflow occurs. The overflow this time is into the eighth (MSB) bit.

The problem of an overflow can occur much sooner with signed numbers. Moreover, the fact that a micro can work with both signed and unsigned numbers can lead to ambiguities. Take, for example, the following problem.

Problem: Find the sum of +125 and +5.

$$
\begin{aligned}
125_{10} &= 01111101 \\
+5_{10} &= \underline{00000101} \\
\text{Sum } 130_{10} &= 10000010
\end{aligned}
$$

In our system the above sum is a negative number, namely, −126, (see Table 32-1, or figure it out, using the 2's complement of the sum to obtain the absolute value). Note that in unsigned arithmetic the sum is 130_{10}. But we cannot switch back and forth in the system we have chosen, since we have agreed that all numbers with a value greater than 127 are negative numbers. The result, in our system, is not valid because an overflow has occurred from the seventh bit to the eighth bit (MSB). When this type of overflow occurs, the V bit of the CC register is *set,* that is, V = 1. By checking the V bit, the programmer can learn that an overflow has occurred. In working with signed numbers, the V bit must be checked, after arithmetic operations, to make certain that the result is valid.

Program for Subtracting Two Numbers

There are quite a few ways to write a program for subtracting two numbers. The actual steps depend on the method chosen. However, they all make use of the basic scheme of converting the subtrahend to its 2's complement and adding the 2's complement to the minuend. First we will consider the manual method of obtaining the result.

Problem: From $+12_{10}$ subtract $+5_{10}$.

We will solve this problem by converting +5 to a negative number −5, and then by adding +12 to −5. The answer, of course, should be +7.

To convert +5 to −5, find the 2's complement of +5. First write +5 in binary form, namely 00000101. To find the 2's complement, change each 0 to a 1 and each 1 to a 0, in the binary, and then add 1 to the LSB. The result is 11111011. Now write +12 in binary form, namely, 00001100. Then add the two binaries.

$$
\begin{aligned}
+12_{10} &= 00001100_2 = 0C_{16} \\
-\ 5_{10} &= \underline{11111011_2} = \underline{FB_{16}} \\
\text{Sum}+7_{10} &= \underline{1}\ 00000111 = 1\ 07_{16}
\end{aligned}
$$

carry ↑ ↑ carry

Ignoring the carry, the result is +7 after the binary sum is converted to its decimal equivalent.

To program this problem for the ET3400, use hex numbers. Note then that $+12 = 0C_{16}$ and $-5 = FB_{16}$. The flow diagram for solving this problem is shown in Fig. 32-1.The program which will accomplish the desired result is Program 32-1.

OC (+12) is stored in address 0090 and FB (−5) in address 0091. After the program is entered and executed, checking accumulator A will reveal the answer, 07. Note that this program is like the elementary program for adding two numbers in Exp. 31.

Now, what is the result of subtracting +12 from +5? Using the above method, the problem resolves itself into adding +5 and −12. So it is necessary to convert +12 into −12 by the 2's complement method. This is left to the student. The hex equivalents of $+5_{10}$ and -12_{10} are respectively 05 and F4. Storing these values in addresses 0090 and 0091 and executing Program 32-1, the result, found in accumulator A, should read F9. It is clear that the result should be a negative number, but to be certain we check the N bit of the condition code (after the program is executed) by pressing the CC key and noting that N = 1. Hence the result is negative. Now, how can the absolute value of F9 be determined? The answer is to write the number $F9_{16}$ in binary form and take the 2's complement of the binary number:

$$
\begin{aligned}
F9_{16} &= 11111001_2 \\
\text{2's complement} &= 00000111_2 = 7
\end{aligned}
$$

The solution to our problem therefore is 7 and $F9_{16} = -7$.

PROGRAM 32-1.

Address	*OP Code*	*Mnemonic*
0000	01	NOP
0001	96	LDAA (DIRECT)
0002	90	operand
0003	9B	ADDA (DIRECT)
0004	91	operand
0005	3E	WAI

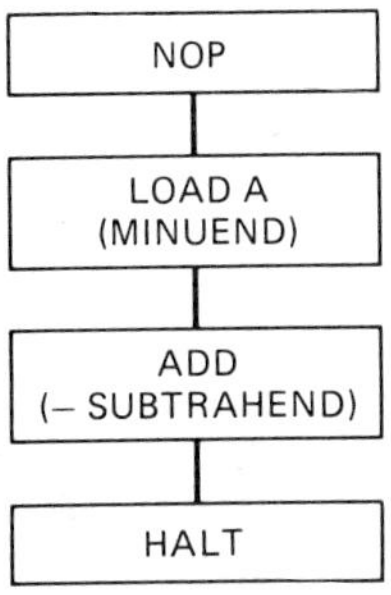

Fig. 32-1. Subtracting two numbers, a simple program.

NEG; NEGA; NEGB

The instruction set of the 6800 has an instruction NEG which automatically finds the 2's complement of a number that is stored in some memory location. NEGA is the instruction which finds the 2's complement of a number stored in accumulator A, and NEGB does the same for a number in accumulator B. The OP code for 2's complementing the number in A is 40 (INHERENT); in B it is 50 (INHERENT) (see Appendix A). Program 32-1 can now be modified to show the absolute value of a *negative* result by making the following changes in the program:

Address	*OP Code*	*Mnemonic*
0005	40	NEGA
0006	3E	WAI

In the case of the preceding problem, subtract 12 from 5. If we modify Program 32-1 as shown, the result read in accumulator A, after the program is executed, is 07, and the answer is, of course, −7.

The NEGA instruction can also be used earlier in the program to convert the subtrahend into a negative number. Now if the subtrahend is larger than the minuend, the final answer should be a negative number. This, too, would require NEG(ating) if we wish the absolute value of the result to appear in the accumulator. Program 32-2 offers another way to effect subtraction. The result is found in accumulator A. The flowchart for Program 32-2 is shown in Fig. 32-2. Enter minuend in 0090 and subtrahend in 0091. Both minuend and subtrahend must be written in hex.

Program 32-2 requires loading the minuend in accumulator B and the subtrahend in accumulator A. Instruction 40 in address 0005 negates the subtrahend (converts it to the 2's complement). Instruction 1B at address 0006 adds accumulator B to A, thus adding the minuend and the (−) subtrahend. The WAI instruction at address 0007 halts the program. The result is found in accumulator A when the program is executed.

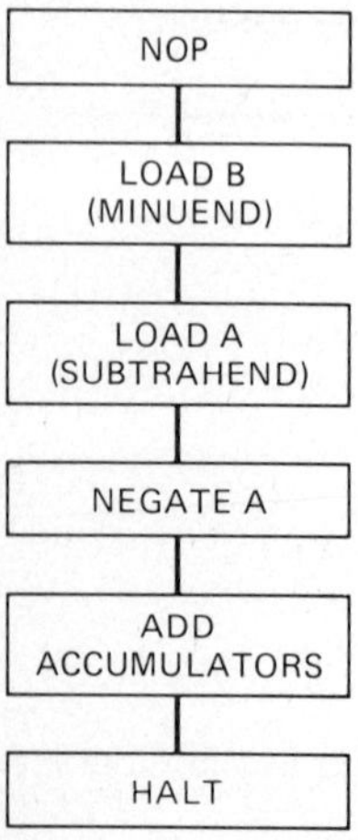

Fig. 32-2. Subtracting two numbers, another approach.

To solve the problem, subtract +12 from +5, we would enter 05 into address 0090 and 0C into address 0091. After entering and executing Program 32-2, the result in accumulator A would read F9. The decimal value of F9, of course, is −7.

PROGRAM 32-2. Subtrahend Larger than Minuend

Address	*OP Code*	*Mnemonic*
0000	01	NOP
0001	D6	LDAB (DIRECT)
0002	90	operand
0003	96	LDAA (DIRECT)
0004	91	operand
0005	40	NEGA (INHERENT)
0006	1B	ABA (INHERENT)
0007	3E	WAI

Program to Subtract with Automatic Signed Readout

There are several difficulties with Programs 32-1 and 32-2. One problem is that the absolute value of the result must be calculated by using the 2's complement if the result of the subtraction is negative. Another problem is that there is no automatic display of the result. Moreover, since the result is either + or −, it would be nice to have a *signed* answer in the result. How can these additional requirements be met? There is a program (Program 32-3) which will solve these problems and provide a signed number readout, which is automatic.

This program will also introduce a new instruction, SUB(tract) and a new routine OUTCH (refer to Appendix A). The SUB(tract) operation instructs the arithmetic-logic unit in the MPU to subtract the subtrahend from the minuend which has been loaded into an accumulator. Since the 6800 has two accumulators, A and B, the mnemonic specifies the accumulator which is involved. Thus SUBA means subtract the subtrahend from the minuend in accumulator A, SUBB from the minuend in accumulator B. When the subtract instruction is executed, the ET3400 automatically takes the 2's complement of the subtrahend and adds it to the minuend, just as we did manually in Programs 32-1 and 32-2. In the process, several programming steps are saved.

OUTCH

This routine is used to display a minus (−) sign when it is needed. Actually OUTCH will display any combination of LED segments, including the decimal point (DP), from the chart shown in Fig. 32-3. The hex code for the character to be displayed must be in accumulator A, when the OUTCH routine is entered. The chart

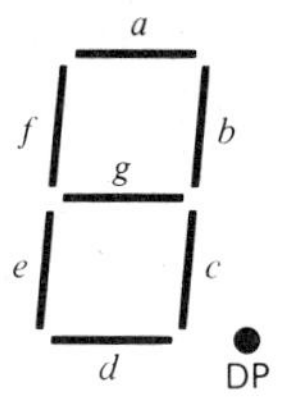

Display	Hex OP Code	Accumulator Positions							
		D_7	D_6	D_5	D_4	D_3	D_2	D_1	D_0
		LED Segment							
		DP	*a*	*b*	*c*	*d*	*e*	*f*	*g*
C	4E	0	1	0	0	1	1	1	0
c	OD	0	0	0	0	1	1	0	1

Fig. 32-3. Segment identification for OUTCH subroutine. (Heath)

in Fig. 32-3 identifies the segment of the LED, whether it is the DP or segments *a, b,* and so on, together with the corresponding segment positions in the 8 bits of accumulator A. A 1 entered in a bit will cause the corresponding segment to light, while a 0 will keep it OFF. The chart shows how the code works. For example, if it is desired to display the capital letter C, segments *a, f, e,* and *d* must be lit. Entering a 1 in the position of each of these segments and a 0 in the remaining positions gives the binary number 01001110. This number must be changed to its equivalent hex value before it can be entered in the trainer. The hex code for the above binary is 4E. So if 4E is entered in accumulator A, followed by the OUTCH routine whose starting address is FE3A, the capital letter C will be displayed. Now, to output the sign (−), segment *g* must be lit while all other segments remain dark. The binary code for this is 00000001, and the corresponding hex code is 01.

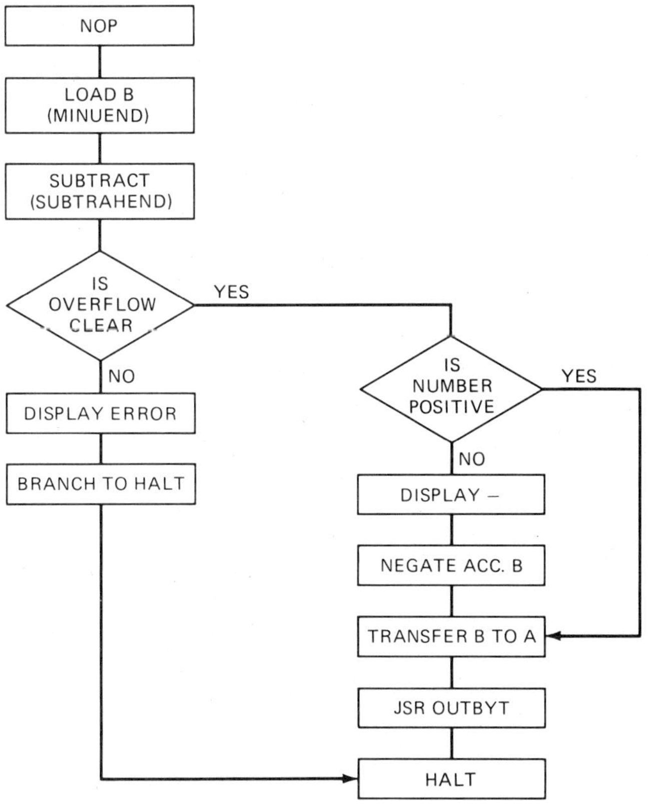

Fig. 32-4. Improved subtract program displays signed output.

Let us examine the flowchart Fig. 32-4 for Program 32-3. Note that the minuend is loaded into accumulator B and the subtrahend is subtracted from it. The next block is a decision-making branch, BVC, which tests to determine if the overflow is clear, that is, if the V bit is 1 or 0. If it is not clear, that is, if V = 1, we know that the result of the subtraction will be an error. So we program the trainer to display the letter E, meaning error, and then we branch to halt. If, on the other hand, there is no overflow, the answer to "Is overflow clear?" is YES, and the program branches to another decision-making block, "Is number positive?," which tests the N bit to determine if the result of the subtraction is a positive number. If the answer is NO (N = 1), that is, if the resultant number in accumulator B is negative, we instruct the trainer to display a (−) sign before the result. Then the NEGATE ACCB block causes the trainer to calculate (2's complement) the absolute

PROGRAM 32-3. Subtract with Signed Display

Address	*OP Code*	*Mnemonic*
0000	01	NOP
0001	D6	LDAB
0002	90	operand
0003	DO	SUBB (DIRECT)
0004	91	operand
0005	28	BVC
0006	— —	relative address
0007	86	LDAA (IMMEDIATE)
0008	0E	operand
0009	BD	JSR (EXTENDED)
000A	FE	operand
000B	28	operand
000C	20	BRA
000D	— —	relative address
000E	2A	BPL
000F	— —	relative address
0010	86	LDAA (IMMEDIATE)
0011	01	operand
0012	BD	JSR (EXTENDED)
0013	FE	operand
0014	3A	operand
0015	50	NEGB
0016	17	TBA
0017	BD	JSR (EXTENDED)
0018	FE	operand
0019	20	operand
001A	3E	WAI

We store the minuend at address 0090 and the subtrahend at address 0091.

value of the negative answer. We next transfer accumulator B to A, call OUTBYT, and halt the program. The OUTBYT routine causes the answer to be displayed as an LED readout.

If, on the other hand, the result of the subtraction is a positive number (YES), the program branches to transfer the contents of accumulator B to A, thereby skipping the *Display*—and *Negate ACCB,* which are not needed for a positive number. The program is then completed as shown.

The flowchart translates into mnemonics and addressing modes as follows: NOP, LDAB (DIRECT), SUBB (DIRECT), BVC, and the letter E for the error message which is displayed with OUTHEX. The mnemonics and addressing modes for displaying E are: LDAA (IMMEDIATE), E, JSR, OUTHEX (FE28), and BRA to end this section. If the overflow was clear, we branch to continue with the program: BPL, LDAA (IMMEDIATE), 01 (to light the minus sign), JSR (EXTENDED), OUTCH (FE3A), NEGB, TBA (this is the address we will have to branch to if the number is positive), JSR (EXTENDED), OUTBYT (FE20), and WAI.

The mnemonics translate into code in Program 32-3.

Notes on Program 32-3

- BVC—means *branch if overflow is clear,* that is, when V = 0.
- BRA—means branch always.
- BPL—means branch if plus, that is, if N = 0.

The addressing modes for each of these branch instructions is relative. So each relative address, that is, the address following each branch instruction, is left empty until the entire program is written. If the answer to BVC is NO (V = 1), steps 0007 and 0008 load the letter E into accumulator A, and steps 0009, 000A, and 000B instruct the trainer to display the letter E (error). Step 000C then instructs the trainer to branch to halt the program.

If the answer to BVC is YES (V = 0), instruction 2A in address 000E activates the BPL instruction. A NO (N = 1) answer to BPL means that the number is negative and requires a minus sign. So address 0010 loads accumulator A with the hex code 01 (address 0011), which will generate a minus sign when the trainer jumps to subroutine OUTCH (addresses 0013 and 0014). At this point the negative number (result of subtraction) in accumulator B must be negated so that its absolute value will finally appear on the LED display. This is accomplished by the NEGB instruction in address 0015. TBA (address 0016) transfers the contents of accumulator B to A. The instructions in addresses 0017, 0018, and 0019 cause the contents of accumulator A to be displayed. WAI (address 001A) ends the program.

A YES answer (N = 0) to BPL means the number is positive. The program therefore skips the *Display*—and *NEGB* instructions and branches to TBA, after which it continues and ends as previously (steps 0016 through 001A).

Now to calculate the relative addresses of the three branch instructions. First note that all branches are *forward,* so that each relative address is a *positive* hex number. These can be calculated as in Program E in Exp. 31, or in the following manner which requires no counting: The first branch is at 0006 and we branch to 000E. Now consider what happens when the trainer has executed step 0006, where the relative address is. At that time the PC is actually at 0007. So the *initiating* address from which we branch is actually 0007. The *destination* address is 000E. Therefore, to find the relative address for step 0006, simply subtract the initiating address from the destination address. Now 000E − 0007 = 0007. Therefore, the relative address is 07, which must be entered in step 0006. Similarly, the relative addresses for steps 000D and 000F are respectively 0C and 06.

SUMMARY

1. In the system of numbers adopted for the 6800, the MSB of the 8-bit binary becomes the *sign* bit.
2. A 0 in the MSB position identifies a positive number.
3. A 1 in the MSB position identifies a negative number.
4. The seven remaining bits can hold any of these positive numbers, *0*0000000 to *0*1111111. The decimal equivalents of this range of numbers is 0 to $+127_{10}$. The hex equivalents are 00 to 7F.
5. The seven remaining bits can hold the following range of negative numbers: *1*0000000 to *1*1111111, which is -128_{10} to -1_{10}. The hex equivalents are 80_{16} to FF_{16}, respectively.
6. To convert a positive number into a negative hex number, write the positive number in its binary form and convert it to its 2's complement. Then write the number in hexadecimal form which the ET3400 uses.
7. To convert a negative number into a positive one, take the 2's complement of the negative number written in binary form. The 2's complement gives the absolute value of the resulting positive number.
8. The 2's complement of a binary number is obtained by changing every 0 to a 1 and every 1 to a 0 and adding a 1 to the LSB.
9. Subtraction of two numbers, say, Y from X, is the equivalent of adding the two numbers X and −Y.

10. The V bit of the condition code determines whether overflow has occurred from the seventh bit to the eighth bit (MSB) of a binary number, in an arithmetic operation.

11. If V = 1, there is overflow, and the results of our computation are invalid.

12. The N bit of the condition code identifies whether the number is negative (N = 1) or positive (N = 0).

SELF-TEST

Check your understanding by answering these questions.

1. Identify the positive and negative numbers in this group of numbers: (*a*) 00010111 (*b*) 10111000 (*c*) 11001110 (*d*) 01100101. Positive __________ negative __________ .

2. In question 1, the decimal value of *a* is __________ ; the hex value is __________ .

3. In question 1, the decimal value of *b* is __________ ; the hex value is __________ .

4. If the two numbers $+127_{10}$ and $+7_{10}$ are added on the ET3400 and the CC register is checked, N will read __________ ; V will read __________ .

5. Assume that we are interpreting the results of the addition in question 4, in terms of signed numbers. The answer to the problem will be (valid/not valid).

6. The instruction __________ changes a positive number in accumulator A to a negative number.

7. The instruction __________ automatically takes the 2's complement of the number specified by the operand and adds it to the number in accumulator B.

8. The +sign __________ (can/cannot) be formed from the segments of the eight-segment LED in Fig. 32-3.

9. The starting address for the __________ subroutine is FE3A.

10. In Program 32-3, if the result of the BPL instruction is NO, a __________ number will be displayed on the LEDs.

MATERIALS REQUIRED

- Equipment: Microprocessor-microcomputer trainer, Heath ET3400, plus operating instructions

PROCEDURE

1. Complete Table 32-2, entering the binary, hexadecimal, and the equivalent 2's complement of the decimal numbers listed.

Simple Subtraction

2. Enter Program 32-1 in the trainer and check it. In steps 3 and 5, enter the 2's complement (in hex) of the subtrahend in address 0091.

3. Using Program 32-1, subtract 11_{10} from 12_{10}. Indicate in Table 32-3 the hex values entered in addresses 0090 and 0091, respectively. Check and enter in Table 32-3 the value found in accumulator A. Enter also the decimal value of the result.

4. For step 3, check the condition code and enter the V and N values in Table 32-3. Indicate if the result

TABLE 32-2. Number Equivalents

Decimal Number	*Binary Number*	*Hexadecimal Number*	*2's Complement*	
			Binary	*Hex*
3				
9				
11				
−12				
15				
−23				
29				
35				
121				

TABLE 32-3. Subtraction with Program 32-1

Decimal Minuend	Decimal Subtrahend	Hex Number in Address		Result		Condition Code		Result	
		0090	0091	Decimal	Hex	N	V	Sign	Valid?
12	11								
3	9								

TABLE 32-4. Subtraction with Program 32-2

Decimal Minuend	Decimal Subtrahend	Hex Number in Address		Result		Condition Code		Result	
		0090	0091	Decimal	Hex	N	V	Sign	Valid?
3	9								
23	15								

TABLE 32-5. Subtraction with Program 32-3

Decimal Minuend	Decimal Subtrahend	Hex Number in Address		ACCB	Condition Code		Final Read-out	Decimal Value	Is Answer Valid?
		0090	0091		N	V			
23	15								
15	23								
29	35								
121	35								
121	−3								
121	−29								

is a positive or a negative number and also if it is valid.

5. Subtract 9_{10} from 3_{10}, repeating steps 3 and 4.

Subtraction—Another Method

6. Enter Program 32-2. We are going to subtract 9_{10} from 3_{10}. Check and execute the program by single-stepping it. Enter the required results in Table 32-4. Record also if the result is a positive or a negative number and if the answer is valid.

7. Repeat step 6 for the problem: subtract 15_{10} from 23_{10}.

Subtraction with Signed Readout

8. Enter Program 32-3, including the relative addresses. We are going to subtract 15_{10} from 23_{10}.

9. Set the PC on 0000 and single-step the program up to address 0005. Press ACCB and record the result in Table 32-5. Press CC and enter the values of N and V. Now press DO and CHAN and enter address 0005. Record the result under FINAL READOUT. Calculate and record the decimal value of the result and indicate if the result is valid.

10. Repeat step 9 for each of the problems listed in Table 32-5.

To Output a Two-Digit Hex Number

11. On a separate sheet of paper draw a flowchart for a program to display a two-digit hex number. Label it, "Flowchart to Output a Two-Digit Hex Number."

12. Write and record the program to achieve this result as Program 32-4 starting at address 0020. Store the number to be displayed in address 0092.

13. Display the hex number 32.

14. Display the hex number F3.

Program 32-4. To Output a Two-Digit Hex Number

Address	*OP Code*	*Mnemonic*
0020		

To Output an LED Segment or Segments

15. On a separate sheet of paper draw a flowchart for a program to display any combination of LED segments. Label it, "Flowchart to Output a Segment or Combination of LED Segments."

16. Write and record a program as Program 32-5 starting at address 0030. Use address 0093 to store the code of the segment(s) to be displayed.

Program 32-5. To Output a Combination of LED Segments

Address	*OP Code*	*Mnemonic*
0030		

17. Output the combination of LED segments which will form the symbol –. Record the binary and hex code in Table 32-6.

TABLE 32-6. Segment Code

Symbol	*Binary Code*	*Hex Code*
–		
H		
J		
8		

18. Output the combination of LED segments which will form the letter H. Record the binary and hex code in Table 32-6.

19. Output the combination of LED segments which will form the letter J. Record the binary and hex codes in Table 32-6.

20. Output the combination of LED segments which will form the number 8. Record the binary and hex codes in Table 32-6.

To Output a Hex Number with a Minus (–) Sign Before It

21. On a separate sheet of paper draw a flowchart for a program to display a two-digit hex number with a minus sign before it. Label it, "Flowchart to Display Any Two-Digit Hex Number with a Minus (–) Sign Before It."

22. Write and record a program as Program 32-6 starting at address 0040. Use address 0094 to store the hex number.

Program 32-6. To Display a Two-Digit Hex Number with a Minus (–) Sign Before It

Address	*OP Code*	*Mnemonic*
0040		

23. Display the number $+27_{16}$, with a (–) before it.

24. Display the number $+33_{16}$, with a (–) before it.

25. Display the number $+3F_{16}$, with a (–) before it.

To Output the Word "Error"

26. On a separate sheet of paper draw a flowchart for a program to display the word "Error." Label it, "Flowchart to Display the Word 'Error'."

27. Write and record a program as Program 32-7 starting at address 0050.

28. Enter the program, check it, and display the word "Error."

Program 32.7. To Display the Word "Error"

Address	*OP Code*	*Mnemonic*
0050		

QUESTIONS

1. How did you convert the decimal numbers −12 and −23 to their binary and hexadecimal forms, respectively?
2. What does the 2's complement of a number accomplish?
3. What are the limitations, if any, of using Program 32-1 to subtract two numbers?
4. What are the limitations, if any, of using Program 32-2 to subtract two numbers?
5. What does step 0006 in Program 32-2 accomplish?
6. What does Program 32-3 accomplish which Programs 32-1 and 32-2 do not?
7. Which subtraction, if any, in Table 32-5 would have given an invalid result? Why?
8. How can you justify the results obtained in subtracting −3 from 121 in Table 32-5?
9. Which subroutine do you use to (*a*) output a two-digit hex number? (*b*) To display a minus (−) sign?
10. Refer to Table 32-4. How can you explain the values of N and V for both problems?

Answers to Self-Test

1. Positive: *a, d;* negative: *b, c*
2. 23_{10}; 17_{16}
3. −72; B8
4. 1; 1
5. not valid
6. NEGA
7. SUBB
8. cannot
9. OUTCH
10. negative

EXPERIMENT

33

PROGRAMMING THE MICROPROCESSOR—PART 3

OBJECTIVES

1. To learn a new programming technique—looping
2. To examine several methods for multiplying two unsigned numbers
3. To debug a program

BASIC INFORMATION

Multiplication by Successive Addition—Looping

The 6800 does not have a multiply instruction. So if it is required to use this micro to solve problems involving multiplication, some means must be found to program it for multiplication.

The simplest way to multiply integral unsigned numbers is by a process known as *successive addition*. For example, the problem 8_{10} multiplied by 5_{10} can be solved by adding 8 five times. Thus

$$8_{10} \times 5_{10} = 8 + 8 + 8 + 8 + 8 = 40_{10} = 28_{16}$$

In this problem, 8_{10} is called the *multiplicand,* 5_{10} is called the *multiplier,* and 40_{10} (or 28_{16}) is the product or the result.

The method of successive addition can be implemented on the ET3400 by a program technique called *looping*. The flowchart in Fig.33-1 illustrates one way to program by successive addition using looping. As will be seen, this program is very limited, but it does have the advantage of simplicity.

The idea involved in Fig. 33-1 is to store the multiplicand in some location such as memory address 0092 and the multiplier at another address such as 0093. Now accumulator A is cleared, because successive addition will take place in accumulator A. The first 8 is added to accumulator A, and at the same time the multiplier in location 0093 is decremented; that is, it is reduced in value by 1, leaving a new multiplier whose value is 4.

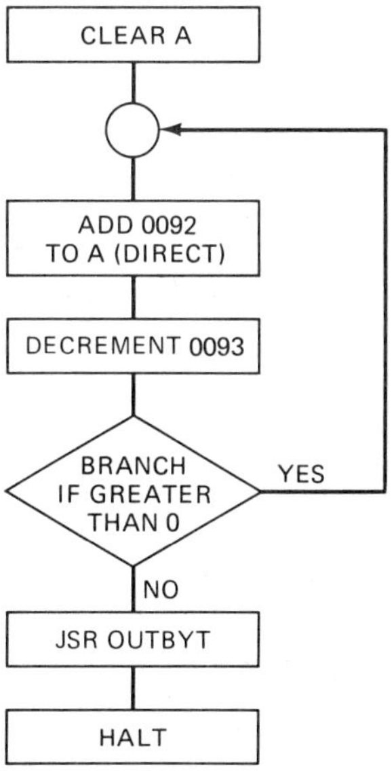

Fig. 33-1. Flowchart for multiplication by successive, without carry.

What was accomplished so far is that the first of five 8s has been added to accumulator A. But four more 8s must be added before the problem is completed.

At this point the looping technique is employed. The trainer is told to branch back (loop) and add a second 8 to accumulator 8, while the multiplier is again decremented, thus reducing its value to 3. Obviously, five complete loops will be required to complete this particular problem. But how will the trainer know when it is to stop looping? There is a 6800 conditional branch instruction (see Appendix A) called Branch if Greater than Zero (BGT) whose mnemonic is BGT. This decision-making instruction checks the value of the decremented multiplier. As long as the multiplier remains greater than zero (YES), the program continues looping. When the multiplier in address 0093 becomes zero, after the fifth loop is completed, the answer to BGT is NO, and the program goes on to output the product held in accumulator A. In the case of $8_{10} \times 5_{10}$, the product, in hex, in the ET3400 will be 28_{16}, which is the equivalent of 40_{10}. You will recall that the ET3400 processes hexadecimal numbers and that the result will be in hex.

Before leaving the flowchart, it must be noted that the loop branching technique involves a *backward* branch. In programming, this must be accounted for by making the relative address following the condition branch instruction a *negative* number.

Translation of the flowchart into code results in Program 33-1.

The OP codes for the instructions are found in Appendix A. The 6800 instructions used here for the first time are CLRA (Clear Accumulator), DEC (Decrement), and BGT (Branch if Greater than Zero).

Of interest in this program is the use of the extended addressing mode for the decrement instruction (see Appendix A). The DEC instruction operates on the multiplier in address 0093. The extended mode uses 16 bits (2 bytes). Hence in programming the OP codes for the operands following the DEC instruction, we write 00 and 93.

Calculating Relative Address for Backward Branch

As in previous programs, the relative address for the BGT instruction was left blank when the program was first written. The reason is that the relative address must be calculated from the program steps. To calculate the relative address, proceed as follows:

1. Determine the *destination* address and the originating address.
2. Subtract the originating address from the destination address.
3. Enter the relative address in the empty space following the BGT instruction.

Now the relative address must be a negative number, written in hex. It will be negative since the value of the originating address is greater than the value of the destination address. Moreover, since the originating and destination address are written in hex in our program, in calculating their difference, the result should be in hex. For example, in Program 33-1 the originating address is 0028_{16} and the destination address is 0021_{16}. To substract these numbers, note that 21_{16} is the minuend and 28_{16} is the subtrahend. Using the 2's complement method for subtraction, first convert the subtrahend into its 2's complement and add this to the minuend. The 2's complement of 28_{16} is D8. Therefore, the problem may be written as

$$\begin{array}{r} 21_{16} \\ +\ D8_{16} \\ \hline \text{result } F9_{16} \end{array}$$

Therefore, F9 is the relative address, which now can be written into Program 33-1.

The ET3400 can be used to determine the relative address by entering the hex values of the subtrahend and the minuend in Program 32-2 (Exp. 32). The resulting number found in accumulator A is the required relative address.

This fact must be borne in mind in using the ET3400. *All numbers entered are hex numbers. The result of an arithmetic operation is also a hex number.* For example, the product of 7_{16} and 4_{16} will not be 28_{16}, but $1C_{16}$. To understand why, the hex numbers must first be converted to their decimal equivalents and the decimal equivalents multiplied. The decimal product, converted into a hexadecimal number, will be the true product of the two numbers. Thus $7_{16} = 7_{10}$ and $4_{16} = 4_{10}$. Now $7_{16} \times 4_{16} = 7_{10} \times 4_{10} = 28_{10}$. Converting 28_{10} into a hex number, $28_{10} = 1C_{16}$. So if the numbers 07 and 04 are loaded into addresses 0092 and 0093, respectively, and if Program 33-1 is entered and executed, the LED readout will be 1C, as it should be.

Now, what if the problem is $7_{10} \times 5_{10}$? Since $7_{10} = 7_{16}$ and $5_{10} = 5_{16}$, if 07 and 05 are loaded into address 0092 and 0093 and Program 33-1 is executed, the trainer will read 23_{16}. Of course, this hex number is the equivalent of 35_{10}.

What will be the product if 12_{16} is loaded into address 0092 and 11_{16} into 0093 and if Program 33-1 is executed? It can readily be shown that the product should be 132_{16}, but when Program 33-1 is tried, the result displayed on the LEDs reads 32. Why? Because Program 33-1 neglected to consider any *carry* resulting from the product. And so Program 33-1 must be modified to take a carry into consideration.

PROGRAM 33-1. Multiplying by Successive Addition

Address	*OP Code*	*Mnemonic*
0020	4F	CLRA (INHERENT)
0021	9B	ADDA (DIRECT)
0022	92	operand
0023	7A	DEC (EXTENDED)
0024	00	operand
0025	93	operand
0026	2E	BGT (RELATIVE)
0027	——	operand
0028	BD	JSR (EXTENDED)
0029	FE	operand } OUTBYT
002A	20	operand } OUTBYT
002B	3E	WAI
0092	——	Store multiplicand
0093	——	Store multiplier

Modified Multiply Program

There is a 6800 instruction, ADCA, meaning Add with Carry to Accumulator A, and if immediate addressing is used and 0 is added, the only addition to A will be the carry. The original addition can be performed in accumulator B, all the carrys accumulated in A, and then by displaying A, transferring B to A, and displaying the result, a four-place answer is possible. Program 33-1 overlooked the fact that the product of two two-digit numbers can be as large as a four-digit number. A solution is therefore required which will be able to output a four-place answer while taking a carry into account. The flowchart in Fig. 33-2 seems to offer a solution. Whether it will solve all the problems remains to be seen.

The modified flowchart in Fig. 33-2 translates into code in Program 33-2*A*.

To determine the relative address, note that the originating address is 002B and the destination address is 0022. Adding the 2's complement of 2B (which is D5) to 22 gives the relative address as F7, which is now written into address 002A.

Now Program 33-2*A* can be entered into the trainer, checked, and used.

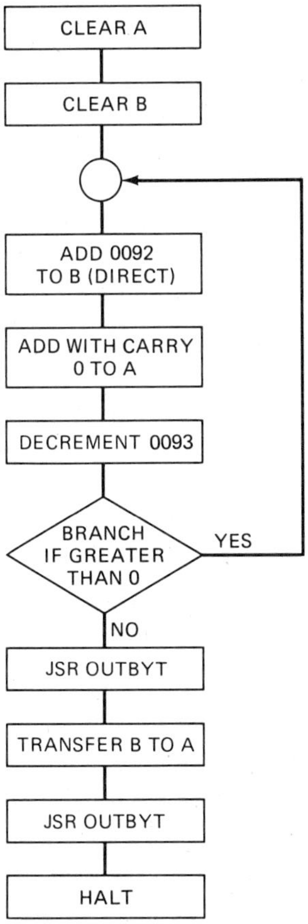

Fig. 33-2. Flowchart for multiplication by successive addition, with carry.

PROGRAM 33-2*A*. Multiply Program with Carry

Address	*OP Code*	*Mnemonic*
0020	4F	CLRA
0021	5F	CLRB
0022	DB	ADDB (DIRECT)
0023	92	operand
0024	89	ADCA (IMMEDIATE)
0025	00	operand
0026	7A	DEC
0027	00	operand
0028	93	operand
0029	2E	BGT (RELATIVE)
002A	— —	operand
002B	BD	JSR (EXTENDED)
002C	FE	operand
002D	20	operand
002E	17	TBA
002F	BD	JSR
0030	FE	operand
0031	20	operand
0032	3E	WAI
0092	— —	Store multiplicand
0093	— —	Store multiplier

What is the result of multiplying 12_{16} by 11_{16} using Program 33-2*A?* If 12 is loaded into address 0092 and 11 into address 0093 and the program is executed, the display will read 0132 as expected.

Will the modified program still produce the same result if the numbers are interchanged, that is, if 11 becomes the multiplicand and 12 the multiplier? By the commutative law, the result should be the same. Entering 11 into memory location 0092 and 12 into 0093 and executing Program 33-2*A* *does* produce the proper result, 0132.

Will the program succeed with other combinations? One way to tell is to try. First, trying the product of 04 and 02 gives the result 0008, and the product of 02 and 04 gives the same result. This would suggest that Program 33-2*A* works with small numbers. What about large numbers, say, the product of 97_{16} and 54_{16}? Entering 97 into 0092 and 54 into 0093 and executing the program results in the readout of 318C. Is this the correct answer? One way to check the answer requires changing the hex numbers into their decimal equivalents, multiplying the decimal numbers together, and finally converting the decimal product back to hex:

$$\begin{array}{rr} 97_{16} = 9 \times 16 + 7 = & 151_{10} \\ \times\ 54_{16} = 5 \times 16 + 4 = & \times\ \ 84_{10} \\ \hline & \text{product } 12{,}684_{10} \end{array}$$

To convert the decimal into hex, it is necessary to recall the decimal values of the appropriate powers of 16. Thus $16^4 = 65{,}536$, $16^3 = 4096$, $16^2 = 256$, $16^1 = 16$, $16^0 = 1$. These values are then set up in ascending powers of 16, reading from right to left as follows:

16^4	16^3	16^2	16^1	16^0
65,536	4096	256	16	1
	3	1	8	C = $12{,}684_{10}$

The largest power of 16 which is smaller than 12,684 (our product) is 3, and 16^3 or 4096 will divide into 12,684, giving a quotient of 3 and a remainder of 396. Dividing 396 by 256 results in a quotient of 1 and a remainder of 140. Dividing 140 by 16 gives a quotient of 8 and a final remainder of 12_{10} or C in hex. If the quotients and the final remainder are entered in the proper columns, the hex equivalent of $12{,}684_{10}$ is $318C_{16}$. So Program 33-2*A* gave a proper answer to the product of 97_{16} and 54_{16}.

What about the product of 54_{16} and 97_{16}? Reversing the numbers in memory locations 0092 and 0093 results *not* in 318C, but in 0054! Is that 0054? Well, we checked the numbers in memory locations 0092 and 0093. They were correctly entered. Program 33-2*A* was again executed, and again the answer 0054! Something obviously was wrong. The result should have been 318C, if no error were made in the program or in entering the program into the trainer.

Debugging a Program

On consideration, it hardly seems likely that an error was made in entering the program because the products 11 × 12, 12 × 11, 4 × 2, 2 × 4, and 97 × 54 were all correct. The problem occurred after 54 was entered in address 0092 and 97 in address 0093. These two addresses were subsequently checked and their contents found correct. It would seem, therefore, that there is a *bug* in Program 33-2*A*, since it does not produce accurate results for the products of all two two-digit numbers. Finding that bug requires a full understanding of each program step and the effect of each instruction as it is executed. *Single-stepping* is the method used for checking the effect of each instruction. But before single-stepping Program 33-2*A*, an analysis of the effect of each instruction on a simple program which is known to be working, is in order.

An analysis of Program 33-2*A* for the product 02 × 04 indicates that as the program is single-stepped for the first time up to the BGT instruction in address 0029, accumulator B should contain 02, and the multiplier in address 0093 should have been reduced by 1 to 03. Accumulator A should be clear (0) because no carry had as yet occurred. Now, when instruction 2E (BGT) is executed, the trainer should loop back to address 0022. Three more single steps should take the trainer to address 0029 again, and the following changes should have taken place as a result of this second loop: accumulator B should now contain the number 04 (the result of adding the number 2 twice), and address 0093 should contain 02 because the multiplier has been decremented twice. Continuing to single-step through the third loop, up to address 0029, will bring the count in accumulator B to 06, in address 0093 to 01. The final and fourth loop to address 0029 will result in a product of 08 in accumulator B, and the multiplier in location 0093 should have been reduced to 00. Accumulator A will still contain 00 because no carry has occurred in this series of calculations. Since the multiplier is no longer greater than zero, looping should stop.

It will not be possible to single-step to completion the program as it was written, because the OUTBYT subroutine entered into ROM at address FE20. But it is not really necessary to single-step through this subroutine, which we assume to be correct. The execution of the program, including display of the result, can now be achieved by pressing DO and 0029.

Now, let us check our analysis of Program 33-2*A* by entering 02 in address 0092 and 04 in address 0093, pressing PC and CHAN, and entering 0020. The program in the trainer may now be single stepped and the effects of each instruction analyzed. Note that single-stepping must begin with a *valid instruction,* not an operand. Step 0020 does contain a valid instruction, Clear Accumulator A. Note also that each time SS is pressed, some change must take place. First, after SS is pressed, the instruction in the display is executed and the next instruction and its address are displayed. *Operands are skipped.* After each SS is pressed, the contents of accumulator A, accumulator B, and address 0093 are checked and noted. This routine is followed with the execution of each instruction.

The record of each check as it was made is shown in Table 33-1.

It is not necessary to continue single-stepping because of the OUTBYT subroutine starting at address FE20. The program is completed by pressing DO, CHAN, and 0029. The display shows the product as 0008.

What does Table 33-1 reveal? It shows that *four* loops were actually completed to produce the product. It shows further that accumulator A read 00 throughout because no carry resulted from multiplying these two numbers. Accumulator B changed after each ADDB (address 0022) instruction, going from 00 to 02, to 04, to 06, and to 08. Accumulator B held whatever number it contained after ADDB was executed, until the next loop began. The multiplier in 0093 held its value until the DEC instruction was executed. Then its value was reduced by 1, which it held until the next DEC instruction was executed. The value of the multiplier in 0093 went to 00 after the fourth DEC instruction was executed. On the next single step the program went to address 002B, showing that BGT was no longer greater than zero, and that looping was therefore finished.

Table 33-1 will be used as a reference for compari-

TABLE 33-1. Effect of Each Instruction in Program 33-2*A* (2 × 4)

Address	*Instruction*		*Accumulator A*	*Accumulator B*	*0093*
0021	5F		00	XX	04
0022	DB	#1	00	00	04
0024	89		00	02	04
0026	7A		00	02	04
0029	2E		00	02	03
0022	DB	#2	00	02	03
0024	89		00	04	03
0026	7A		00	04	03
0029	2E		00	04	02
0022	DB	#3	00	04	02
0024	89		00	06	02
0026	7A		00	06	02
0029	2E		00	06	01
0022	DB	#4	00	06	01
0024	89		00	08	01
0026	7A		00	08	01
0029	2E		00	08	00
002B	BD		00	08	00
FE20	36		00	08	00

son when Program 33-2*A* is used for single-stepping through the program for 54 × 97.

After entering 54 into address 0092 and 97 into address 0093 and setting PC on 0020, we again start to execute the program by single-stepping. The results are shown in Table 33-2.

Now we press RESET, DO 0020, and the display reads 0054. Checking accumulator A, we find 54, and accumulator B shows 54, while address 0093 shows 96.

The results in Table 33-2 clearly show that the *program did not loop when the BGT instruction was executed*. Everything up to that point (address 0029) was normal. Why didn't the BGT instruction work? The reason lies in the branch instruction we chose. The micro works on 2's complement arithmetic and therefore considers a *negative* number *smaller* than zero. Note that the number 97_{16} is a negative number (= 10010111_2) in the 2's complement system and the branch instruction looking at that number in address 0093 interprets it as a number less than zero. And so when the branch instruction is executed, the number in accumulator B (54) is transferred to A and appears in the display as 0054.

The BGT is obviously the wrong branch instruction. Is there a right branch instruction? Referring to Appendix A, we note that there is an instruction BNE (Branch if Not Equal to Zero) which should branch for numbers greater than zero or less than zero, *but not for zero*. This instruction would appear to be the solution to the problem, and if it is, the branch will be executed until the contents of address 0093 are reduced to 00. Hence Program 33-2*A* must be modified by changing the instruction at address 0029 from 2E to 26 (the OP code for BNE). Now the modified program will be checked on the product 54 × 97 to determine if it works.

After 26 was entered in address 0029, 54 in address 0092, and 97 in address 0093, the program was executed and the resultant LED display read 318C. The numbers reversed (97 in 0092 and 54 in 0093) gave the same result, 318C. And so, modified Program 33-2*B* performed as expected and may be used to find the product of two unsigned two-digit numbers, positive or negative.

TABLE 33-2. Effect of Each Instruction in Program 33-2*A* (54 × 97)

Address	*Instruction*	*Accumulator A*	*Accumulator B*	*0093*
0021	5F	00	XX	97
0022	DB	00	00	97
0024	89	00	54	97
0026	7A	00	54	97
0029	2E	00	54	96
002B	BD	00	54	96
FE20	36	00	54	96

PROGRAM 33-2*B*. Modified Multiply with Carry

Address	*OP Code*	*Mnemonic*
0020	4F	CLRA
0021	5F	CLRB
0022	DB	ADDB (DIRECT)
0023	92	operand
0024	89	ADCA (IMMEDIATE)
0025	00	operand
0026	7A	DEC
0027	00	operand
0028	93	operand
0029	26	BNE (RELATIVE)
002A	— —	operand
002B	BD	JSR (EXTENDED)
002C	FE	operand
002D	20	operand
002E	17	TBA
002F	BD	JSR
0030	FE	operand
0031	20	operand
0032	3E	WAI
0092	— —	Store multiplicand
0093	— —	Store multiplier

Multiplication by Shift and Add

Program 33-2*B* is code-efficient; that is, it uses few instructions and therefore takes little room in memory. However, it is quite slow, a fact which is not too evident in the trainer because multiplication in the ET3400 is performed by an 8-bit number. If the time it takes to execute a program is calculated, this fact appears: the time to complete a loop is 15 machine cycles (Add Direct is three cycles, Add with Carry Immediate is two cycles, Decrement is six cycles, and the Branch instruction takes 4 cycles). At 1-MHz clock frequency, cycle time would be 1 μs, and a multiplication with a large multiplier would take 15 μs times the multiplier. Even when the multiplier is only 200, this time is 3000 μs, or 3 ms, a time which is too slow for many real-time applications.

There is another method for multiplication which is faster, although it does require more programming steps. This method employs a different algorithm, that is, a different set of rules or steps to solve the problem. This method might well be called *multiplication by shift and add.*

Ordinary manual multiplication is not done by successive addition of the multiplicand because it would take too long. Instead, the familiar method illustrated below is used. Take, for example, the problem $23_{10} \times 34_{10}$. This is normally written and calculated as follows:

```
   23
 × 34
 ----
   92
  69
 ----
  782₁₀
```

The first operation involves multiplying the multiplicand (23) by the unit's digit (4) of the multiplier and setting this partial product (92) down. Then the ten's digit (3) of the multiplier multiplies the multiplicand and the second partial product (69) is placed beneath the 92, but shifted one digit to the left of the 2 in 92. The two partial products are then added and the result is 782. Why is the 69 shifted one digit to the left? Because it is the result of multiplying the 23 by 3, and the 3 is the next higher power of 10 (and therefore is really 30) as compared to the first multiplier digit, the 4. So the 69 is really 690 and if it were written as 690, would appear as

```
   92
+ 690
-----
  782
```

This same technique may be used in multiplying binary numbers except the process is simpler because binary numbers use just two digits, 0 and 1. The rules for binary multiplication are

$1 \times 1 = 1$
$1 \times 0 = 0$
$0 \times 1 = 0$
$0 \times 0 = 0$

Written in 8-bit binary,

$23_{10} = 0001\ 0111_2$		
$\times\ 34_{10} = 0010\ 0010_2$		
0000 0000	multiply	by 0
0 0010 111	"	by 1
00 0000 00	"	by 0
000 0000 0	"	by 0
0000 0000	"	by 0
0 0010 111	"	by 1
Product = 782_{10} 0 0011 0000 1110		

An analysis of the binary multiplication shows that the first partial product is the result of multiplying the multiplicand by the units digit, namely, 0. The second partial product is the result of multiplying the multiplicand by the 2's digit, 1 in this case; the third multiplier is 0, the fourth is 0, the fifth is 0, and the sixth is 1. Note that each partial product is shifted 1 bit to the left, to account for the *weight* of the multiplier. Observe also that in binary multiplication the partial product is either all zeros (when the multiplier bit is 0) or is the multiplicand itself (when the multiplier is 1) shifted 1 bit to the left.

The array of partial products could have been obtained by shifting the multiplicand 1 bit to the left for each multiplier bit and by writing the partial products in an array, moving toward the left. Zero rows can be totally eliminated, but the shifted partial products must take into account the number of shifts caused by 0 multiplier bits. So for the problem 0001 0111 × 0010 0010, we can write

	0000 0010 1110	partial product due to first 1
	0010 1110 *0000*	partial product due to second 1
Product	0011 0000 1110	= sum of partial products

For the shift-and-add method of multiplication, these facts must be noted:

1. The multiplicand must be shifted 1 bit to the left for each multiplier bit.
2. Each LSB in the multiplicand that is left vacant as a result of a shift to the left must be filled with a 0.
3. The shifted multiplicand becomes a partial product after it is multiplied by an operational multiplier bit.
4. If the operational multiplier bit is a 0, the shifted multiplicand is ignored.
5. If the operational multiplier bit is a 1, the partial product is the shifted multiplicand.
6. As the MSB of the multiplicand is shifted to the left, it must not be lost, so that it may be accounted for in the final summing of the partial products.

When multiplication by shift and add is done manually, the eye and the brain keep track of the multiplier bit which is being used. The micro uses a simple strategy to recognize each multiplier bit as it becomes operational. As each multiplier bit is used, it is shifted to the right, out of the register, as a carry bit, as in Fig. 33-3. The micro can then easily determine whether there is a 0 or a 1 in the carry (by instruction BCC—Branch if Carry is Clear). Since the 0 multiplier bits cause a 0 partial product, which will not affect the sum of the other partial products, the micro can be instructed to ignore the 0 partial product when it recognizes a 0 multiplier bit, but to shift the multiplicand 1 bit to the left, while the multiplier is shifted 1 bit to the right. As soon as a 1 multiplier bit is shifted into the carry and is recognized, the micro can be instructed to add the shifted multiplicand as a partial product to the sum of the other partial products. When the last 1 is shifted out of the multiplier and the multiplicand added, the micro can be instructed to output the sum of the partial products as the final *product*.

In this system, then, for every shift right of the multiplier, there is a corresponding shift left of the multiplicand. Each LSB shifted right out of the multiplier is tested to determine whether it is a 0 or 1. Each MSB shifted left out of the multiplicand is saved so that it can be added as part of the final product of the numbers being multiplied.

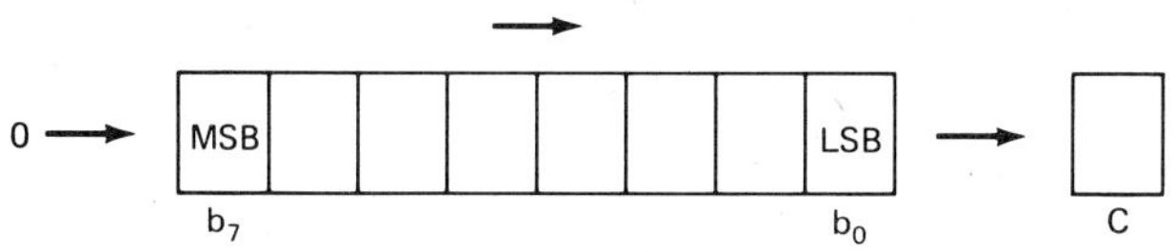

Fig. 33-3. Multiplier shifted to the right. LSB is shifted into carry. Zero is shifted into MSB.

How can the ET3400 multiply the two decimal numbers 23 and 34? These numbers are entered into the system in hex and are recognized as the binary numbers 0001 0111 (23_{10}) and 0010 0010 (34_{10}). The LSB of the multiplier is shifted *right* into the carry. The micro checks the number in the carry. It is a 0. So the multiplicand is shifted *left,* the multiplier is shifted *right,* and the next bit of the multiplier is checked in the carry. It is a 1. So the shifted multiplicand (call it the *partial product*) is added to the other partial product (0 in this case). Let us call the sum of these partial products the *first partial sum*. The process continues then until completion.

Schematically, the process can be shown as follows:

Shifted Multiplicand	*Partial Sum*	*Multiplier*
0000 0000 0001 0111	0000 0000 0000 0000	0010 0010
0000 0000 0010 1110	0000 0000 0010 1110	0001 0001
0000 0000 0101 1100	0000 0000 0010 1110	0000 1000
0000 0000 1011 1000	0000 0000 0010 1110	0000 0100
0000 0001 0111 0000	0000 0000 0010 1110	0000 0010
0000 0010 1110 0000	0000 0000 0010 1110	0000 0001
	+ 0000 0010 1110 0000	
	0000 0011 0000 1110	

Note that the array of bits in the multiplicand and in the partial sum indicates that 2 bytes (16 bits) are needed for both the partial sum and the multiplicand.

Flowchart—Multiply by Shift and Add

Figure 33-4 is a flowchart for this process, and an examination of this chart shows that at the start, the memory locations which are to be used for storing and adding the shifted numbers and carrys are cleared, so that the result will not be affected by any random number in them. For any number in these locations, other than 0, would give erroneous results. The full multiplicand, which was stored in address 0090, is loaded into accumulator B, and the multiplier (in address 0091) is shifted to the right. In this process the LSB of the multiplier is shifted into the carry and examined. If the carry is clear (C = 0), the program branches to shift the multiplicand in address 0090 left. In this shift the MSB of the multiplicand is shifted into the carry, and the carry is then rotated into address 0094 (see Fig. 33-5). The multiplier bits remaining in the register are tested to determine if there are 1s left in the multiplier. If there are (YES), the previously shifted multiplicand is loaded into accumulator B, and the multiplier in address 0091 is again shifted to the right. The new carry

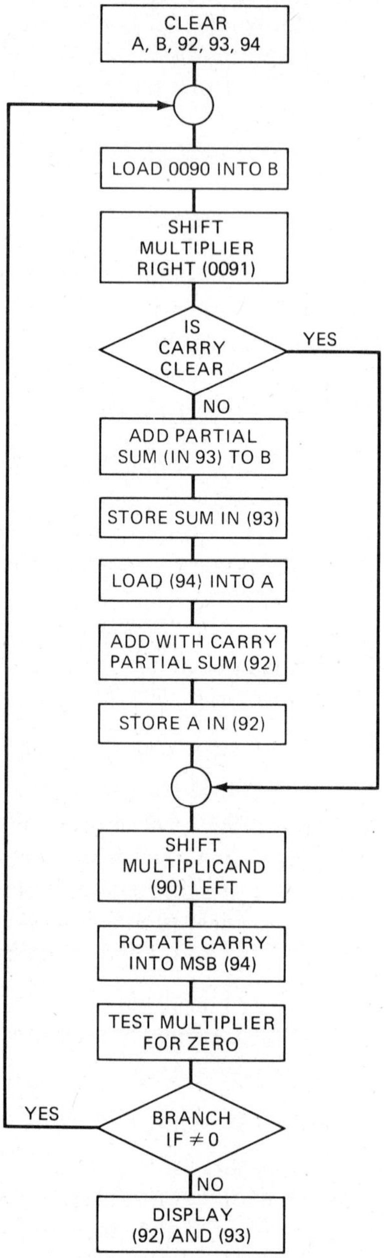

Fig. 33-4. Flowchart for multiplication by shift and add.

(which is the next multiplier bit) is tested. If it is a 1 (NO, carry is not clear), there is no branch, and the number in address 0093 is added to accumulator B, and the result is stored in address 0093. (NOTE: In our problem, since the first multiplier bit is 0 and the second is a 1, the shifted multiplicand 0010 1110 is now stored in address 0093). The number in address 0094 (this is the MSB that was shifted out of the multiplicand and rotated into 0094) is now loaded into accumulator A. Accumulator A is then stored in address 0092. The program now causes the multiplicand to be shifted left again and continues as before. The program comes to an end only when all the bits in the shifted

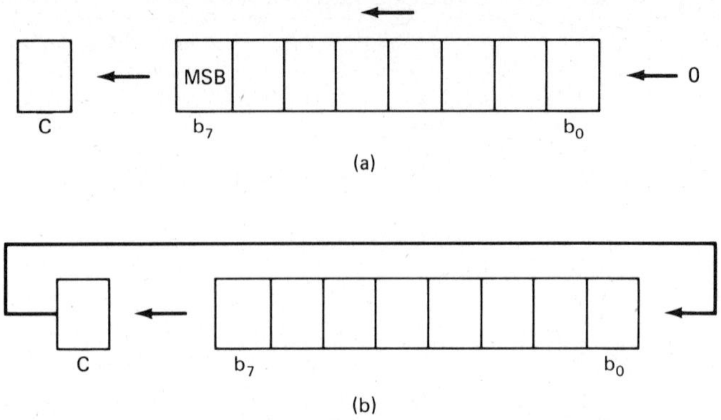

Fig. 33-5. (*a*) Multiplicand shifted to left; MSB is shifted into carry; (*b*) MSB is rotated into address 0094.

multiplier test 0. Then the numbers in addresses 0092 and 0093 are displayed as the final answer.

What are the numbers in addresses 0092 and 0093? Well, 0092 holds the sum of the MSBs shifted out of the multiplicand. If they are all 0s, the display of address 0092 will read 00. In our problem, address 0092 will display the number 03. Address 0093 holds the sum of the shifted (partial products) multiplicands. In our problem, this sum is 0E. So the total display will be 030E which equals 782_{10}.

Before translating the flowchart in Fig. 33-4 into code, it should be stressed that the instructions used in the program may not be available in every micro instruction set. Therefore, for micros other than the 6800, even the flowchart may need to be modified, in order to accomplish the task before us. As a matter of fact, some advanced micros have an instruction "multiply," and then the entire program is superfluous.

And now Program 33-3, based on the flowchart in Fig. 33-4, follows, with starting address 0040.

Comparing Programs

A comparison of the advantages and disadvantages of Programs 33-2*B* and 33-3 (exclusive of the output functions) yields the following:

1. The successive addition program (Program 33-2*B*) requires 11 memory locations, while the shift-and-add program (Program 33-3) requires 35.
2. The time required to complete a multiplication by a multiplier equal to 200 takes about 3 ms with Program 33-2*B*. In the worst case, with Program 33-3, 0.43 ms would be required (assuming a 1-μs cycle time).
3. Program 33-3 is therefore an order of magnitude faster than Program 33-2*B*.

It should be noted that there are hardware multipliers available which can be connected into a microprocessor system, and some of these devices can perform

PROGRAM 33-3. Multiply by Shift and Add

Address	*OP Code*	*Mnemonic*
0040	5F	CLRB
0041	4F	CLRA
0042	97	STAA (DIRECT)
0043	92	operand
0044	97	STAA
0045	93	operand
0046	97	STAA
0047	94	operand
0048	D6	LDAB (DIRECT)
0049	90	operand
004A	74	LSR (EXTENDED)
004B	00	operand
004C	91	operand
004D	24	BCC (RELATIVE)
004E	— —	operand
004F	DB	ADDB (DIRECT)
0050	93	operand
0051	D7	STAB (DIRECT)
0052	93	operand
0053	96	LDAA (DIRECT)
0054	94	operand
0055	99	ADCA (DIRECT)
0056	92	operand
0057	97	STAA (DIRECT)
0058	92	operand
0059	78	ASL (EXTENDED)
005A	00	operand
005B	90	operand
005C	79	ROL (EXTENDED)
005D	00	operand
005E	94	operand
005F	7D	TST (EXTENDED)
0060	00	operand
0061	91	operand
0062	26	BNE (RELATIVE)
0063	— —	operand
0064	96	LDAA (DIRECT)
0065	92	operand
0066	BD	JSR (EXTENDED)
0067	FE	operand
0068	20	operand
0069	96	LDAA (DIRECT)
006A	93	operand
006B	BD	JSR (EXTENDED)
006C	FE	operand
006D	20	operand
006E	3E	WAI
0090	— —	Store multiplicand
0091	— —	Store multiplier

an 8 × 8 multiply in 1 μs. Which of these alternatives is used in any given system depends on how quickly the multiplication must be performed and how much room is available in the program memory.

SUMMARY

1. One way of multiplying two unsigned integers is by the method of successive addition.
2. Successive addition requires adding the multiplicand as many times as the value of the multiplier. For example, $8_{10} \times 5_{10}$ requires adding 8 *five* times.
3. In programming this repetitive process for a microprocessor, a technique called *looping* is employed.
4. Looping involves backward branching over the steps that must be repeated.
5. A backward branch is indicated to the 6800 by a *negative* relative address.
6. To determine the relative address of a backward-branch instruction, subtract the originating address from the destination address. The originating address is the address *after* that of the relative address. For the 6800, all numbers are in hex, as are all numbers entered.
7. A *bug* in a program means an error in programming.
8. A useful technique in debugging a program for the ET3400 is to execute the instructions, one instruction at a time (single-stepping) and observe the effects of each instruction as it is executed. Trouble is pin-pointed when an instruction does not cause the expected result.
9. An algorithm is a set of rules to solve a problem.
10. Multiplication by shift and add requires a longer program, but it takes less time to run than does multiplication by successive addition.
11. When binary numbers are multiplied by the shift-and-add method, the partial product resulting from each multiplier bit is either a row of zeros (when the multiplier bit is a 0) or the multiplicand shifted 1 bit to the left (when the multiplier bit is a 1).
12. The MSBs of the multiplicand must be saved, as they are shifted out of the register, so that they may be added to the final sum of partial products to make up the final product.

SELF-TEST

Check your understanding by answering these questions.

1. The shift-and-add method of multiplication is faster than the successive addition method. ____________ (true/false)
2. The product of $9_{10} \times 7_{10}$ will be displayed as the hex number ____________ on the ET3400.

3. A backward programming loop requires a ____________ ____________ address.

4. A multiplication program requires a backward loop from address 004A to 0043. The relative address for the loop instruction is ____________ .

5. In Program 33-2*B* the OP code for the instruction which tells the micro to branch as long as the multiplier is not equal to zero is ____________ .

6. In Program 33-2*B* the OP code of the instruction which saves the carrys resulting from successive addition of the multiplicand is ____________ .

7. Single-stepping in the ET3400 must begin with an operand. ____________ (true/false)

8. In Program 33-3 the OP code of the instruction which tells the micro to shift the multiplier to the right is ____________ . The mnemonic for this instruction is ____________ .

9. The mnemonic of the instruction in Program 33-3 which tests the multiplier to determine if there are any remaining 1s is ____________ and its OP code is ____________ .

10. In Program 33-3, memory address ____________ contains the sum of the multiplicand carrys.

MATERIALS REQUIRED

- Equipment: Microprocessor-microcomputer trainer, Heath Company, ET3400 and instruction manual

PROCEDURE

Multiplication without Carry

1. Enter Program 33-1 in the trainer and check it.
2. In Table 33-3, convert each decimal number into its equivalent hex number and record it in the appropriate space in the column labeled "Hex Numbers."
3. Using Program 33-1, find the product of $3_{10} \times 2_{10}$ and record the hex readout in Table 33-3. Also record the decimal equivalent of the product. Indicate whether the readout value is correct. (NOTE: Remember the hex value of the decimal numbers to be multiplied must be entered in addresses 0092 and 0093.)
4. Repeat step 3 for each of the problems listed in Table 33-3.

Multiplication with Carry—Program 33-2*A*

5. Enter Program 33-2*A* in trainer and check it.

TABLE 33-3. Multiplication with Program 33-1

Multiplication Problem		*Product*		
Decimal Number Problem	*Hex Numbers*	*Hex Readout*	*Decimal Equivalent*	*Is Readout Correct?*
3 × 2				
2 × 3				
8 × 6				
6 × 8				
9 × 11				
11 × 9				
13 × 15				
15 × 13				
15 × 17				
17 × 15				
16 × 16				
16 × 17				
17 × 18				

6. In Table 33-4, convert each of the decimal numbers into its equivalent hex number and record it in the appropriate column labeled "Hex Numbers."

7. Using Program 33-2*A*, find the product of $4_{10} \times 3_{10}$ and record the hex readout in Table 33-4. Also record the decimal equivalent of the product (readout). Indicate whether the readout value is correct.

8. Repeat step 7 for each of the problems listed in Table 33-4.

Checking Each Instruction as It Is Executed

9. Single-step Program 33-2*A* for the product 4 × 3, starting at address 0020. After each instruction is executed record in Table 33-5 the address and the OP code of the instruction (following) which appears in the readout. Also check and record in Table 33-5 the value in accumulator A (ACCA), accumulator B (ACCB), and addresses 0092 and 0093. Stop when the address FE20 appears in the readout.

10. Repeat step 9 for the product $127_{10} \times 128_{10}$. Record in Table 33-6.

TABLE 33-5. Single-Stepping Program 33-2*A*: 4 × 3

Address	*OP Code*	*ACCA*	*ACCB*	*0092*	*0093*
0021					

Multiplication with Carry—Program 33-2*B*

11. In Program 33-2*A*, change the OP code of the instruction at address 0029 from 2E to 26. Leave unchanged the remainder of the program. The modified program is Program 33-2*B*.

TABLE 33-4. Multiplication with Program 33-2*A*

Multiplication Problem		*Product*		*Is Readout Correct?*
Decimal Number Problem	*Hex Numbers*	*Hex Readout*	*Decimal Equivalent*	
4 × 3				
9 × 11				
11 × 9				
15 × 17				
17 × 15				
16 × 16				
17 × 18				
18 × 17				
126 × 126				
127 × 127				
1 × 128				
128 × 1				
127 × 128				
128 × 127				
127 × 255				
255 × 127				

TABLE 33-6. Single-Stepping Program 33-2*A*: ($127_{10} \times 128_{10}$)

Address	*OP Code*	*ACCA*	*ACCB*	*0092*	*0093*
0021					

12. In Table 33-7, convert each of the decimal numbers to its equivalent hex value and record it in the appropriate space in the column labeled "Hex Numbers."

13. Using Program 33-2*B*, find the product of 4 × 3 and record the hex readout in Table 33-7. Also record the decimal equivalent of the hex readout. Indicate whether the readout value is the correct product.

14. Repeat step 13 for each of the problems listed in Table 33-7.

Multiplication with Carry—Program 33-3

15. In Table 33-8, convert each of the decimal numbers into its equivalent hex value and record it in the appropriate space in the column labeled "Hex Numbers."

16. Enter Program 33-3 in the trainer starting at address 0040 and check it.

17. Using Program 33-3, find the product of 7 × 3 and record the readout in Table 33-8. Record also the decimal equivalent of the readout. Indicate whether the readout value is correct.

18. Examine and record in Table 33-9 the contents of addresses 0090, 0091, 0092, 0093, 0094, and ACCA and ACCB.

19. Repeat step 18 for each of the products in Table 33-8.

Single-Stepping Program 33-3

20. Enter the numbers $6D_{16}$ in address 0090 and 03 in 0091 for Program 33-3. What is the decimal equivalent of this problem? __________ × __________ What is the product of the above in hex numbers? __________

21. Now set program counter (PC) to 0040.

22. Record in Table 33-10 on page 248 the OP code of the instruction at this address as it appears in the readout. Also check and record the contents of addresses 0090, 0091, 0092, 0093, 0094, ACCA, and ACCB.

TABLE 33-7. Multiplication with Program 33-2*B*

Multiplication Problem		*Product*		*Is Readout Correct?*
Decimal Number Problem	*Hex Numbers*	*Hex Readout*	*Decimal Equivalent*	
4 × 3				
15 × 17				
17 × 15				
17 × 18				
18 × 17				
127 × 127				
1 × 128				
127 × 128				
128 × 127				
127 × 255				
255 × 127				
255 × 255				

TABLE 33-8. Multiplication with Program 33-3

Multiplication Problem		*Product*		*Is Readout Correct?*
Decimal Number Problem	*Hex Numbers*	*Hex Readout*	*Decimal Equivalent*	
7 × 3				
3 × 7				
17 × 18				
18 × 17				
127 × 127				
127 × 128				
128 × 127				
127 × 255				
255 × 127				
255 × 255				

TABLE 33-9. Effects of Program 33-3 on Accumulators and Addresses

Problem in Decimal Numbers	*Address*					*ACCA*	*ACCB*
	0090	*0091*	*0092*	*0093*	*0094*		
7 × 3							
3 × 7							
17 × 18							
18 × 17							
127 × 127							
127 × 128							
128 × 127							
127 × 255							
255 × 127							
255 × 255							

23. Press SS and repeat step 22, continuing to single-step the program until you reach the readout FE2036.

24. Now press DO FE20. The product as it appears in the readout is ____________ . Is it correct? NOTE: You will use Table 33-10 as a reference in debugging your program in the steps that follow.

Debugging Program 33-3

25. Have your instructor put a bug into Program 33-3 in your trainer. Then use the bugged program to multiply the numbers in Table 33-8. List all products and circle those that are incorrect. If any product is incorrect, debug the program, remove the bug, and repeat the multiplications.

26. Do as many bugs as time will permit.

TABLE 33-10. Single-Stepping Program 33-3 (6D × 03)

				Address				
Address	*OP Code*	*ACCA*	*ACCB*	*0090*	*0091*	*0092*	*0093*	*0094*
0040	5F			6D	03			

QUESTIONS

1. In Table 33-3, using Program 33-1, which products, if any, were incorrect? Why were they incorrect?

2. Refer to Table 33-4. List those products, if any, for which Program 33-2*A* gave incorrect answers. Why were they incorrect?

3. Refer to Tables 33-5 and 33-6. Did the value of the multiplier change in exactly the same way for the product 4 × 3 as for 127 × 128? If it did not, why was there a difference?

4. What did Program 33-2*A* provide for the product that Program 33-1 did not?

5. Refer to Table 33-7. Which products, if any, were incorrect? How did Program 33-2*B* correct the difficulty with Program 33-2*A?*

6. What is the maximum value of unsigned decimal numbers which Program 33-2*B* can multiply? Why is this the maximum value?

7. Refer to Table 33-8. Which products, if any, were incorrect? What advantages does Program 33-3 have over Program 33-2*B?* Disadvantages?

8. How did you compute the relative address for address 004E in Program 33-3? Be specific.

9. Refer to Table 33-9. Which address(es) contain(s) the 2 bytes of the resultant product?

10. Refer to Table 33-9. What happened to each of the multipliers? Why?

11. Refer to Table 33-10. For the product 6D × 03, how many backward branches were there in Program 33-3? Why?

12. Refer to Table 33-10. How many forward branches, if any, were there in Program 33-3 for the product 6D × 03? Why?

Answers to Self-Test

1. true
2. 3F
3. negative relative
4. F9
5. 26
6. 89
7. false
8. 74; LSR
9. TST; 7D
10. 0092

EXPERIMENT

34

MOTORS AND MOTOR CONTROLS

OBJECTIVES

1. To study the operation of a dc shunt motor
2. To learn how the speed and direction of rotation of a dc motor may be controlled
3. To become familiar with some industrial control circuit schematic symbols and typical control circuits
4. To demonstrate the operation of a dc shunt motor

BASIC INFORMATION

Characteristics of a DC Shunt Motor

Most ac motors tend to rotate at a single speed determined by their construction and the frequency of the applied voltage. They are used to turn exhaust fans and drive machinery which does not require continuously variable speeds.

DC motors, on the other hand, are used extensively in industrial electronic control systems because of their desirable speed and torque characteristics: their speed can be easily varied and they have high starting torque. The dc shunt motor can be operated as a variable-speed motor with good speed regulation and good torque characteristics. Electronic control units are available for use with these motors thus making possible automatic operation over a wide range of speeds.

Good speed regulation means that the speed of the motor is constant (or nearly so) with changes in the mechanical load on the motor. High starting torque is desirable to accelerate a load which is not moving or to change its direction of rotation.

DC Motor Construction

A motor rotates because of the interaction of two magnetic fields. These magnetic fields are usually created by passing current through two windings, one stationary and the other rotating. The nonmoving winding is called the *field winding,* and the rotating winding is called the *armature*. In some modern dc motors, the field winding is replaced by a permanent magnet.

The dc shunt-wound motor employs a rotating armature which is slotted to accommodate the armature winding. These windings are connected to segments of a cylindrical *commutator,* a part of the armature, which carries current to the armature through brushes which make a sliding contact with the commutator. The field winding is connected in parallel, or in *shunt,* with the armature winding. The name of the motor is derived from this connection.

If the field and armature winding are connected in series, we have the *series* motor. This connection provides very high starting torque but very poor speed regulation. The series motor is not as widely used as the shunt dc motor. A modified form of the shunt motor, the *compound motor,* has both a shunt and a series field winding. It is a common motor type.

Figure 34-1 shows how a shunt motor may be connected with external rheostats to control the operation of the motor. The resistance of the field winding is relatively high and limits the field current to an acceptable value when it is connected across the line. The resistance of the armature winding, however, is small and would permit a large current to flow when the motor is not rotating. (This high current is responsible for

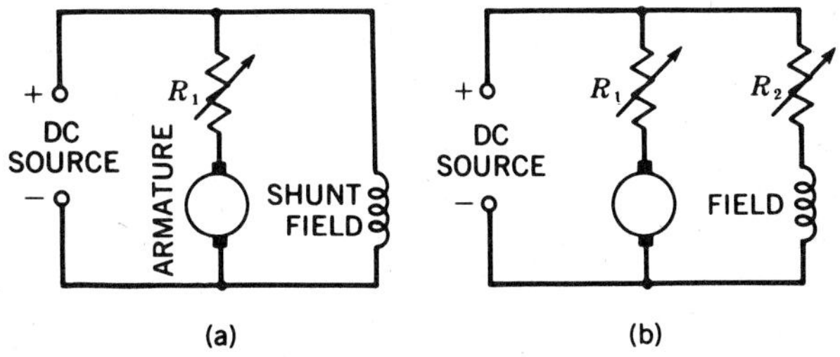

Fig. 34-1. Armature and field connections to dc shunt motor.

the high starting torque of the motor.) Low-power dc shunt motors are started by connecting them directly across the dc power source, but motors with high horsepower ratings use a starting resistance in series with the armature to limit the starting current and starting torque to acceptable values.

When power is first applied to the motor armature, only the small ohmic resistance of the winding would limit the current through it if R_1 were not present (Fig. 34-1*a*). As the armature begins to rotate, however, its conductors cut the magnetic lines of force established by the field winding. This action induces in the armature winding a voltage called a *counter emf,* which increases with speed and acts in opposition to the dc voltage applied to the armature. This reduces the current through the armature as it speeds up since it is the difference between the applied voltage and the counter emf which causes current flow through the armature. At full speed and no load, the counter emf is nearly equal to the externally applied source voltage and the armature current is small. As the load increases (tending to slow the motor), the counter emf decreases. The armature current then increases to provide the needed torque, thus causing the motor to run at nearly a constant speed as the load changes.

The armature current may be calculated from:

$$I(\text{armature}) = \frac{(V_{dc} - \text{cemf})}{R_t}$$

where I = armature current

V_{dc} = dc source voltage

cemf = counter emf generated in the armature as it rotates

R_t = ohmic resistance of armature plus resistance of armature rheostat, if present

When starting the motor (or when it is stalled), the counter emf is zero and the armature current is maximum.

Shunt Motor Speed Control

The speed of a dc shunt motor can be changed by changing the field or armature current. Figure 34-1*b* shows how rheostats may be connected in series with the field and armature windings to control the motor. Within its normal range of operation, we will find that the speed of the motor may be increased by increasing the armature current or by decreasing the field current. Let us see how these ideas apply to a specific motor.

A manufacturer's specification sheet lists the characteristics of a motor, including its "base speed." This is the speed at which the motor operates under load with rated voltages applied to it. Figure 34-2 shows that the base speed of a particular motor is 1800 revolutions per minute (rpm) when 220 V is applied to

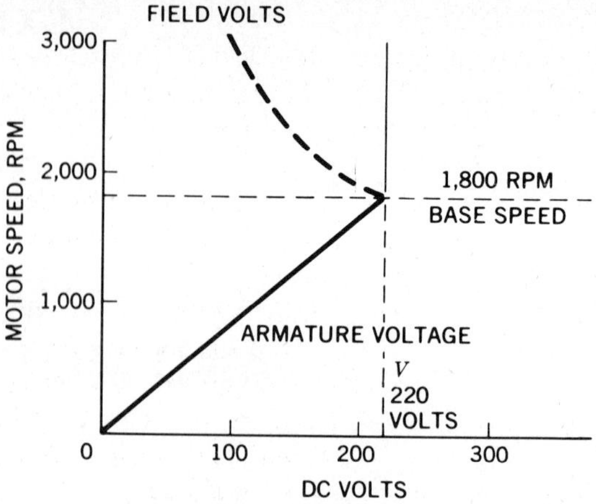

Fig. 34-2. DC motor speed control by varying armature or field excitation.

both field and armature windings. The graph indicates that we may increase the motor speed beyond the base speed by reducing the field voltage while keeping the armature voltage constant or that we can decrease the motor speed below the base speed by reducing the armature voltage while keeping the field voltage constant.

Use of a rheostat in series with the armature to adjust the speed of a motor is limited by the *IR* drop across the rheostat and its effect on the speed regulation of the motor. As the load on the motor increases, armature current increases. This increased current through the rheostat causes an increase in the *IR* drop across it, thereby reducing the voltage applied to the armature. Therefore, the motor tends to slow down due to the increased load and due to the decreased voltage applied to the armature. The speed regulation (SR) of a motor may be calculated in percent as:

$$\text{SR\%} = \frac{100\ (\text{no-load rpm} - \text{loaded rpm})}{\text{loaded rpm}}$$

For example, the no-load speed of a motor is 1800 rpm. With full load its speed is 1725 rpm. Calculate its speed regulation.

$$\text{SR} = \frac{100(1800 - 1725)}{1725}$$

$$= \frac{100(75)}{1725} = 4.35\%$$

In Fig. 34-2 notice that speed increases as the field voltage decreases. To prevent excessive and possibly damaging motor speed, the field voltage must be maintained within specified limits.

In this experiment, and in the illustrations, there is a single dc source. However, in practice, the design of a motor may be such that it has different armature voltage and field voltage ratings and therefore separate sources for the windings are required.

Operating Precautions

Shunt motors should be started with full field voltage but with low armature voltage. Armature voltage is increased slowly to increase motor speed. This is accomplished by starting the motor with a "starting resistance" connected in series with the armature; the amount of resistance is manually or automatically reduced as the motor speeds up.

If a large motor were started without a starting resistance in the armature circuit, it would take a very large starting current, possibly affecting other equipment connected to the same power source; it could also develop such a large torque that mechanical equipment connected to it could be damaged.

When stopping a dc shunt motor with separate field and armature sources, power should be removed from the armature circuit first. This will prevent a potentially hazardous increase in speed which might otherwise occur if armature voltage is present without field excitation.

Measuring the Counter EMF of a Motor

Figure 34-3 describes how the counter (or "back") emf generated in the armature of a dc shunt motor may be measured. The switch S_1 is used either to apply power to the armature winding or to remove it. A voltmeter across the winding measures the counter emf in position 2.

With S_1 in position 1 and S_2 closed, the motor starts and its terminal voltage is measured by the voltmeter. (This is *not* the counter emf). After the motor has achieved a steady speed, S_1 is placed in position 2, and due to the spinning armature conductors cutting the magnetic lines of the field winding, the counter emf may be measured. The back emf decreases as the motor slows down. If the back emf is plotted as a function of the motor speed as measured with a tachometer, the result should be a straight line with a positive slope.

Reversing Direction of Rotation

Since the rotation of the dc shunt motor is due to the interaction of two magnetic fields, it is necessary to reverse the north-south polarity of one of them to reverse the direction of rotation of the motor. For example, if the south poles of two electromagnets were brought near each other, there would be a force of repulsion between them. If the direction of current flow through the coils were now reversed, the south poles would become north poles and there would still be a force of repulsion between them. On the other hand, if the direction of current through one coil were reversed, its south pole would become a north pole, but the south pole of the other coil would be unaffected and there now would be a force of attraction between the two electromagnets. Therefore, to change the direction of rotation of a dc shunt-wound motor, the direction of the current flow through either the armature or the field must be reversed. Figure 34-4 shows how a DPDT switch may be connected to reverse the connection of the field with respect to the armature. If the motor rotated clockwise with S_1 in position 1, it will rotate counterclockwise in position 2.

Dynamic Braking

Because of inertia, a motor will not stop instantly when its power source is removed. If armature voltage is disconnected and the motor continues to turn with its field energized, it will act as a generator. A resistance placed across the armature will dissipate the stored energy in the form of heat. Because a generator is harder to turn when it is electrically loaded, the motor will slow down as if a brake had been applied.

In position 1 of Fig. 34-5, normal operating voltage is applied to the motor. In position 2, however, voltage is removed from the armature and a load R is placed across it. The smaller this resistor is, the more rapidly the motor will stop. Maximum braking current flows at the instant the brake is applied. As the motor slows down, the counter emf decreases, the braking current is less, and the braking effect is reduced.

An idea similar to dynamic braking or "damping" is used to prevent excessive deflections of meter movements during shipment or handling: a shorting wire is placed across the movement. When the movement moves, it also acts as a generator. The current which flows through the shorting wire acts as a dynamic brake. This prevents excessive movement of the needle which might affect the calibration of the meter.

Industrial Control Symbols and Circuits

Table 34-1 shows some common industrial control diagram symbols. Normally closed (NC) and normally open (NO) push-button switches are commonly used

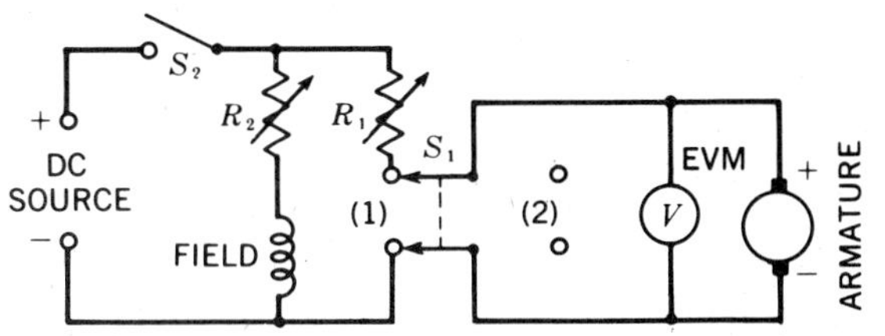

Fig. 34-3. Circuit for measuring counter emf of a motor.

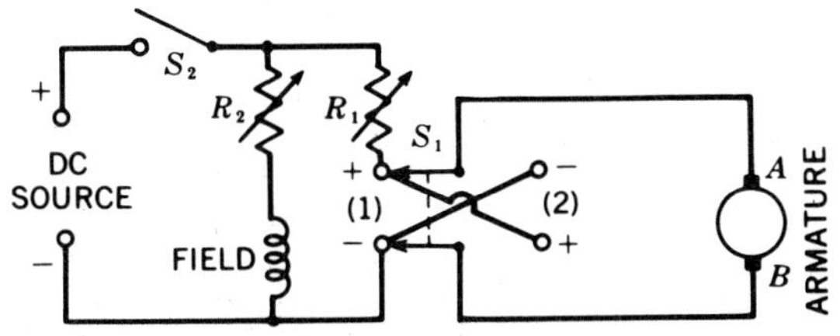

Fig. 34-4. Reversing the direction of rotation of a dc shunt motor.

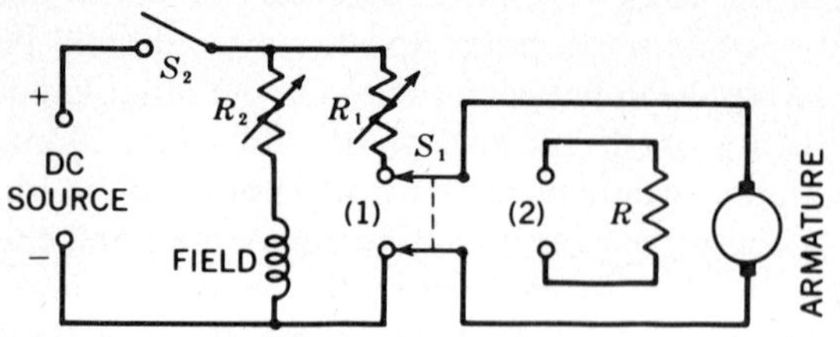

Fig. 34-5. Method of dynamic braking.

TABLE 34-1 Common Industrial Control Systems

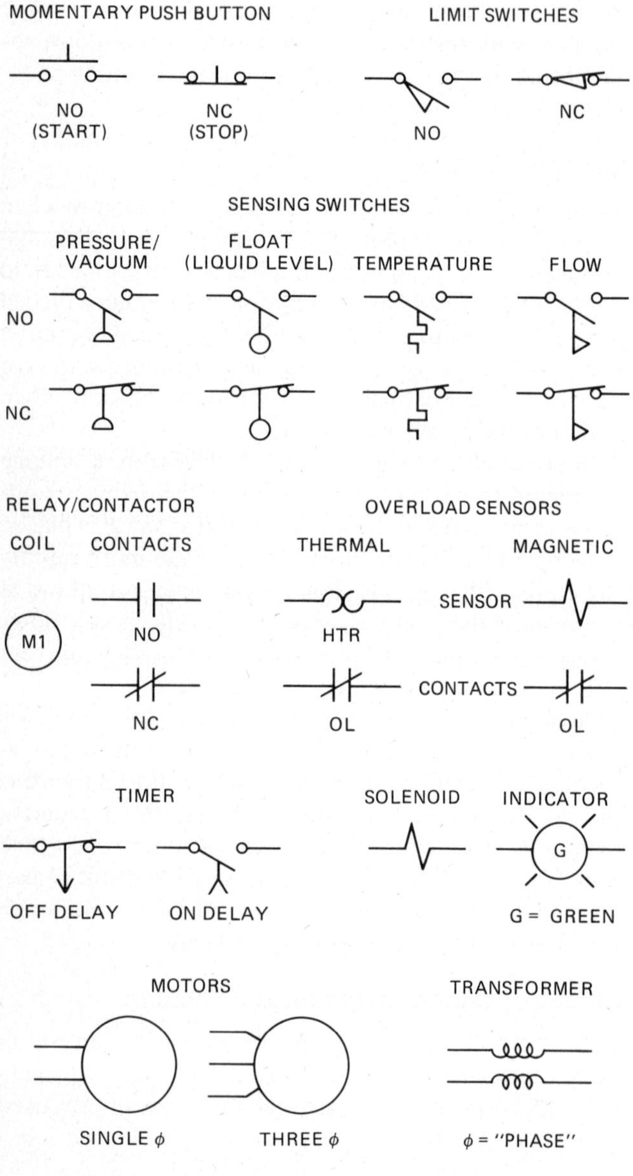

in the circuits which STOP or START a motor, machine, or process. Limit switches are often used to "limit" the travel of a machine by interrupting power to its drive motor, for example. Many industrial processes involve control of a machine by sensing whether or not a required pressure, liquid level, temperature, or flow value is present.

Motors and control relays are almost always included as a part of an industrial machine or process. When a motor is controlled by a latching control relay, it is possible to introduce automatic means of stopping the motor if the electrical or mechanical load on the motor is too great. This is accomplished by sensing the current entering the motor either by the heat it generates or the magnetic field it causes. In either case, a set of overload contacts opens the motor control relay circuit as if the STOP button had been pressed.

Transformers, interval timers, solenoids, and indicator lamps are also very common components of industrial control systems and their symbols are included in Table 34-1.

In low-power applications, such as those in the home, motors are started by applying full line voltage to them through a simple switch. But in industrial applications, with motors of very large horsepower, with the need for remote operation, to meet safety and reliability requirements, more complex methods are required.

Figure 34-6 shows how a motor may be started remotely by briefly pressing the START button and stopped by briefly pressing the STOP button. Motor starter relay M_1 has two sets of contacts: one rated at values of voltage and current sufficient to reliably start and stop the motor, and a second set of contacts used to latch the relay when the start button is pressed.

The operation of this circuit may be understood by analyzing the *ladder diagram*, Fig. 34-6. If we assume that relay M_1 is not energized, the motor is not turning because contacts M_1 are normally open. Pressing the START button PB_1 will complete the path from L_1 of the voltage source through PB_2 to the coil of M_1 and the other side of the line voltage L_2. M_1 now pulls in, closing the M_1 contacts in series with the motor, causing it to start. Another set of contacts on M_1 are in parallel with the START button and maintain power to the coil of M_1 after PB_1 is released. M_1 has been "latched" through a set of its own contacts.

If the STOP button is now pressed, the coil of M_1 is deenergized and the M_1 contacts around PB_1 open, thus "unlatching" M_1. The set of M_1 contacts controlling the motor also open and the motor stops.

A single-phase motor and control circuit have been

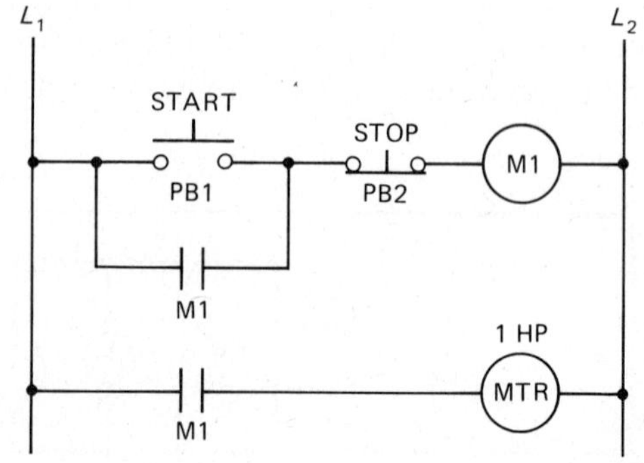

Fig. 34-6. START STOP motor.

shown thus far to illustrate the general idea. Industrial systems often use three-phase power and higher voltages, with low-voltage control circuits. Figure 34-7 illustrates how a three-phase motor may be started "across the line" and how overload-sensing devices can be added to remove voltage from the motor if the line current in any leg becomes excessive.

M_1 is a three-phase, high-voltage, high-current, "contactor" with a set of auxiliary contacts for use with the control circuits. Heaters 1, 2, and 3 respond to the amount of current entering the motor by getting warm in proportion to the current. When excessive current flows for a long enough time, the heaters cause a low-temperature alloy to melt in the overload sensors (OL_1, OL_2, OL_3). Because they are in series with the coil of the control relay, it is deenergized, and the power to the motor is interrupted as if the STOP button had been pressed.

Safety

If the motor drives a conveyor system with belts, chains, and moving parts, these hazardous components might be enclosed behind panels which are removable for maintenance or repair. To prevent injury, these panels often operate microswitches, or interlocks, so that the machine cannot be turned ON when the panels are removed for maintenance. The interlock switches, which are closed when the panels are in place, are connected in series with the STOP button. Removing a panel stops the machine.

Where equipment damage may result from jamming of parts on a conveyor line, additional normally closed switches with sensing arms may be connected in series with the STOP button and may be placed at appropriate places along the conveyor to stop the machine. The likelihood of human injury may also be minimized if additional EMERGENCY STOP buttons, connected in series with the main STOP button, are installed at several well-chosen points around the equipment.

SUMMARY

1. A shunt-wound motor has two windings, the armature and field windings, which are connected in parallel, or "shunt."
2. DC shunt motors are used in industrial control systems because they have good speed and torque characteristics.
3. The speed of a dc shunt motor is variable and easily controlled electronically.
4. The ohmic resistance of the field winding of a shunt motor is relatively high, but the armature resistance is quite low.
5. Because the armature resistance is low, a dc shunt motor takes a large starting current. To limit this current to a specified value, resistance is inserted in series with the armature during starting.
6. The speed of a shunt-wound dc motor increases with armature voltage and decreases with field voltage.
7. The direction of rotation of a shunt motor is reversed by reversing the direction of current through one of the windings.
8. During normal operation, a dc shunt motor generates within itself a counter emf (or back emf) which subtracts from the externally applied operating voltage, thus limiting the amount of armature current.
9. Industrial control diagrams generally use different symbols and drawing techniques from the electronic diagrams used for TV, computers, etc.

SELF-TEST

Check your understanding by answering these questions.

1. In a simple dc shunt-wound motor, the windings are connected in parallel. __________ (true/false)

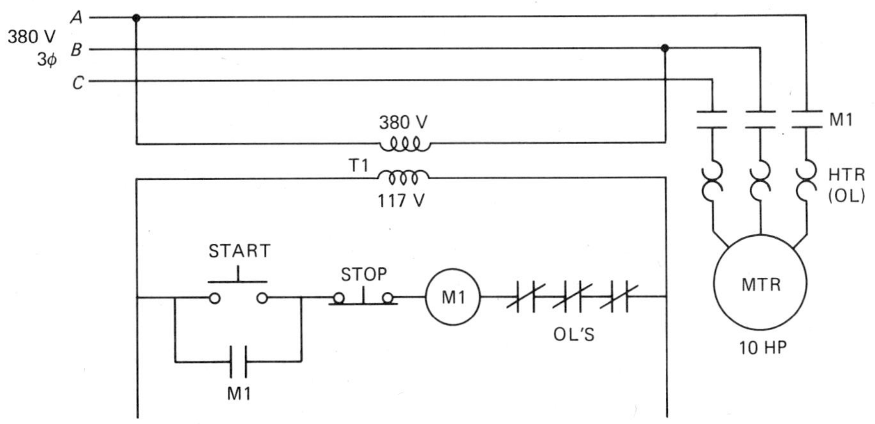

Fig. 34-7. Three phase motor.

2. The segmented cylindrical object which makes a sliding electrical connection to the outside world through brushes is called the ___________ .

3. The windings of a dc shunt motor are called the ___________ and the ___________ windings.

4. The counter emf generated in the armature winding as it rotates has ___________ (the same/opposite) polarity as the source voltage.

5. If the polarity of the voltages applied to both windings of a dc shunt motor are reversed, the direction of rotation will not be affected. (true/false)

6. How well a motor maintains a given speed with varying load is indicated by its ___________ ___________ , in percent.

7. A dc shunt motor should be started by applying full voltage to the ___________ winding and a low voltage to the ___________ winding.

8. When a relay is maintained in the energized condition through a set of its own contacts, it is said to be ___________ .

MATERIALS REQUIRED

- Power supply: 120-V 60-Hz source; variable isolation transformer (RCA isotap or the equivalent)
- Equipment: 0- to 5-A ammeter; EVM (or 20,000 Ω/V VOM); tachometer capable of showing speed changes up to 25 rpm, in range 50 to 5000 rpm (approximately)
- Motor: ⅙ horsepower dc shunt field, 100 V, 1.8-A armature 90 V, rated at 1725 rpm, GE-type 5BCD56BA6 or 5BCD56BD258 or the equivalent
- Semiconductors: Rectifier diodes, four 1N5625 or the equivalent
- Miscellaneous: Adjustable wire-wound resistors: 0 to 100 Ω at 15 W, and 0 to 25 Ω at 75 W; two SPST switches; one DPDT switch; 100-W test lamp

PROCEDURE

NOTE: The dc motor you will use in this experiment is designed for use on full-wave power supplies. Full-wave rectification will be achieved by using a bridge rectifier, as in Fig. 34-8. The field of the ⅙-horsepower motor is rated at 100 V and the armature, at 90 V. The rated speed of the motor is 1725 rpm.

The pulsating dc voltage which appears at the output of the bridge rectifier is applied through rheostats R_1 and R_2 respectively to the armature and field windings of the motor.

1. You will receive a shunt-wound fractional-horsepower dc motor with four terminals. Two of the terminals are connected to the armature winding. The other two are the terminals of the field winding.

   By a continuity check at the terminals, determine and record in Table 34-2 the connections associated with each of the windings. Measure the resistance of the armature and field windings, and record in Table 34-2.

2. Connect the circuit shown in Fig. 34-8. S_1 and S_2 are open. Meter A is a 0- to 5-A ammeter. It is used to measure armature current.

   Set R_1 for maximum resistance in the circuit. Set R_2 at minimum resistance, so that the full dc voltage is applied to the field. *Do not* vary R_2 until instructed to do so. There is no load on the motor at this time.

3. Close S_1, applying voltage to the field winding. Now close S_2, applying power to the armature. Observe, measure, and record in Table 34-2 the

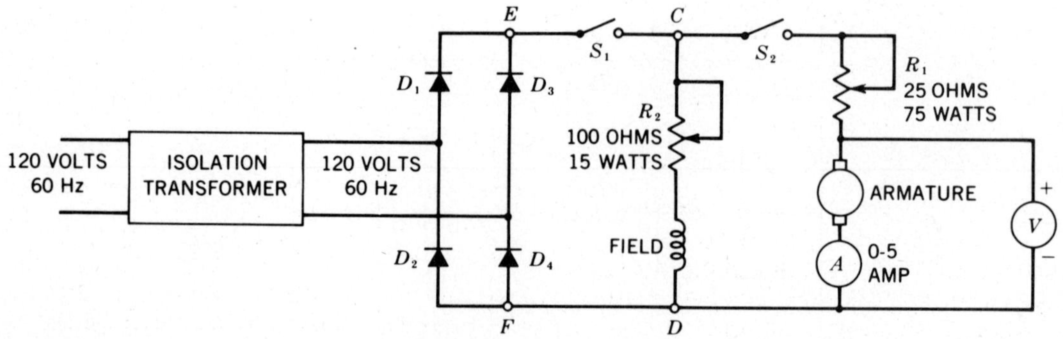

Fig. 34-8. Experimental motor control circuit and supply.

TABLE 34-2. DC Shunt-Motor Characteristics Data (No Load)

	Armature		*Fielding Winding*	
Step	*Terminals*	*Resistance, Ω*	*Terminals*	*Resistance, Ω*
1				
	Armature Current, A	*Voltage across Armature, V*	*Motor Speed, rpm*	*Effect on Motor Speed*
3				
4, 5, 6				
7, 8		90		
9				
10		90		
11				
14	Counter emf	V		
16	Direction of rotation			
17	Direction of rotation			
20	Time required for motor to stop (without dynamic braking)		s	
21	Time required for motor to stop (with dynamic braking)		s	

armature current at the instant S_2 is closed. Note that armature current is highest at the instant power is applied. However, this current is reduced as the motor speed stabilizes.

4. After the motor speed has stabilized, measure and record armature current.

5. Measure the voltage across the armature winding, and record in Table 34-2.

6. Measure and record the speed of the motor in revolutions per minute.

7. Adjust R_1 until the armature voltage measures 90 V. Measure and record the armature current. Indicate in Table 34-2 the effect of increased armature voltage on motor speed.

8. Measure and record the motor speed.

9. Adjust R_1 for minimum resistance, thus applying the full voltage across the armature winding. Measure and record armature voltage and armature current, still without load. Indicate in Table 34-2 the effect on motor speed of increase in armature voltage. Measure and record motor speed.

10. Set R_1 for 90 V across the armature winding. Observe and record the effect on motor speed as the field rheostat is varied from minimum resistance to maximum, that is, as the field voltage (and current) is reduced from maximum to minimum.

    Measure and record armature current under conditions of 90-V voltage and minimum field voltage, without load.

    Measure and record motor speed under conditions of maximum armature voltage and minimum field voltage.

11. Adjust R_1 for minimum armature voltage, and set R_2 for maximum field voltage. Record armature voltage and current, and motor speed without load.

12. Open S_2. Now there is no power applied to the armature circuit. Field circuit is still ON.

13. Connect a DPDT switch S_3 in the armature circuit as shown in Fig 34-9. This circuit arrangement will make it possible for an EVM to measure the counter emf developed in the armature. There is still no load on the motor.

14. Close switches S_2 and S_3, applying power to the armature. Adjust R_1 for maximum speed. When motor speed has stabilized, open S_3. Observe and

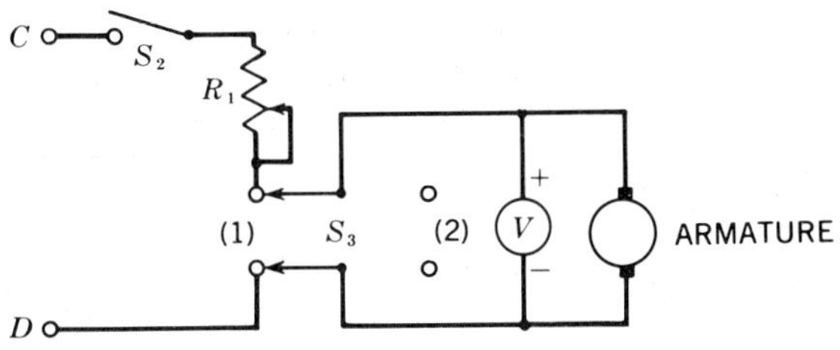

Fig. 34-9. Measuring counter emf of motor.

record in Table 34-2, the counter emf as measured by V at the moment switch S_3 is opened. Why does the voltage, measured by V, decline as the motor slows down?

15. Open switch S_2, removing power from the armature circuit. Field current is still ON. Connect the armature circuit as shown in Fig. 34-10. R_1 is the same rheostat that was previously used. This circuit arrangement will make it possible to demonstrate how reversal of motor rotation is accomplished.

16. Close S_2, applying power to the armature. Adjust R_1 for minimum speed. Switch S_3 is in position 1. Observe direction (clockwise or counterclockwise) of rotation of the motor, and record in Table 34-2.

17. Throw S_3 from position 1 to position 2. Observe and record in Table 34-2 the effect on direction of motor rotation.

18. Open switch S_2, removing power from the armature circuit. Field current is still ON. Connect the armature circuit as shown in Fig. 34-11. L is a 100-W lamp. This circuit arrangement will make it possible to show the effect of dynamic braking.

19. Close S_2, applying power to the armature. Adjust R_1 for maximum speed. Switch S_3 is in position 1. Permit motor speed to stabilize.

20. Open switch S_2. Measure the length of time it takes the motor to stop, using a stopwatch or timer. Record this time in Table 34-2.

21. Close S_2. Permit motor speed to stabilize. Now throw S_3 from position 1 to position 2. Measure and record the length of time it takes the motor to stop.

22. Open switch S_2. (Armature power is removed first when stopping this dc shunt motor with separately excited armature and field windings.) Reconnect the armature circuit shown in Fig. 34-8.

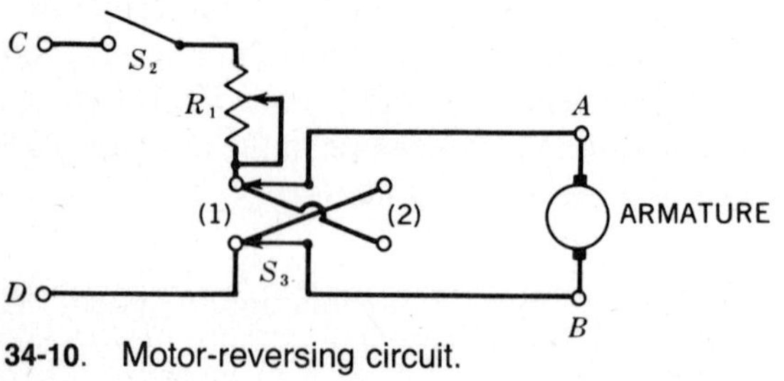

Fig. 34-10. Motor-reversing circuit.

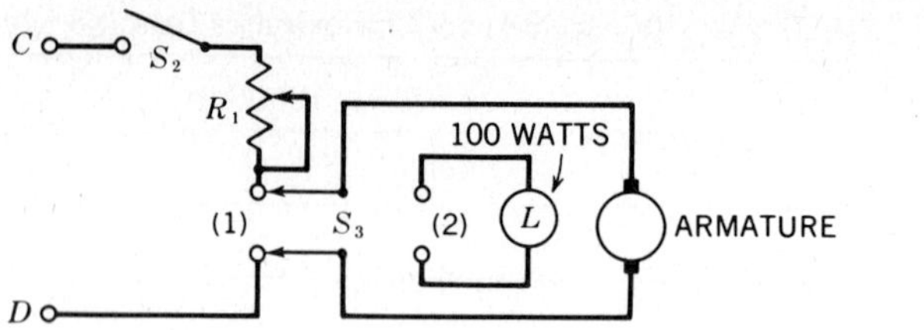

Fig. 34-11. Circuit to demonstrate dynamic braking.

23. Open switch S_1. Connect a variable load to the motor. A generator, or any other device whose loading effects may be varied, may be used as the load.

24. Close switch S_1, applying power to the field. Adjust R_2 for maximum field current.

25. Close switch S_2, applying power to the armature circuit. Adjust R_1 for rated speed. Adjust load so that armature draws 1.5 A (approximately) of current.

    Measure armature voltage and current, and record in Table 34-3. Measure and record motor speed.

26. Reduce load current so that armature current is 0.9 A (approximately). Do not change settings of R_1 and R_2. Measure and record armature voltage, current, and motor speed.

27. Adjust R_1 for minimum speed, and set load so that there is about 1.5 A of current in armature circuit. Measure and record the armature voltage, armature current, and motor speed.

28. Do not change setting of R_1 and R_2. Reduce motor load so that there is 0.9 A (approximately) of armature current. Measure and record the armature voltage, armature current, and motor speed.

Extra Credit

Obtain the following additional materials and construct a motor-starting circuit similar to Fig. 34-6:

1 normally open push-button switch (momentary)

1 normally closed push-button switch (momentary)

1 DPST or DPDT relay

1 motor

QUESTIONS

NOTE: *Refer to the data in Tables 34-2 and 34-3, where applicable, to substantiate the answers to the questions.*

1. Explain the construction of a dc shunt motor.

2. What is the effect on motor speed of increasing armature voltage? What is the effect on armature current of increasing the load on the motor?

TABLE 34-3. DC Shunt-Motor Characteristics (with Load)

Step	*Armature Current,* A	*Armature Voltage,* V	*Motor Speed,* rpm
25	1.5		
26	0.9		
27	1.5		
28	0.9		

3. What is meant by counter emf in a motor? Where is the counter emf generated?
4. What is the effect on counter emf of an increase in motor speed?
5. What is a disadvantage of using a rheostat in the armature circuit to control the motor speed of a dc shunt motor? How was this disadvantage demonstrated in this experiment?
6. What is the effect on motor speed of decreasing the field current of a dc shunt motor?
7. How may the direction of rotation of a dc shunt motor be changed?
8. What is the purpose of a dynamic brake? Explain the principle of operation of a dynamic brake.
9. If the armature resistance of a 28-V dc motor is 0.1 Ω, what is the maximum current through the armature when 28 V direct current is applied?
10. How much resistance should be connected in series with the armature of the motor above to limit the peak-starting current to 100 A?
11. If the speed of a certain motor is 3500 rpm under load and its percent speed regulation is 2.86 percent, what is its approximate no-load speed?

If a 117-V ac appliance motor is used, be sure to use a relay with the same ac voltage and current ratings. *Observe safety precautions.*

Answers to Self-Test

1. true
2. commutator
3. field; armature
4. same
5. true
6. speed regulation
7. field; armature
8. latched

EXPERIMENT

35

SCR AUTOMATIC CONTROL OF THE SPEED OF A DC MOTOR

OBJECTIVE

To study the operation of an SCR automatic-speed-control (ASC) circuit

BASIC INFORMATION

The arrangement of a rheostat in series with the armature winding for regulating the speed of a motor has drawbacks. One limitation is poor speed regulation with variations in load. An increase in load results in an increase in load current and hence in an increased voltage drop across the rheostat in series with the armature winding. The increased voltage across the rheostat subtracts from the dc output of the rectifier and reduces the voltage applied to the armature. As a result the motor tends to slow down with an increase in load. There is also an I^2R power loss across the rheostat.

Controlling Motor Speed by Controlling Firing Point of an SCR

The limitations that a rheostat imposes on speed control of a motor can be overcome by use of electronic motor control circuits. There are many types of electronic controls. The block diagram for one arrangement

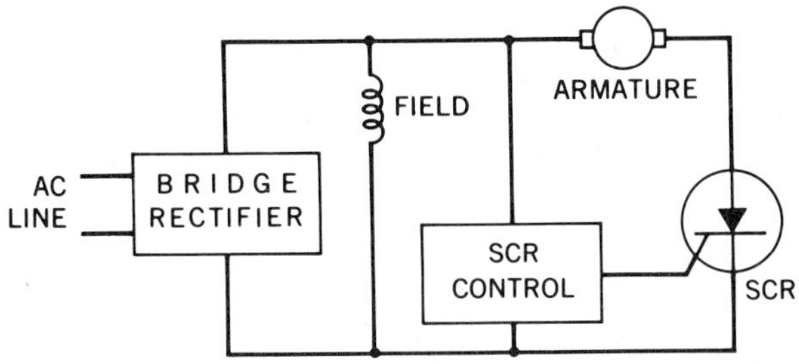

Fig. 35-1. Block diagram of an SCR motor speed control circuit.

is shown in Fig. 35-1. This utilizes an SCR in series with the armature winding. The SCR gate-control circuit (not shown) is used to set the triggering point of the SCR. The output pulsating dc voltage which appears across the armature depends on the point at which the SCR is triggered. Thus the maximum output voltage occurs when there is no delay in triggering time. The output voltage decreases as the triggering delay increases. We can thus vary motor speed by varying the triggering time of the SCR. With this arrangement, an increase in load does not appreciably affect the dc voltage supplied to the armature by the SCR. We will find that the control circuit may be so designed that we obtain not only motor speed control, but also automatic speed regulation. That is, the speed of the motor may be made relatively independent of load changes and variations in line voltage, within a specified range. This is accomplished by feeding back speed information to the SCR gate-control circuit.

Figure 35-2 shows solid-state control for a shunt-wound dc motor. A full-wave bridge rectifier supplies pulsating dc voltage to the motor field, which is connected across the rectifier output. Armature voltage is also derived from the rectifier, but the level of armature voltage is controlled by the SCR, whose anode-cathode circuit is connected in series with the armature winding. The gate-control circuit determines the various points in each half-cycle when the SCR is turned ON, thus fixing the level of armature voltage and hence motor speed. The SCR turns OFF at the end of each half-cycle. Rectifier diode D_3, in parallel with the armature winding, provides a discharge path for the energy stored in the inductance of the armature winding when the SCR turns OFF, at the end of each cycle. Without D_3, the inductive "kick" would develop a high voltage across the armature, which would keep the SCR conducting at the end of each half-cycle, thus defeating the effect of the gate-control (also speed-control) circuit.

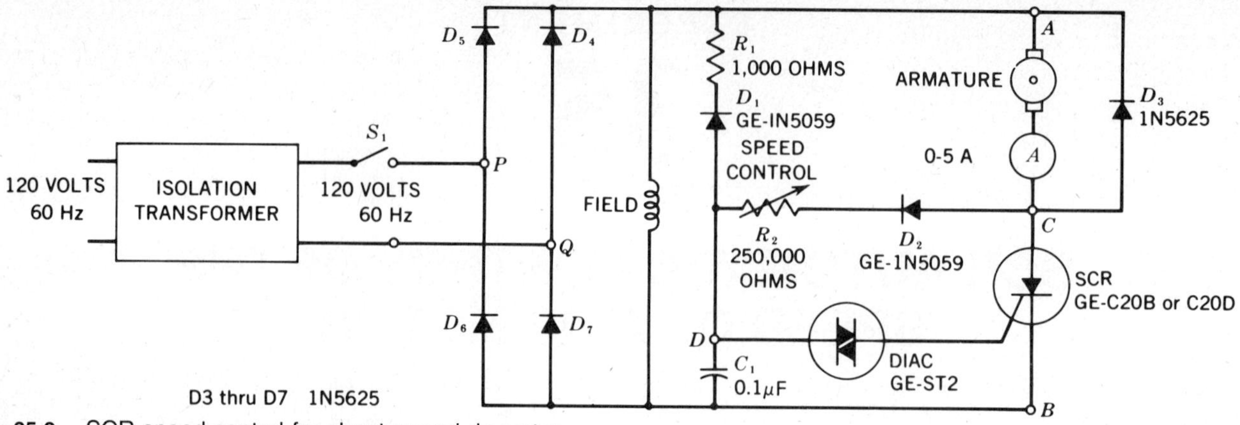

Fig. 35-2. SCR speed control for shunt-wound dc motor.

SCR Gate-Speed-Control Circuit

In earlier experiments, phase-shift control and unijunction transistor (UJT) timing circuits were used to trigger the gate and turn the SCR ON. In this experiment, a diac is used in conjunction with a capacitor C_1 charge-discharge circuit to trigger the gate in the circuit of Fig. 35-2, in a manner similar to the charge-discharge triggering circuit of the UJT, studied in an earlier experiment. A diac is a two-terminal, three-layer, bidirectional avalanche diode, which can be switched from OFF to ON by the proper level of positive or negative voltage across it (see Experiment 6). The diac conducts when the voltage across it reaches V_{BR}, the breakover level. The way in which this is achieved is as follows: The full-wave bridge rectifier delivers two positive-going pulsating dc half-cycles of voltage for each cycle of input ac line voltage. At the end of each half-cycle, capacitor C_1 is completely discharged. C_1 begins to charge at the start of each half-cycle, through the resistance R_2 (speed control), diode D_2, and the armature winding. When C_1 has charged to the forward breakover voltage of the ST2 diac (about 32 V), a trigger pulse is delivered by the diac to the SCR gate, thus turning ON the SCR and completing the armature circuit. The pulsating voltage is then applied to the armature for the remainder of the half-cycle. The charge time for C_1 determines the phase angle at which the SCR is turned ON.

The SCR is turned OFF practically at the end of each half-cycle, when SCR anode current drops below the holding level. It is at this point that the magnetic field about the armature winding collapses and is discharged by diode D_3.

At the end of the half-cycle, capacitor C_1 discharges through the field winding, current-limiting resistor R_1, and diode D_1. The discharge time constant for C_1 must be made short enough to assure that C_1 is properly reset at the start of the next half-cycle. However, the time constant cannot be so short that the discharge current becomes greater than the current of the field winding. Proper balance is achieved by using a 1000-Ω resistor R_1 in the circuit in Fig. 35-2.

The length of time that it takes C_1 to charge to the forward breakover voltage of the diac is controlled by the resistance of R_2. The lower the value of R_2, that is, the shorter the charging time constant, the faster is the charge time, thus permitting the SCR to be triggered early during the half-cycle. The power delivered to the armature lasts during the SCR conduction interval. Therefore, the longer the power interval, the faster the motor revolves. A low R_2 resistance, therefore, increases motor speed; a high R_2 resistance decreases it. Accordingly R_2 is labeled the speed control.

Speed Regulation with Changes in Load

The time required for C_1 to charge to the diode breakover voltage depends not only on the resistance of R_2, but also on the level of the voltage across the SCR, to which C_1 is charging. The voltage across the SCR is the difference between the output voltage of the bridge rectifier and the counter emf of the armature. Since counter emf depends on motor speed, the voltage across the SCR is governed partially by motor speed. Therefore, the charging rate of C_1 is governed by counter emf and hence by motor speed. At low motor speeds, when counter emf is low, voltage across the SCR is high and C_1 can charge to the forward breakover voltage of the diac more rapidly. At higher motor speeds, counter emf is higher, voltage across the SCR lower, and C_1 charging takes longer to reach the forward breakover voltage.

Motor speed, as reflected in counter emf, therefore provides feedback which is sensed by C_1. This feedback is the basis for the ASC characteristic of the circuit, thus maintaining the motor at a relatively constant speed, despite changes in load. To illustrate how this is accomplished, let us assume that R_2 is set to operate the motor at a specified speed, say, 1500 rpm for a particular load. Assume now that the load is increased, tending to slow down the motor to 1400 rpm. The reduced speed results in a lower counter emf and accordingly in a higher voltage across the SCR. Hence C_1 charges more rapidly and turns the SCR ON somewhat earlier than before the increase in load. Increased SCR conduction time increases power delivered to the armature, speeding up the motor to 1500 rpm. Similarly, a reduction in load, which would tend to speed up the motor, results in feedback in such a direction as to reduce motor speed to its initial level. In this manner, automatic regulation of speed is accomplished for load changes.

In addition to the regulation which results from feedback from motor speed, regulation is also dependent on feedback from armature current. As we have noted before, the energy stored in the inductance of the armature winding causes current to flow in rectifier D_3 at the start of each half-cycle. During this conduction interval, the rectifier acts as a short circuit across the armature winding and no counter emf can appear across the armature. As a result, the voltage across the SCR to which C_1 is charging during this time is the full voltage delivered by the bridge rectifier. How long it will take the inductive energy in the armature to dissipate, and hence for the current through D_3 to die out, depends both on armature current and motor speed. D_3 remains conducting for a longer interval at the start of each half-cycle for higher armature currents and lower motor speeds. Capacitor C_1 can therefore charge faster because it is charging longer toward a higher voltage, thus regulating motor speed from feedback resulting from *both* armature current *and* motor speed.

Speed Regulation with Changes in Supply Voltage

The circuit in Fig. 35-2 provides partial compensation for the changes in supply voltage which would otherwise directly affect motor speed. Thus an increase in supply voltage affects both the field and armature circuit. Consider first the armature circuit. Since C_1 charges to the voltage across the SCR, which is the voltage difference between the bridge-rectifier output and the counter emf across the armature, an increase in line voltage will result in a higher charging voltage and the motor will tend to speed up. This tendency is partially counteracted by the fact that the increased voltage appears also across the field winding, thus increasing field current. An increase in field current will tend to slow down motor speed. Therefore, there is a balancing between armature and field circuits.

When R_2 is set for maximum speed, that is, when $R_2 = 0\ \Omega$ the time delay introduced by the charging of C_1 disappears, because the C_1 charging time constant is now very short. Accordingly, full power is applied to the armature because the feedback sensing circuit, which helped provide speed regulation, is now eliminated.

SUMMARY

1. Control of the speed of a dc shunt motor may be achieved by connecting an SCR in series with the armature of the motor and varying the conduction angle of the SCR.
2. Proper design of the series SCR system assures control of motor speed *and* makes the speed of the motor relatively independent (within limits) of load changes and changes in line voltage.
3. The SCR gate supply circuit in Fig. 35-2 can be adjusted to set the motor speed by increasing or decreasing the resistance of R_2. Varying R_2 changes the phase of the sine wave applied to the diac which triggers the SCR.
4. In the SCR speed control circuit in Fig. 35-2, automatic feedback to maintain a relatively constant motor speed results from the counter emf developed across the armature winding and from the level of current in the armature.
5. Feedback in an ASC circuit is of such a nature that it tends to counteract either an increase or a decrease in motor speed, thus maintaining the speed at a relatively constant level.

SELF-TEST

Check your understanding by answering these questions.

1. In the ASC circuit in Fig. 35-2, the anode-cathode circuit of the SCR is connected in series with the armature winding of the motor. __________ (true/false)
2. In an electronic ASC circuit, automatic control of motor speed requires __________ from the motor to the speed-controlling circuit.
3. The purpose of D_3 in Fig. 35-2 is to discharge the inductance of the armature winding when the SCR turns __________ (ON/OFF).
4. Increasing the resistance of R_2 in Fig. 35-2 __________ the speed of the motor.
5. The speed of the motor in Fig. 35-2 is __________ proportional to the length of time that the SCR is conducting.
6. In Fig. 35-2 the voltage applied to the armature is (*a*) pure dc; (*b*) half-wave pulsating dc; (*c*) full-wave pulsating dc. __________
7. In the circuit in Fig. 35-2, the field winding receives a __________ voltage than the armature winding.
8. In the circuit in Fig. 35-2, if C_1 were short-circuited, it would still be possible to set motor speed by set-

ting the resistance of R_2, but automatic control of motor speed would be lost.
__________ (true/false)

MATERIALS REQUIRED

- Power supply: 120-V 60-Hz source; variable isolation transformer (RCA isotap or the equivalent)
- Equipment: Oscilloscope; EVM or VOM; 0- to 5-A ammeter; tachometer capable of showing speed changes up to 25 rpm in range 50 rpm to 5000 rpm
- Motor: Same as in Experiment 35 with device for loading motor
- Resistor: ½-W 1000-Ω
- Capacitor: 0.1-μF 200-V
- Semiconductors: Rectifier diodes, five 1N5625, two 1N5059 or the equivalent; SCR, GE C20B, C20D or C220D, or the equivalent; diac, GE-ST2 or the equivalent
- Miscellaneous: SPST switch (2-A); 250,000-Ω 2-W potentiometer

PROCEDURE

CAUTION: *Armature current must not exceed 1.5 A.*

No-Load Characteristics

1. Connect the circuit shown in Fig. 35-2. D_1 and D_2 are 1N5059 diodes. D_3 and D_5 through D_7 are 1N5625 rectifier diodes. Set the output of the variable isolation transformer at 120 V. There is no load on the motor.
2. Close S_1. Power ON. Adjust speed control R_2 for maximum motor speed. Measure and record in Table 35-1 the dc voltage across the shunt field *AB* and across the armature. Also measure and record the armature current and motor speed.
3. Set the output of the isolation transformer at 110 V rms. Measure and record motor speed.
   Also measure and record motor speed with the output of the isolation transformer set for 130 V rms.
4. Set R_2 for minimum motor speed and repeat the measurements in steps 2 and 3.
5. Set the output of the isolation transformer at 120 V rms. Adjust the speed control until the motor reaches its rated speed, 1725 rpm.
6. Connect the vertical input leads of an oscilloscope set for dc input across the line input to the bridge rectifier, hot lead to *P*, ground lead to *Q*. Set sync selector to line sync. Adjust the controls of the oscilloscope until the reference waveform shown in Table 35-2 is stationary on the screen, and center the waveform.
7. Observe, measure, and record in Table 35-2, in proper time phase with the reference waveform, two half-cycles of the output waveform of the bridge rectifier, points *A* to *B*. Also observe, measure, and record, in proper time phase with the reference, the waveform across the armature, points *A* to *C*, across the SCR, points *C* to *B*, and the charge waveform across capacitor C_1, points *D* to *B*.

TABLE 35-1. Voltage, Armature Current, and Motor Speed—Motor Has No Load

Step	*Condition*	*DC, V*		*DC Armature Current,* A	*Motor Speed,* rpm *at Indicated Line Voltage*		
		Field	*Armature*		*110* V	*120* V	*130* V
2, 3	R_2 set for maximum motor speed, 120-V line						
4	R_2 set for minimum motor speed, 120-V line						

TABLE 35-2. Electronic Motor Control Waveforms

Step	*Motor Speed*	*Test Points*		*Waveform*	*Voltage,* p-p
6, 7	1725 rpm	*P* to *Q*		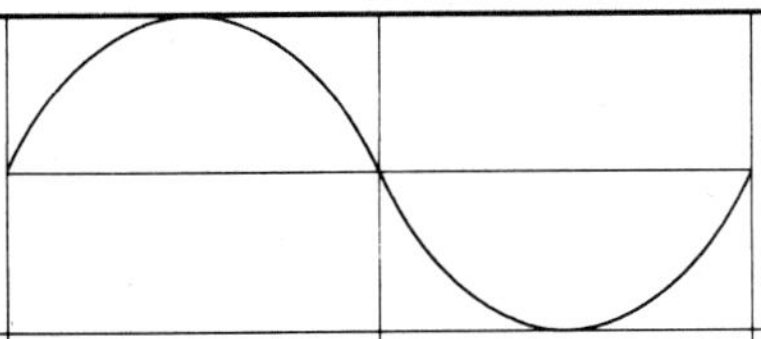	
		A to *B*			
		A to *C*			
		C to *B*			
		D to *B*			
8	1200 rpm	*A* to *C*			
		C to *B*			
		D to *B*			

8. Reset R_2, reducing the motor speed to 1200 rpm. Observe, measure, and record in Table 35-2, in proper time phase with the reference, the voltage across the armature, across the SCR, and across C_1.

Load Characteristics

9. Isolation transformer output is still 120 V. Set no-load motor speed at 1000 rpm and adjust the motor load until the armature current measures 1.5 A (no higher). Measure and record in Table 35-3 motor speed for this load condition.

10. Set the no-load motor speed at 1725 rpm. Adjust the motor load for 1.5 A of armature current. Measure and record in Table 35-3 motor speed for this load condition.

TABLE 35-3. Electronic Motor Control (Fig. 35-2) Load Characteristics

Step	*No-Load Speed Setting of R_2*, rpm	*Armature Current under Load*, A	*Motor Speed under Load*, rpm
9	1000	1.5	
10	1725	1.5	

QUESTIONS

1. Compare the load speed control characteristic of the circuit in Fig. 35-2 with that of the control circuit in Exp. 34. Refer to your data in Table 35-3 and to the equivalent data in Exp. 34.
2. Explain how the SCR in Fig. 35-2 is triggered.
3. In Fig. 35-2 how does the setting of R_2 affect motor speed? Refer specifically to the waveforms in Table 35-2 to illustrate your answer.
4. Is no-load motor speed affected by line voltage variations in Fig. 35-2?
5. What specific characteristic of the diac was the determining factor in its use in Fig. 35-2?
6. (a) How does armature voltage vary with motor speed?
   (b) How is this variation used to achieve electronic speed regulation in this experiment?
7. Would a large electrolytic filter capacitor, for example 1000 μF, connected across the output of the bridge rectifier, have helped or hindered the regulation capabilities of the system in Fig. 35-2? Explain why.
8. What is the advantage of using a bridge rectifier over a half-wave rectifier in the electronic motor control system used in this experiment? Explain.

Answers to Self-Test

1. true
2. feedback
3. OFF
4. reduces
5. directly
6. c
7. higher
8. false

EXPERIMENT

36

PROGRAMMABLE CONTROLLERS

OBJECTIVES

1. To study the programmable controller and learn why it is useful
2. To learn experimentally how a programmable controller is programmed and demonstrate its use

BASIC INFORMATION

In previous experiments we have learned about relays, time delays, logic, microprocessors, motor controls, and other devices which are commonly used in industrial electronics. The programmable controller is a device which is increasingly being used in industry to perform the operations previously accomplished with those discrete devices. The programmable controller is extremely flexible and may be modified quickly to perform a new or different machine-control task.

The programmable controller was introduced to replace electromechanical relays in sequential control systems, in which one event follows another in a prescribed way to complete a task.

An industrial machine or process may include pushbuttons, limit switches, timers, interlocks, etc., to control motors, valves, solenoids, and other output devices. In the past, these items have been hard-wired together into a "dedicated" machine which could not easily be modified to perform different tasks or accommodate a change in the process. With a programmable controller, however, all that is required to delete or change a timed interval, for example, is to change the controller's program.

The Personal Computer versus the Programmable Controller

A program could be written to use a personal computer as a programmable controller, but a controller cannot be used as a general-purpose computer. What are the similarities and differences between them?

The controller is electrically and physically designed for use in a harsh industrial environment involving extremes of temperature, corrosive vapors, dirt, shock, and vibration. Although it is flexible, it is intended for a specific application. The personal computer is intended for use in a clean, temperature-controlled environment by skilled operators. Its applications are limited only by the capabilities of the programs written for it and by the types of devices with which it is interfaced.

The personal computer may be programmed in a variety of languages: Fortran, Cobol, Basic, Pascal, etc. The controller is designed to be programmed with relay, schematic, or ladder diagrams in mind. An electrician who is familiar with relay and logic diagrams can be trained in a short time to program a controller, but programming a personal computer requires more skill and training.

Programmable Controller System Components

The basic elements of the programmable controller are the processor, which is a special-purpose computer; I/O modules; a power supply; and a programming panel (Fig. 36-1). The processor and its associated memory are loaded with a program which accepts switch closures as inputs, performs logical operations on these inputs, and causes appropriate loads to be energized at the proper time and under the prescribed conditions.

The operator utilizes a keyboard to enter the logical relations given in the ladder diagram for the control circuits of the machine. A liquid crystal display (LCD) or cathode-ray tube (CRT) displays the symbols for the relay coils, contacts, switches, and so forth which the programmer has entered by the keyboard. The starting place, then, for programming the controller is preparation of a ladder diagram using standard symbols as if the control circuits were going to be hard-wired. A typical main programming terminal and a hand-held terminal for program modification are shown in Fig. 36-2.

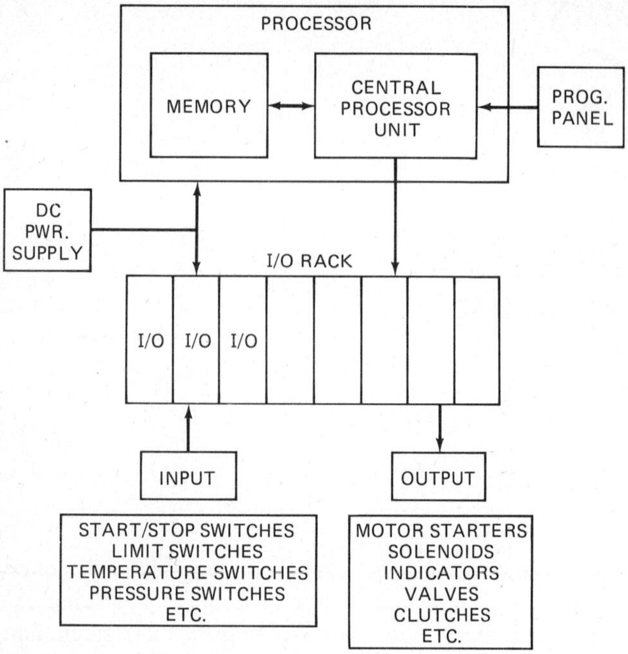

Fig. 36-1. Block diagram, programmable controller.

relationships between them, and causes specified outputs (switch closures) to occur in the proper sequence and at the required time.

Programming the Controller

The main program may be entered from a personal computer which uses a special program (software) and a special interface card (hardware) to place instructions into the memory of the controller. It is more common, however, to utilize a specially designed industrial terminal for this purpose. Once the main program is entered into the controller, modifications to the program may be made on the factory floor by an operator using a small hand-held unit.

Developing the Program

Let us assume that an industrial process involves filling three containers with a fluid, one after the other, after a START button is pressed. When the last container is full, the pump causing the fluid to flow should stop. Pressing the STOP button at any time should interrupt the process.

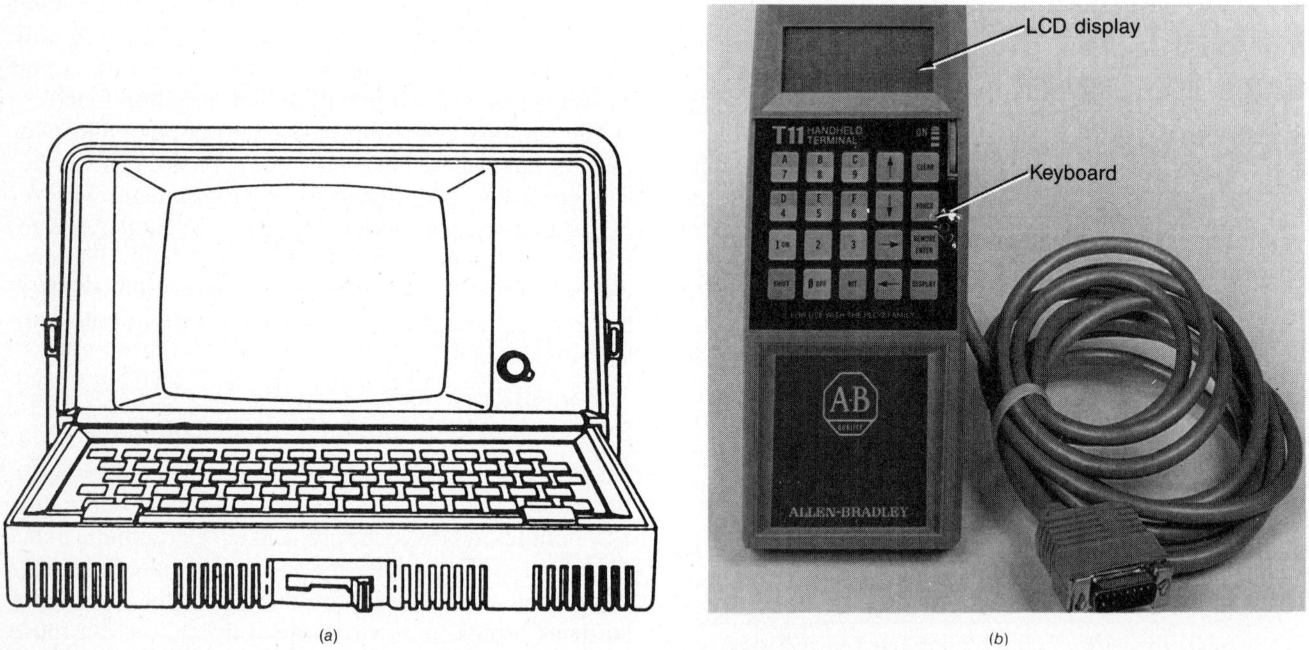

Fig. 36-2. (*a*) Programming terminal; (*b*) Hand held terminal (*Allen-Bradley Company*)

The input circuits of the I/O modules often use optoisolators to prevent spikes and other noise in the industrial environment from entering or damaging the low-voltage, solid-state components within the processor. Similarly, the output module may employ an optically isolated solid-state relay where the internal TTL signals of the controller control a triac serving as the "contacts" of an output relay. These I/O modules are mounted in a rack as in Fig. 36-3.

The program in the processor causes all of the input signals to be scanned continuously, solves the logical

This verbal description is amplified pictorially in Fig. 36-4*a*, and the corresponding electrical ladder diagram is shown in Fig. 36-4*b*. The level of the fluid in each container is sensed by a float switch. Fluid flow into each container is controlled by a solenoid valve.

When the START switch is briefly depressed, motor control relay M_1 pulls in and latches in the ON condition through a set of its own contacts. Another set of M_1 contacts starts the pump motor.

Assuming container 3 is not full, switch S_3 is closed, solenoid 3 causes the valve for container 3 to operate,

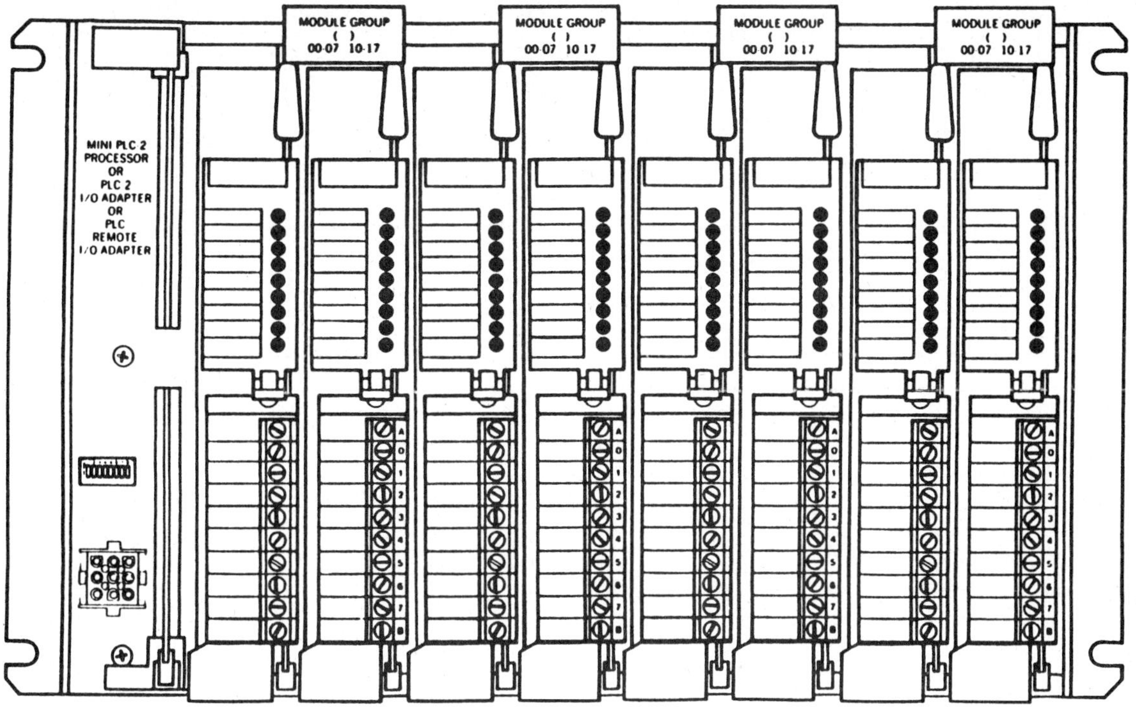

Fig. 36-3. I/O rack. (*Allen-Bradley Company*)

and the filling process begins. When container 3 is full, float switch S_3 opens. A set of normally open contacts on S_3 closes, causing valve 2 to operate and filling of container 2 to begin. When it is full, float switch S_2, closed when the level is low, opens, and filling of container 2 stops. A set of S_2 contacts, which are open until the high level is reached, closes, and filling of container 1 begins.

When container 1 is filled, its float switch S_1 breaks the path to solenoid 1 and filling stops. Another set of S_1 contacts interrupts the motor control circuit and the pump stops. If the STOP switch is pressed at any time, the process is stopped.

Connecting the I/O Devices to the Controller

Figure 36-5 shows how the input and output devices for the filling system are connected to the controller. Each of the input switches completes a path between L_1 and L_2 of the power source through the input mod-

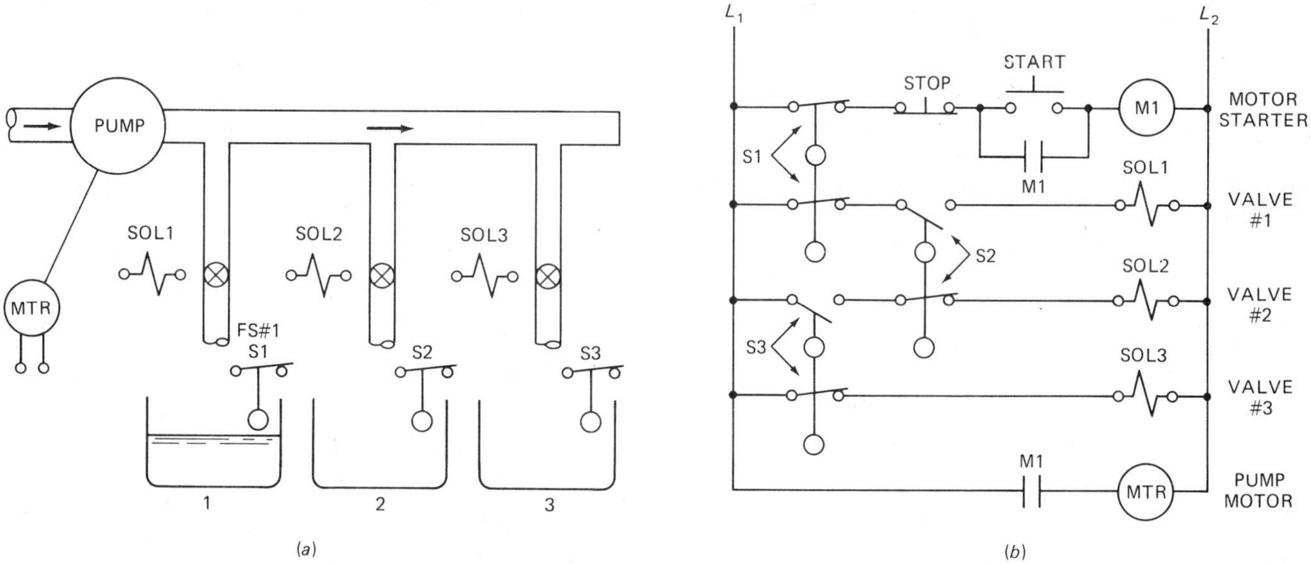

Fig. 36-4. (*a*) Pictorial diagram of filling system; (*b*) Ladder diagram of filling system.

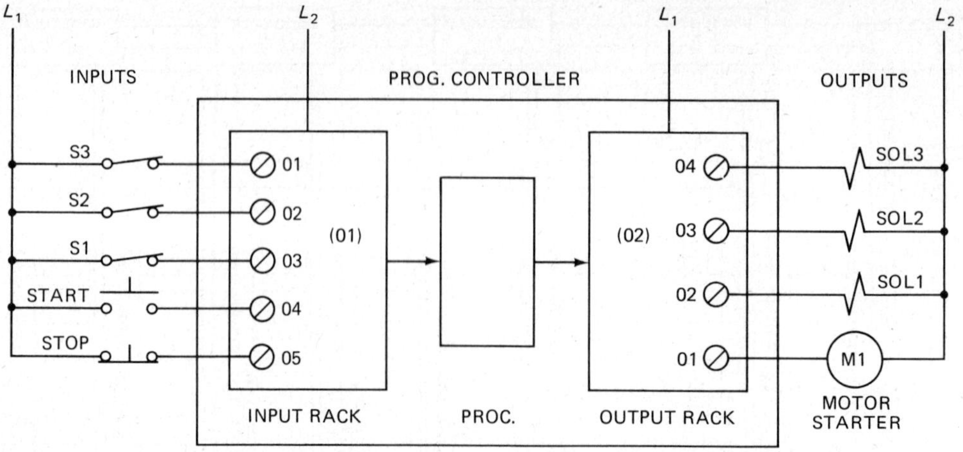

Fig. 36-5. Input output connection to the controller.

ule, thus providing an input to the CPU. An output device is connected across L_1 and L_2 when the appropriate output module is energized in accordance with the program stored in the memory of the CPU.

In the system discussed here, a four-digit number identifies each input or output by the rack number and terminal number of the module. For example, a normally closed set of contacts on S_3 connected to input module 01, terminal 01, would have the address 0101. Solenoid valve SOL_3 connected to rack 02, terminal 04, would have the address 0204. The ladder diagram may be revised to show these address assignments, and the resulting relationships may be entered through a hand-held programming keyboard.

The hand-held programming keyboard includes keys corresponding to the type of symbol to be entered from the ladder diagram, and a display which shows the operator reference information, addresses, and symbols for the devices entered in the program.

SUMMARY

1. A programmable controller uses computer technology to replace discrete components such as timers and relays in the control of machines.
2. Whereas a machine or process controller using discrete components is "dedicated" to that machine and is difficult to modify, the controller easily can be reprogrammed to change a process or the sequence of events. A controller can be removed from one machine and used to control an entirely different machine or process.
3. Programmable controllers are programmed in terms of the ladder diagrams which are used to describe the electrical connections and operation of industrial machines.

SELF-TEST

Check your understanding by answering these questions.

1. An electrician must know Basic programming to program a programmable controller. __________ (true/false)
2. The starting point for programming a programmable controller is the __________ diagram which shows the connections between inputs, such as switches, and outputs, such as motors, solenoids, heaters, etc.
3. A machine controller made up of relays, timers, and other discrete parts is called a __________ controller.
4. A machine or process controlled by a programmable controller may be readily modified to incorporate new steps or to change the sequence of events. __________ (true/false)
5. A programmable controller can be programmed by a hand-held unit, a larger industrial terminal, or a __________ __________ with appropriate software and hardware.

MATERIALS REQUIRED

- Equipment: Allen-Bradley SLC-100 Demonstrator, catalog no. 1745-DEMO1; includes wired SLC-100, pocket programmer, EEPROM, power cord; user manual, publication no. 1745-800; self-teach manual, publication no. 1745-800A; SLC-100 demonstrator exercises, publication no. 1745-802
- Programming the SLC-100 can also be performed by using an IBM-compatible personal computer and Allen-Bradley Personal Computer Interface Kit, catalog no. 1745-PCK
- Miscellaneous switches, lamps, motors, relays, etc., as required

PROCEDURE

CAUTION: In the following procedures, hazardous voltage is present. Observe safety precautions. Have the work checked by your instructor before applying power.

1. Complete the demonstration exercises given in Allen-Bradley publication no. 1745-802. This will familiarize you with the SLC-100, its use, capabilities, and programming method.

2. One of the most common industrial control circuits is control of a motor by momentarily depressing a START or STOP button, as previously studied. Set up the circuit in Fig. 36-6 with a control relay and push-button switches to demonstrate its operation. Notice that momentarily pressing the START button causes the motor to operate and momentarily pressing the STOP button causes it to stop.

   Additional push-buttons may be connected in parallel with the START button so that the motor can be started from several locations. Overload, interlock, and emergency stop buttons may be connected in series with the STOP button so that the motor can be automatically stopped when an overload occurs, when safety interlock switches are open, or when an emergency occurs.

3. The motor may be controlled in a similar way by the SLC-100. Inputs from the START and STOP buttons are connected to the input strip, and the motor is controlled by a set of contacts on the output strip of the SLC-100.

NOTE: The following procedures require that connections to the input and output terminals of the SLC-100 be temporarily removed so that it can be used with external devices (switches and motor).

4. With power disconnected from the SLC-100, carefully remove and insulate all of the numbered wires connected to the input and output strips.

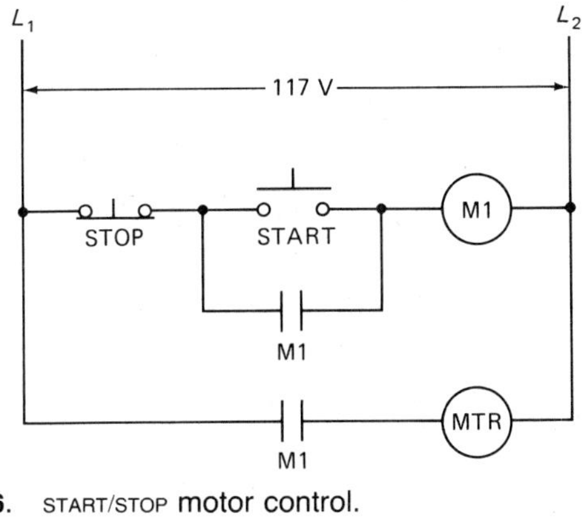

Fig. 36-6. START/STOP motor control.

NOTE: The outputs of the SLC-100 are relay contacts and are isolated from each other. The contacts are rated to switch 120-V ac. Do not use the outputs to switch inductive loads unless proper transient suppression across the load is provided. For long life, do not use the SLC-100 to control loads greater than 1 A continuously. Short-term peak current on "make" should not exceed 15 A.

5. Figure 36-7*a* shows the interconnections between the SLC-100 and the other components. Connect the external START and STOP buttons to the input terminals defined in the program of step 5 (below), and control the motor by the terminals on the output strip defined in the program. To keep within the current rating of the output relay, use a small 117-V ac fan motor (or other small motor) rated at 1 A or less.

6. Using the programming methods supplied with the programmable controller, enter a program based on the elements of the ladder diagram for the START/STOP motor controller (Fig. 36-6).

7. With power ON, observe that the operation of the programmable controller circuit is the same as that with the control relay: pressing the START button causes the motor to start, and pressing the STOP button causes it to stop.

For additional programming experience, use the ladder diagram of Fig. 36-4*b* as a basis to write and enter a program which will simulate the operation of the container filling machine previously discussed.

For S_1, S_2, and S_3, use DPDT switches. Simulate the motor and solenoids with low-wattage light bulbs. The sequence of events will be demonstrated by first pressing the START button and then operating S_3, S_2, and S_1 in turn. All switches should initially be in the positions shown in Fig. 36-4*b*.

Figure 36-7b is a suggested diagram for the interconnections between the SLC-100 and the external switches, indicators and control relay.

For a permanent accessory mount the switches, etc., in a small box or on a panel with appropriate labels together with a copy of the ladder diagram. The power cord to the accessory should be fused.

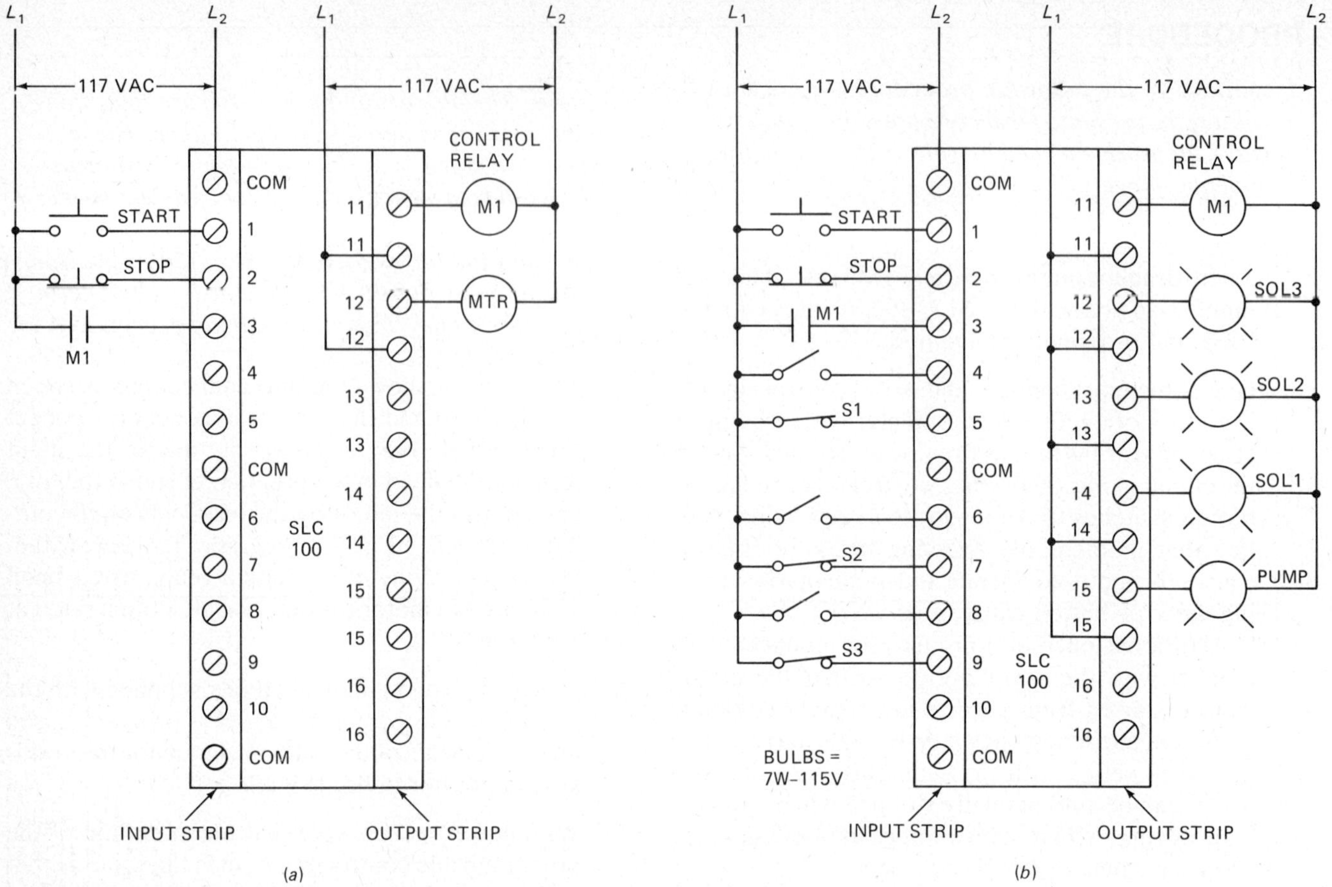

Fig. 36-7. (*a*) Interconnections between the SLC-100 and other components; (*b*) Interconnections between the SLC-100 and external switches, indicators, and relays.

QUESTIONS

1. Does an electrician have to be a "computer programmer" in order to program a programmable controller?
2. What types of diagrams should a person be able to interpret if he or she wishes to use a programmable controller effectively?
3. What is the difference between a "dedicated" machine controller and a programmable controller?
4. Should a programmable controller always be preferred over a machine controller using discrete parts? What is probably the deciding factor?
5. Why should an industrial technician or electrician be familiar with programmable controllers?
6. Why must a programmable controller be more rugged than a personal computer?

Answers to Self-Test

1. false
2. ladder
3. dedicated
4. true
5. personal computer

APPENDIX A

6800 INSTRUCTION SET

ACCUMULATOR AND MEMORY		ADDRESSING MODES															BOOLEAN/ARITHMETIC OPERATION (All register labels refer to contents)	COND. CODE REG.					
		IMMED			DIRECT			INDEX			EXTND			INHER				5	4	3	2	1	0
OPERATIONS	MNEMONIC	OP	~	#	OP	~	#	OP	~	#	OP	~	#	OP	~	#		H	I	N	Z	V	C
Add	ADDA	8B	2	2	9B	3	2	AB	5	2	BB	4	3				A + M → A	↕	●	↕	↕	↕	↕
	ADDB	CB	2	2	DB	3	2	EB	5	2	FB	4	3				B + M → B	↕	●	↕	↕	↕	↕
Add Acmltrs	ABA													1B	2	1	A + B → A	↕	●	↕	↕	↕	↕
Add with Carry	ADCA	89	2	2	99	3	2	A9	5	2	B9	4	3				A + M + C → A	↕	●	↕	↕	↕	↕
	ADCB	C9	2	2	D9	3	2	E9	5	2	F9	4	3				B + M + C → B	↕	●	↕	↕	↕	↕
And	ANDA	84	2	2	94	3	2	A4	5	2	B4	4	3				A • M → A	●	●	↕	↕	R	●
	ANDB	C4	2	2	D4	3	2	E4	5	2	F4	4	3				B • M → B	●	●	↕	↕	R	●
Bit Test	BITA	85	2	2	95	3	2	A5	5	2	B5	4	3				A • M	●	●	↕	↕	R	●
	BITB	C5	2	2	D5	3	2	E5	5	2	F5	4	3				B • M	●	●	↕	↕	R	●
Clear	CLR							6F	7	2	7F	6	3				00 → M	●	●	R	S	R	R
	CLRA													4F	2	1	00 → A	●	●	R	S	R	R
	CLRB													5F	2	1	00 → B	●	●	R	S	R	R
Compare	CMPA	81	2	2	91	3	2	A1	5	2	B1	4	3				A − M	●	●	↕	↕	↕	↕
	CMPB	C1	2	2	D1	3	2	E1	5	2	F1	4	3				B − M	●	●	↕	↕	↕	↕
Compare Acmltrs	CBA													11	2	1	A − B	●	●	↕	↕	↕	↕
Complement, 1's	COM							63	7	2	73	6	3				$\bar{M} \rightarrow M$	●	●	↕	↕	R	S
	COMA													43	2	1	$\bar{A} \rightarrow A$	●	●	↕	↕	R	S
	COMB													53	2	1	$\bar{B} \rightarrow B$	●	●	↕	↕	R	S
Complement, 2's (Negate)	NEG							60	7	2	70	6	3				00 − M → M	●	●	↕	↕	①	②
	NEGA													40	2	1	00 − A → A	●	●	↕	↕	①	②
	NEGB													50	2	1	00 − B → B	●	●	↕	↕	①	②
Decimal Adjust, A	DAA													19	2	1	Converts Binary Add. of BCD Characters into BCD Format	●	●	↕	↕	↕	③
Decrement	DEC							6A	7	2	7A	6	3				M − 1 → M	●	●	↕	↕	④	●
	DECA													4A	2	1	A − 1 → A	●	●	↕	↕	④	●
	DECB													5A	2	1	B − 1 → B	●	●	↕	↕	④	●
Exclusive OR	EORA	88	2	2	98	3	2	A8	5	2	B8	4	3				A ⊕ M → A	●	●	↕	↕	R	●
	EORB	C8	2	2	D8	3	2	E8	5	2	F8	4	3				B ⊕ M → B	●	●	↕	↕	R	●
Increment	INC							6C	7	2	7C	6	3				M + 1 → M	●	●	↕	↕	⑤	●
	INCA													4C	2	1	A + 1 → A	●	●	↕	↕	⑤	●
	INCB													5C	2	1	B + 1 → B	●	●	↕	↕	⑤	●
Load Acmltr	LDAA	86	2	2	96	3	2	A6	5	2	B6	4	3				M → A	●	●	↕	↕	R	●
	LDAB	C6	2	2	D6	3	2	E6	5	2	F6	4	3				M → B	●	●	↕	↕	R	●
Or, Inclusive	ORAA	8A	2	2	9A	3	2	AA	5	2	BA	4	3				A + M → A	●	●	↕	↕	R	●
	ORAB	CA	2	2	DA	3	2	EA	5	2	FA	4	3				B + M → B	●	●	↕	↕	R	●
Push Data	PSHA													36	4	1	A → M_{SP}, SP − 1 → SP	●	●	●	●	●	●
	PSHB													37	4	1	B → M_{SP}, SP − 1 → SP	●	●	●	●	●	●
Pull Data	PULA													32	4	1	SP + 1 → SP, M_{SP} → A	●	●	●	●	●	●
	PULB													33	4	1	SP + 1 → SP, M_{SP} → B	●	●	●	●	●	●
Rotate Left	ROL							69	7	2	79	6	3				M	●	●	↕	↕	⑥	↕
	ROLA													49	2	1	A 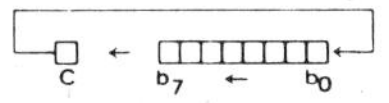	●	●	↕	↕	⑥	↕
	ROLB													59	2	1	B	●	●	↕	↕	⑥	↕
Rotate Right	ROR							66	7	2	76	6	3				M	●	●	↕	↕	⓪	↕
	RORA													46	2	1	A 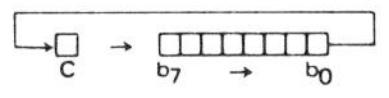	●	●	↕	↕	⑥	↕
	RORB													56	2	1	B	●	●	↕	↕	⑥	↕
Shift Left, Arithmetic	ASL							68	7	2	78	6	3				M	●	●	↕	↕	⑥	↕
	ASLA													48	2	1	A 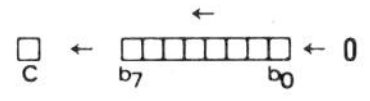	●	●	↕	↕	⑥	↕
	ASLB													58	2	1	B	●	●	↕	↕	⑥	↕
Shift Right, Arithmetic	ASR							67	7	2	77	6	3				M	●	●	↕	↕	⑥	↕
	ASRA													47	2	1	A 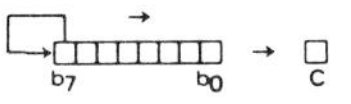	●	●	↕	↕	⑥	↕
	ASRB													57	2	1	B	●	●	↕	↕	⑥	↕
Shift Right, Logic.	LSR							64	7	2	74	6	3				M	●	●	R	↕	⑥	↕
	LSRA													44	2	1	A 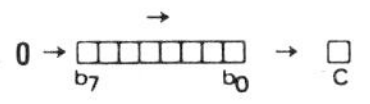	●	●	R	↕	⑥	↕
	LSRB													54	2	1	B	●	●	R	↕	⑥	↕
Store Acmltr.	STAA				97	4	2	A7	6	2	B7	5	3				A → M	●	●	↕	↕	R	●
	STAB				D7	4	2	E7	6	2	F7	5	3				B → M	●	●	↕	↕	R	●
Subtract	SUBA	80	2	2	90	3	2	A0	5	2	B0	4	3				A − M → A	●	●	↕	↕	↕	↕
	SUBB	C0	2	2	D0	3	2	E0	5	2	F0	4	3				B − M → B	●	●	↕	↕	↕	↕
Subract Acmltrs.	SBA													10	2	1	A − B → A	●	●	↕	↕	↕	↕
Subtr. with Carry	SBCA	82	2	2	92	3	2	A2	5	2	B2	4	3				A − M − C → A	●	●	↕	↕	↕	↕
	SBCB	C2	2	2	D2	3	2	E2	5	2	F2	4	3				B − M − C → B	●	●	↕	↕	↕	↕
Transfer Acmltrs	TAB													16	2	1	A → B	●	●	↕	↕	R	●
	TBA													17	2	1	B → A	●	●	↕	↕	R	●
Test, Zero or Minus	TST							6D	7	2	7D	6	3				M − 00	●	●	↕	↕	R	R
	TSTA													4D	2	1	A − 00	●	●	↕	↕	R	R
	TSTB													5D	2	1	B − 00	●	●	↕	↕	R	R

INDEX REGISTER AND STACK POINTER OPERATIONS	MNEMONIC	IMMED OP	IMMED ~	IMMED #	DIRECT OP	DIRECT ~	DIRECT #	INDEX OP	INDEX ~	INDEX #	EXTND OP	EXTND ~	EXTND #	INHER OP	INHER ~	INHER #	BOOLEAN/ARITHMETIC OPERATION	5 H	4 I	3 N	2 Z	1 V	0 C
Compare Index Reg	CPX	8C	3	3	9C	4	2	AC	6	2	BC	5	3				$(X_H/X_L) - (M/M+1)$	•	•	(7)	↕	(8)	•
Decrement Index Reg	DEX													09	4	1	$X - 1 \rightarrow X$	•	•	•	↕	•	•
Decrement Stack Pntr	DES													34	4	1	$SP - 1 \rightarrow SP$	•	•	•	•	•	•
Increment Index Reg	INX													08	4	1	$X + 1 \rightarrow X$	•	•	•	↕	•	•
Increment Stack Pntr	INS													31	4	1	$SP + 1 \rightarrow SP$	•	•	•	•	•	•
Load Index Reg	LDX	CE	3	3	DE	4	2	EE	6	2	FE	5	3				$M \rightarrow X_H, (M+1) \rightarrow X_L$	•	•	(9)	↕	R	•
Load Stack Pntr	LDS	8E	3	3	9E	4	2	AE	6	2	BE	5	3				$M \rightarrow SP_H, (M+1) \rightarrow SP_L$	•	•	(9)	↕	R	•
Store Index Reg	STX				DF	5	2	EF	7	2	FF	6	3				$X_H \rightarrow M, X_L \rightarrow (M+1)$	•	•	(9)	↕	R	•
Store Stack Pntr	STS				9F	5	2	AF	7	2	BF	6	3				$SP_H \rightarrow M, SP_L \rightarrow (M+1)$	•	•	(9)	↕	R	•
Indx Reg → Stack Pntr	TXS													35	4	1	$X - 1 \rightarrow SP$	•	•	•	•	•	•
Stack Pntr → Indx Reg	TSX													30	4	1	$SP + 1 \rightarrow X$	•	•	•	•	•	•

JUMP AND BRANCH OPERATIONS	MNEMONIC	RELATIVE OP	RELATIVE ~	RELATIVE #	INDEX OP	INDEX ~	INDEX #	EXTND OP	EXTND ~	EXTND #	INHER OP	INHER ~	INHER #	BRANCH TEST	5 H	4 I	3 N	2 Z	1 V	0 C
Branch Always	BRA	20	4	2										None	•	•	•	•	•	•
Branch If Carry Clear	BCC	24	4	2										C = 0	•	•	•	•	•	•
Branch If Carry Set	BCS	25	4	2										C = 1	•	•	•	•	•	•
Branch If = Zero	BEQ	27	4	2										Z = 1	•	•	•	•	•	•
Branch If ⩾ Zero	BGE	2C	4	2										$N \oplus V = 0$	•	•	•	•	•	•
Branch If > Zero	BGT	2E	4	2										$Z + (N \oplus V) = 0$	•	•	•	•	•	•
Branch If Higher	BHI	22	4	2										$C + Z = 0$	•	•	•	•	•	•
Branch If ⩽ Zero	BLE	2F	4	2										$Z + (N \oplus V) = 1$	•	•	•	•	•	•
Branch If Lower Or Same	BLS	23	4	2										$C + Z = 1$	•	•	•	•	•	•
Branch If < Zero	BLT	2D	4	2										$N \oplus V = 1$	•	•	•	•	•	•
Branch If Minus	BMI	2B	4	2										N = 1	•	•	•	•	•	•
Branch If Not Equal Zero	BNE	26	4	2										Z = 0	•	•	•	•	•	•
Branch If Overflow Clear	BVC	28	4	2										V = 0	•	•	•	•	•	•
Branch If Overflow Set	BVS	29	4	2										V = 1	•	•	•	•	•	•
Branch If Plus	BPL	2A	4	2										N = 0	•	•	•	•	•	•
Branch To Subroutine	BSR	8D	8	2										}	•	•	•	•	•	•
Jump	JMP				6E	4	2	7E	3	3				} See Special Operations	•	•	•	•	•	•
Jump To Subroutine	JSR				AD	8	2	BD	9	3				}	•	•	•	•	•	•
No Operation	NOP										01	2	1	Advances Prog. Cntr. Only	•	•	•	•	•	•
Return From Interrupt	RTI										3B	10	1		—	—	(10)	—	—	—
Return From Subroutine	RTS										39	5	1	}	•	•	•	•	•	•
Software Interrupt	SWI										3F	12	1	} See special Operations	•	S	•	•	•	•
Wait for Interrupt	WAI										3E	9	1	}	•	(11)	•	•	•	•

CONDITIONS CODE REGISTER OPERATIONS	MNEMONIC	INHER OP	INHER ~	INHER #	BOOLEAN OPERATION	5 H	4 I	3 N	2 Z	1 V	0 C
Clear Carry	CLC	0C	2	1	$0 \rightarrow C$	•	•	•	•	•	R
Clear Interrupt Mask	CLI	0E	2	1	$0 \rightarrow I$	•	R	•	•	•	•
Clear Overflow	CLV	0A	2	1	$0 \rightarrow V$	•	•	•	•	R	•
Set Carry	SEC	0D	2	1	$1 \rightarrow C$	•	•	•	•	•	S
Set Interrupt Mask	SEI	0F	2	1	$1 \rightarrow I$	•	S	•	•	•	•
Set Overflow	SEV	0B	2	1	$1 \rightarrow V$	•	•	•	•	S	•
Acmltr A → CCR	TAP	06	2	1	A → CCR	—	—	(12)	—	—	—
CCR → Acmltr A	TPA	07	2	1	CCR → A	•	•	•	•	•	•

CONDITION CODE REGISTER NOTES:

(Bit set if test is true and cleared otherwise)

1. (Bit V) Test: Result = 10000000?
2. (Bit C) Test: Result = 00000000?
3. (Bit C) Test: Decimal value of most significant BCD Character greater than nine? (Not cleared if previously set.)
4. (Bit V) Test: Operand = 10000000 prior to execution?
5. (Bit V) Test: Operand = 01111111 prior to execution?
6. (Bit V) Test: Set equal to result of $N \oplus C$ after shift has occurred.
7. (Bit N) Test: Sign bit of most significant (MS) byte of result = 1?
8. (Bit V) Test: 2's complement overflow from subtraction of LS bytes?
9. (Bit N) Test: Result less than zero? (Bit 15 = 1)
10. (All) Load Condition Code Register from Stack. (See Special Operations)
11. (Bit I) Set when interrupt occurs. If previously set, a Non-Maskable Interrupt is required to exit the wait state.
12. (ALL) Set according to the contents of Accumulator A.

LEGEND:

OP	Operation Code (Hexadecimal);	00	Byte = Zero;
~	Number of MPU Cycles;	H	Half-carry from bit 3;
#	Number of Program Bytes;	I	Interrupt mask
+	Arithmetic Plus;	N	Negative (sign bit)
−	Arithmetic Minus;	Z	Zero (byte)
•	Boolean AND;	V	Overflow, 2's complement
M_{SP}	Contents of memory location pointed to be Stack Pointer;	C	Carry from bit 7
+	Boolean Inclusive OR;	R	Reset Always
⊕	Boolean Exclusive OR;	S	Set Always
$\overline{M}$	Complement of M;	↕	Test and set if true, cleared otherwise
→	Transfer Into;	•	Not Affected
0	Bit = Zero;	CCR	Condition Code Register
		LS	Least Significant
		MS	Most Significant

APPENDIX B

PARTS REQUIREMENTS

RESISTORS, 1/2-W

(1)	33 Ω	*(3)*	1 kΩ	*(1)*	12 kΩ
(1)	47 Ω	*(1)*	1.2 kΩ	*(4)*	15 kΩ 5%
(1)	56 Ω	*(1)*	1.5 kΩ	*(1)*	22 kΩ
(1)	68 Ω	*(1)*	1.8 kΩ	*(1)*	33 kΩ
(1)	100 Ω	*(1)*	2.2 kΩ	*(1)*	47 kΩ
(1)	150 Ω	*(1)*	2.7 kΩ	*(2)*	100 kΩ
(1)	220 Ω	*(1)*	3.3 kΩ	*(1)*	120 kΩ
(7)	330 Ω	*(1)*	4.7 kΩ	*(1)*	220 kΩ
(1)	470 Ω	*(2)*	6.8 kΩ	*(1)*	1 MΩ
(3)	560 Ω	*(2)*	10 kΩ	*(1)*	12 MΩ 5%
(2)	820 Ω				

RESISTORS, HIGHER WATTAGE

(1)	5 Ω 5 W	*(2)*	1.2 kΩ 20 W
(1)	33 Ω 2 W	*(1)*	2.5 kΩ 10 W
(1)	50 Ω 5 W	*(1)*	5 kΩ 5 W
(1)	250 Ω 5 W	*(1)*	10 kΩ 10 W
(1)	1 kΩ 1 W		

POTENTIOMETERS

One each, 2 W unless otherwise specified.

1 kΩ	100 kΩ	500 kΩ
10 kΩ 4 W	250 kΩ	1 MΩ
50 kΩ 4 W		

CAPACITORS

(1)	47 pF 400 V	*(1)*	1.0 μF 400 V
(1)	250 pF 400 V	*(1)*	25 μF 100 V
(1)	0.001 μF 400 V	*(2)*	50 μF 50 V
(1)	0.0015 μF 25 V	*(1)*	100 μF 15 V
(1)	0.002 μF 400 V	*(1)*	100 μF 50 V
(2)	0.01 μF 200 V	*(1)*	100 μF tantalum 25 V
(2)	0.02 μF 10 V		
(2)	0.1 μF 400 V		

SEMICONDUCTORS

Diodes

(6) 1N5625 *(2)* 1N5059 *(1)* 1N4154

Zener

(1) 18 V/1 W

Transistors

(1) 2N3397 *(2)* 2N6005 *(2)* D40E5 -GE
(2) 2N6004 *(1)* D40D4 -GE

Optoelectronic

(1) BPX38-1 (phototransistor—Litronix)
(2) H11C6 (optoisolator—GE)
(1) CL703 (photocell—Clairex)
(2) Silicon solar cells at 0.5 V (Radio Shack 276–120)
(4) LED-RL2000 (Litronix)
(1) LED-FND360 or DIALCO 745–0017 (seven-segment, common anode)

NOTE: Italic numbers in parentheses indicate quantity.

ICs

(1)	7408	*(1)*	555	*(2)*	7476
(1)	7427	*(1)*	7447	*(1)*	7490
(1)	7432				

SCR

(1) C220D or C20D/C22D (GE)
(1) 2N1596

Diac

(1) ST2 (GE) breakover 30 V

Triac

(1) SC140B (GE)

UJT

(1) 2N2160

Thyrite Varistor

(1) 71D-7000 (GE)

TRANSFORMERS AND CHOKES

(1) Power: Primary 120 V 60 Hz; secondary 180, 160, 0, 160, 180 V at 70 mA (Triad-R22A or equivalent)
(1) Filament: Primary 120 V 60 Hz; secondary 25 V-CT at 1 A
(1) Variable isolation transformer (RCA Isotap or equivalent)
(1) Variable autotransformer
(1) Inductance 8 H at 85 mA

MISCELLANEOUS

Microprocessor—microcomputer trainer —Heathkit ET3400 *(1)* plus operating instructions
Relay (dc): Type RBM 10730–8 or equivalent *(1)* 400-Ω field, 7-mA pickup, 1-A contacts
Switches (toggle): SPST-3, SPDT-3, DPDT-1
Switches (bell type): Two normally open, with spring snap action
Light source: 60-W with reflector
Test light: 10-W with socket, 100-W with socket
Neon lamp: NE2 *(1)*
Pilot lamps and sockets: type 1477 *(2)*
Resistor decade box, variable in 1-Ω steps, 0–100 kΩ
Adjustable wire wound resistors: 0 to 100 Ω/15 W; 0 to 25 Ω/75 W
Fused line cord
Line cord adapter, three-prong to two-prong

NOTE: Rotating components and servo components are for optional experiments. If used, they should be stocked at the rate of 1 per 3 or 4 teams. Servomotor and amplifier components should match and should be matched with synchros.

Motor: 1/6 hp at 1725 rpm, 100-V shunt field, 90-V armature at 1.8 A, GE type 5BCD56BA6 or 5BCD56BD258 with base #161L494AA-G1 (56 frame)
Synchrogenerator
Synchromotor
Synchro differential generator
Synchro control transformer